Advances in

CMP/POLISHING TECHNOLOGIES FOR THE MANUFACTURE OF ELECTRONIC DEVICES

Edited by

PROF. TOSHIRO DOI

PROF. IOAN D. MARINESCU

PROF. SYUHEI KUROKAWA

ELSEVIER

Amsterdam • Boston • Heidelberg • London
New York • Oxford • Paris • San Diego
San Francisco • Singapore • Sydney • Tokyo
William Andrew is an imprint of Elsevier

William Andrew is an imprint of Elsevier
The Boulevard, Langford Lane, Kidlington, Oxford OX5 1GB, UK
225 Wyman Street, Waltham, MA 02451, USA

First edition 2012

British Library Cataloguing in Publication Data
A catalogue record for this book is available from the British Library

Library of Congress Cataloging in Publication Data
A catalog record for this book is available from the Library of Congress

ISBN: 978-1-4377-7859-5

For information on all William Andrew publications
visit our website at books.elsevier.com

Printed and bound in United States of America

11 12 13 14 15   10 9 8 7 6 5 4 3 2 1

CONTENTS

CONTRIBUTORS

Prof. Michio Uneda
Kanazawa Institute of Technology

and

Mr. Yasuhiko Takeno
Global Net Corp.

PREFACE

CMP and polishing are the most precise processes used to finish the surfaces of mechanical and electronic or semiconductor components. *Advances in CMP/Polishing Technologies for the Manufacture of Electronic Devices* is a book dedicated to updates in these processes more than a book about the basics. Even these processes are very precise. There has been very little scientific research undertaken into the study and application of these processes. These processes may be characterized as "more an art than a science." The aim of this book is to present the developments of recent years so that a foundation may be laid to enable the transformation of these operations into more deterministic processes by the involvement of some mechanical, chemical and tribological science.

The "Current situation in ultra-precision technology" (Chapter 3) will give an overview of CMP and polishing. The stock removal mechanisms of CMP and polishing are very different from any other processes, and because both CMP and polishing are free abrasive processes, most of the mechanisms are under a probability percentage. All abrasive processes have an overlap of rubbing, plowing and scratching mechanisms that are functions of a large number of parameters of the process, of the abrasive, and of the work piece. This makes any prediction of the outcomes of these processes very difficult.

Most of the applications of these processes are kept as confidential as possible (proprietary information), and specific details are not seen in professional or technical journals and magazines. This is the reason there are not many books which emphasize these processes.

The authors of this book have put together the latest knowledge concerning these processes in two leading industrial countries: the United States and Japan. Even though the authors are from academia, they all possess extensive experience in both theoretical and application domains.

Due to the high pace of development of the electronics and semiconductor industry, many of the presented processes and applications come from these industries, which are also the engines for the development of these processes. Few people using a computer realize how much CMP and polishing are involved in a computer's components. The most critical components of the disk drive are finished with special nanopolishing techniques, not to mention the CMP of chips, which has already become a standard technology.

Developments in the abrasive industry in recent years, mainly of superabrasives, have generated more challenges for industries that utilize these processes. The reality that day-by-day we get finer diamond and cubic boron nitride (CBN) grits is challenging these industries. It is not unusual today to talk about nano-grit, mainly in the case of diamond. Relatively new technology such as obtaining diamonds by explosion has allowed the development of products with grits as small as 5 nm and even smaller. To use these grits, avoiding the formation of clusters, is a challenge, which has only been partially solved.

Most of the knowledge used in the study of polishing and CMP has been borrowed from tribology, the science of friction, wear and lubrication. A book published in 2004 (*Tribology of Abrasive Processes)* was exclusively dedicated to the application of tribology to abrasive processes, but had more emphasis on the grinding process, which is largely used in industry. Not many researchers from the tribology field are dealing with manufacturing processes, even through this "marriage" is a win-win solution. Lately more people, mainly from academia, have taken this approach, and the results are great.

The audience for this book is very large. The book will be useful for a large category of professionals, starting with technicians and engineers and extending to researchers and academics. The book can also be used as a complementary textbook for undergraduate and graduate studies.

I would like to express special thanks to my colleague at the University of Toledo: Emily Lewandowski, for her help reading and editing the manuscript and putting up with my English, as well as that of my Japanese co-authors.

Finally, I would like to sincerely thank all the co-authors of this book, including their universities and families for allowing them to spend the time required for writing the chapters of this book. I would like to make a special mention of my co-author and very good friend Dr. Doi, who inspired this book and with whom I have a wonderful friendship for many years.

Dr. Ioan Marinescu
White Lake
Michigan
USA

ABOUT THE AUTHORS

Dr. Toshiro K. Doi is currently a Professor in the Department of Mechanical Engineering at Kyushu University, Japan. He earned his PhD in Polishing Technology from the University of Tokyo in 1985. He was a visiting professor at the University of Arizona in USA from 2003 to 2005. His research covers precision processing including CMP technology for the functional materials and its applications. He has published several books and over 130 papers in Japan and abroad. He is the inventor or co-inventor of more than 170 patents. He is a fellow, and a distinguished chairman of the Planarization CMP Committee of JSPE, and was a chairman of the 136th Committee on Future-oriented Machining of JSPE, Electrochemical Society, and a member of other national and international associations.

Dr. Ioan D. Marinescu is a Professor of Mechanical, Industrial and Manufacturing Engineering at the University of Toledo. He is also the Director of the Precision Micro-Machining Center of the College of Engineering (www.eng.utoledo.edu/pmmc) of the same University. He has a PhD in Manufacturing Processes, an Honorary Doctorate from the University of Iashi, Romania, and is a member of numerous international professional organizations: JSPE, SME, ASME, ASPE, CIRP, IDA, ASAT, NAMRI.

In 2010 he was named Honorary Professor of De La Salle University, Manilla Philippines.

Professor Marinescu is author of more than 15 books and over 300 technical and scientific papers. He has given lectures and workshops in more than 40 countries around the world. He is also the Executive Director and Co-Founder of the American Society for Abrasive Technology, ASAT.

Dr. Marinescu founded his own company: Advanced Manufacturing Solutions Co., LLC, a company specializing in consulting, R&D, Manufacturing and Trade.(www.interams.com) He is the President and CEO of this company.

Dr. Syuhei Kurokawa is currently an Associate Professor in the Department of Mechanical Engineering at Kyushu University. He earned his PhD Degree in Production Engineering from Kyushu University in 1992. He acted as a Visiting Professor of Laboratory for Machine Tools and Production Engineering (WZL) of the RWTH Aachen in Germany in 1998. His research fields include measurement and evaluation of gear accuracy, characterization of surface topography, nanomachining of micro machine elements and measuring devices, and planarization CMP technology of device wafers. He is a member of the JSME, JSPE, JSPS and JSAT.

CHAPTER 1

Introduction

Various functional materials, such as semiconductors, glass or dielectric, magnetic, ceramic, metallic or macromolecular materials have been used for recent electronic and optical devices, and have produced high performance and diversification. In order for the unique properties of such functional materials to be fully utilized within electronic/optical devices or applied to base substrates as high-quality films, they need to be effectively and precisely machined to the required geometry/dimensions. Machining processes make use of ultra-precision machining technology to produce a diverse range of high-performance parts. Fabrication processes involve lapping, polishing and ultra-precision polishing. For crystal materials such as electronic/optical devices, lapping and polishing are used as a pre-processing and ultra-precision polishing finishing step. Ultra-precision polishing is the final stage of a fabrication process, and directly affects the quality of the final device.

"Lapping – (mechanical) polishing – ultra-precision polishing" is an abrasive process, illustrated in Figure 1.1. The figure shows a simplified process model that has process unit of depth of cut/chip size on the horizontal axis. When brittle materials like silicon or glass are used in lapping, the material breaks down, accompanied by crack creation as a result of the mechanical action of hard abrasives. Polishing, however, can produce a mirror finish by a micronized mechanical action that causes micro-cracks to accumulate. In the case of ultra-precision polishing, these actions are more micronized; subtle interaction among work pieces, chips and machining liquid cannot be ignored, and the process progresses at the level of atoms and elements.

During ultra-precision polishing, relatively soft micro-particles continuously remove materials in extremely small amounts by elastically plastic deformation, without creating brittle fractures on the surface of a work piece. This produces a smooth, mirror surface with some small roughness in given geometry precision. Polishing the functional materials used in electric/optical devices produces a surface that has no deteriorated layer

Advances in CMP/Polishing Technologies
DOI: 10.1016/B978-1-4377-7859-5.00001-6

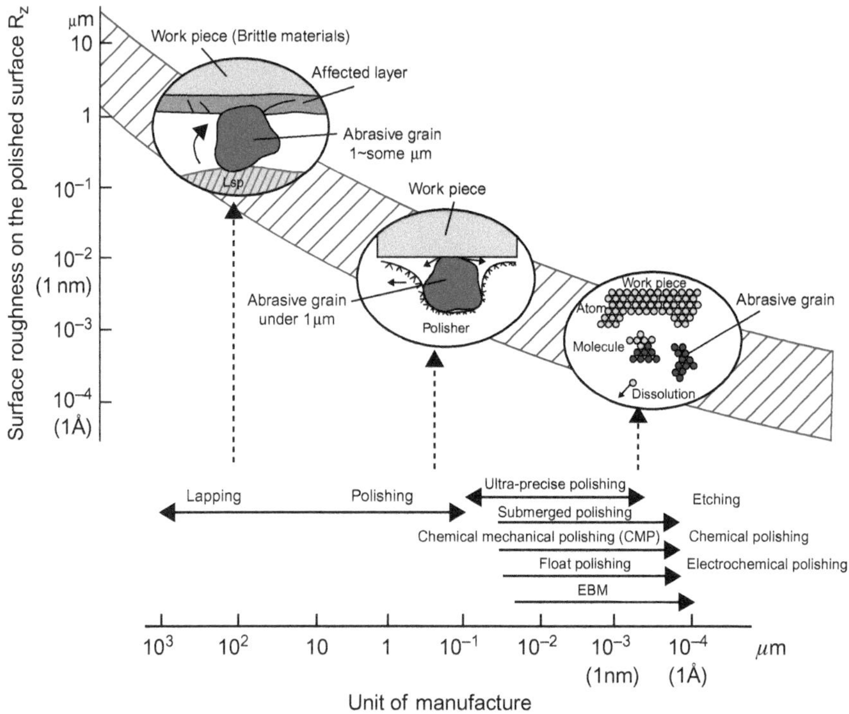

Figure 1.1 Positioning of lapping – (mechanical) polishing – ultra-precision polishing in an abrasive process

at a given geometry and precision. Current optical and metallographic ultra-precision polishing methods are an extension of conventional methods, and can include chemical/physical approaches. This book attempts to explain the ultra-precision process technologies which are indispensable for producing high performance devices and equipment. The fabrication process of semiconductor devices is used as a specific example. Actual ultra-precision technologies will also be introduced, and some examples will be given to show how these can be applied to other technological areas.

REFERENCES

1. T. Doi, Details of Semiconductor CMP Technology (in Japanese), Kogyo-chosaka; Publishing Co., Tokyo, Japan, 2001, pp. 13–38.
2. I.D. Marinescu, E. Uhlman, T.K. Doi, Handbook of Lapping and Polishing, CRC Press (Taylor & Francis Group), NY, 2006, pp. 266–279 and pp. 343–363.

CHAPTER 2

Details of the Fabrication Process for Devices with a Silicon Crystal Substrate

The groundbreaking invention of the transistor marked the beginning of the information technology era in the twentieth century. It is said that the development of the transistor by Shockley et al. at the Bell Institute in the US originated from a phenomenon in which an electric current changed according to the position of the electrode, when one more needle was added to a germanium wave inspector to make it a tripole. The word transistor is a combination of the words transfer and resistor, and suggests amplification of the electric current.

After the advent of the transistor, the progress of solid electronic devices slowed, until in 1959 a functional element that formed the basis of the IC (Integrated Circuit) was designed: a condenser and a resistor were used in a transistor on a single crystal substrate made of Ge (germanium) and Si (silicon) with interconnections assembled together and integrated. Silicon in particular was supported by high-quality crystal growth technology and has contributed to today's ultra-LSI.

In this chapter, the fabrication process for a device with a silicon substrate is introduced. The development process is discussed, and an overview is given of current and future challenges for device process technology.

2.1 HISTORY OF SEMICONDUCTOR DEVICES AND THEIR TYPES

Silicon single crystals are used as semiconductor devices primarily because it is easy to form SiO_2 (oxidation) films on the crystal surface, which have excellent stability and insulation properties. As this technology of utilizing oxidizing film has progressed, silicon IC technology has developed. This technology, called the Planar Process, was initially developed

Advances in CMP/Polishing Technologies
DOI: 10.1016/B978-1-4377-7859-5.00002-8

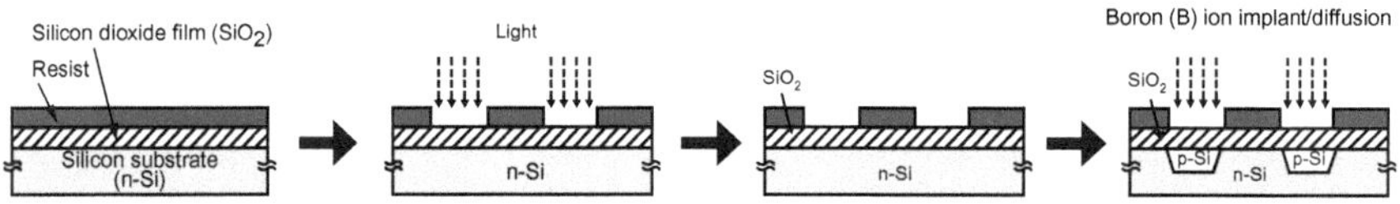

Figure 2.1 Basic scheme showing the formation of a pn junction with lithography and the Planar Process

by Noyce, of Fairchild Co. in the US (a founder of the current Intel) in 1959, and was an extremely important technology that opened the way for IC production, so marking the starting point of the current semiconductor IC. Figure 2.1 shows the basic scheme of the formation of a pn junction by lithography and the planar process. This opens windows in SiO_2 films which have been formed on a semiconductor silicon substrate, and, through these windows, impurities (boron in this case) are injected and diffused. Because SiO_2 films are insulators, electrical conductive films (for example an Al film) can be interconnected.

The first IC was reported in a patent applied for by Jack Kilby of Texas Instrument Co. in 1959. He produced a basic IC, consisting of only two transistors and several resistors on Ge crystals; thin lead was used as interconnectors. This technology was combined with the above-mentioned planar technology and has since progressed into current ICs. J. Kilby was awarded the Nobel Prize in Physics for his "contribution to the invention of the IC" in 2000.

The IC that emerged in 1963 made a debut as a MOS (Metal Oxide Semiconductor) or CMOS (Complementary MOS, MOS transistor having both p/n channels) device (Figure 2.2). Integration technology was enhanced in 1967 when the announcement of CMOS logics (by RCA Co.) was made, and in 1971 Intel developed a microprocessor that could deal with 1 kbit memory DRAM (Dynamic Random Access Memory) and 4 bit data. This opened the door to LSIs (Large Scale Integrated Circuits). ICs gradually came to have several hundred elements and, around 1970, developed into an LSI, having several thousand elements.

At this time, a magnetic memory was used as the main memory in a computer, but DRAM invented later, replaced it. As early as 1976, 64 k DRAM was developed; 1982, 1 M emerged, and the era of Mega began. In 1971, a 4-bit microprocessor was developed, increasing to an 8-bit MPU (Micro Processing Unit) in 1975, and, in 1981, to a 16-bit MPU. LSI was a tremendous breakthrough for the IC, and fierce competition in development between Japan and the US took place in the late 1970s,

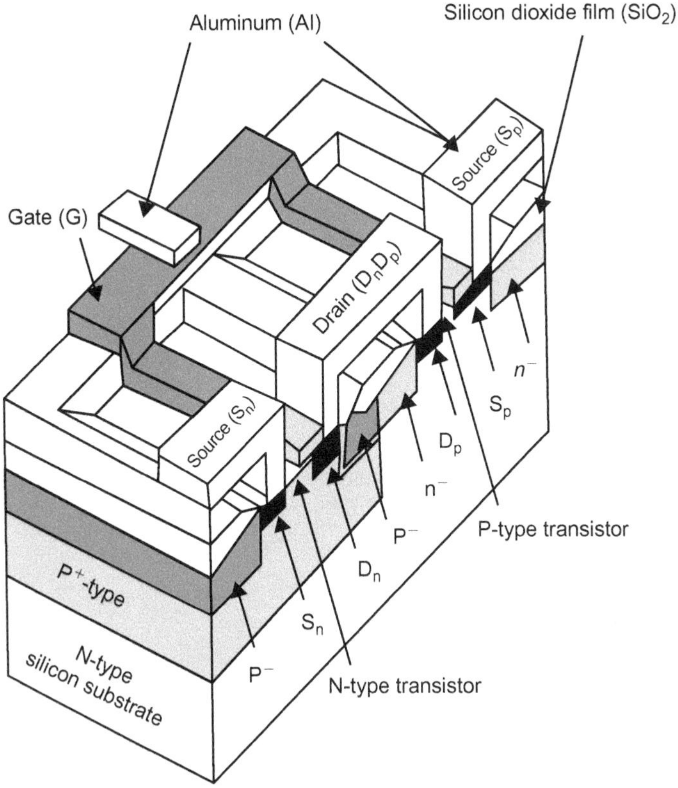

Figure 2.2 Basic structure of CMOS device

symbolized by the 64 kbit memory launched into the US market by Japanese companies. Moore's Law, advocated by Dr. Gordon Moore in his paper in 1965, describes a long-term trend of micro-processors. The paper indicated that the number of transistors on an integrated circuit would double every 2 years. He left Fairchild and founded what is currently Intel, with Noyce. This declaration soon became the goal of the whole semiconductor industry and formed the basis for a road map for a semiconductor technology of SEMATECH in the US.

Meanwhile the LSI cooperative research committee was set up in Japan in 1976, and both public and private sectors worked together for four years to develop an LSI with 1 μm machining technology. In the 1980s, 64 kbit, 256 kbit and 1 Mbit memories occupied more than half of the market, showing the lead gained by Japanese technology, which had excellent mass productivity and reliability.

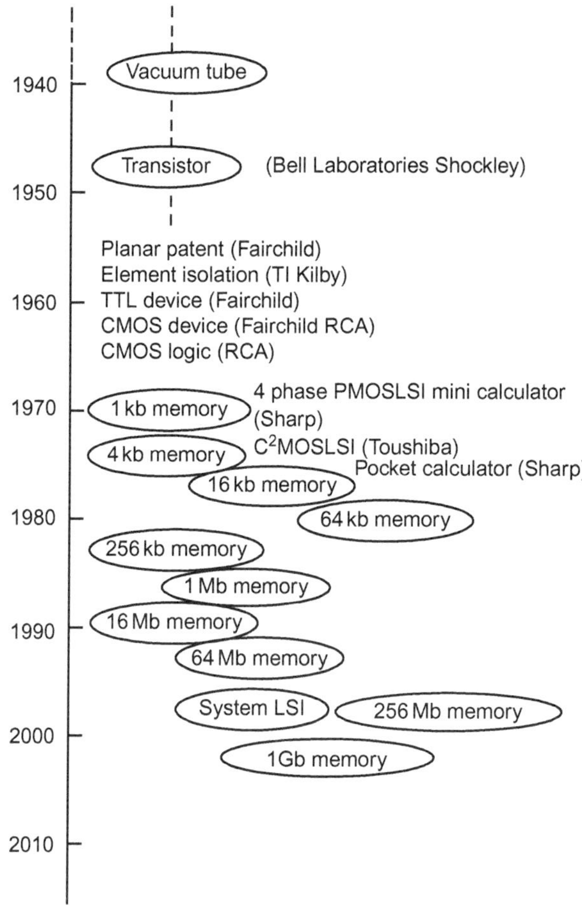

Figure 2.3 Progress of CMOS devices

Figure 2.3 shows the types and progress of IC, classified by its basic integration. Figure 2.4 shows a schematic picture of the development process of a CMOS device, as the basic circuit of an ultra-LSI. Thirty years after the introduction of the LSI in 1970, processor ability has increased by an amazing 1 million times. DRAM has developed just as quickly, from k- to Mbit and then, finally, to Gbit. Factors contributing to Moore's Law have been microprocess technology, circuit technology and the technology of new materials. Various innovative technologies have made it possible to achieve line widths down to 40 μm for today's LSIs, and 32 μm or less is expected to be achievable after 2013.

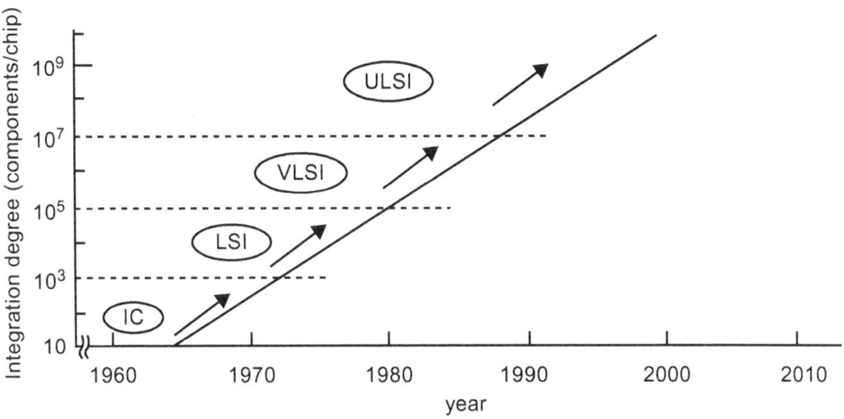

Figure 2.4 Device trend having MOS at the center and the progress

2.2 SEMICONDUCTOR DEVICE PROCESS TECHNOLOGY AND CURRENT SITUATION

Process integration technology makes a given IC by assembling elements (unit or elementary), as seen in Table 2.1. The reference (name) differs between makers and includes consistent process, integrated process, total process, through process and integration process.

In an effort to get as small a tip surface area as possible, this dimension is not fixed, but is continuously getting smaller over time. ULSI process integration needs to be inexpensive but also reliable, stable, and of high performance. In general, the use of amplifying elements in an IC, and the MOS transistor in particular, has been minimized. In the early 1970s a scaling law (a proportional shrinking law) for a micronized MOS transistor was announced. This law calculates circuit performance; the device dimension, electric voltage and current are set at 1/k, and the impurity concentration is k times higher. In this scaling law, because the electric field and current density are consistent, surface density of electric power in an IC is consistent.

However, in an actual product, large scale and high speed are both required. As micronization increased, the electric current density and electric field strength increased. As a result, reliability decreased due to current leaks. Current leaks, particularly in DRAM, lead to operational errors in memory. In order to resolve these problems, a much more complicated structure was introduced. Process steps and production time increased, and, as a result, production costs increased.

Table 2.1 Process Integration Technology by Assembling Element Processes

• Integrated Circuit Substrate	– Single crystal wafer base plate: Intrinsic Gettering – Epitaxial grown – Homo: Si grown on Si – Hetero: a-Si, SiC, Si-Ge – SOI – Fusion Recrystallization – Solid phase grown – SIMOX (Separation by Implanted Oxygen) – Pasting wafer: Direct connection, organic Film integration
• Lithography	– Light transfer process: Same size projection, shrinking projection (stepper, scanner) – Electron beam portrayal: point beam, fairing beam – X-ray portrayal: Same size, Shrinking
• Etching	– Solution technique – Dry etching – RIE (Reactive Ion Etching) – Ion milling – UV exposure, Usher – CMP (Chemical Mechanical Polishing) – Lift off
• Insulating Film Formation	– Oxidation – CVD (Chemical Vapor Deposition) – Sputter – Chemical anode
• Junction Formation	– Ion milling – Solid diffusion – Silicide: Face of Si base plate – Annealing
• Forming Electric Distribution Electrode	– Sputter – Damascene, dual-Damascene – Coat especially Cu trace – MOCVD (Metal-Organic CVD) – Silicide: Gate electrode, trace
• Multilayer Interconnection	– Planarized insulating film – Damascene, dual-Damascene – Implant plug

(*Continued*)

Table 2.1 (Continued)

• Decontamination	– Gettering: P, O, Micro-defect – Clean up – Grit and dust (vacuum, air, submerge)
• Assessment Analysis	– Crystal fault – Pattern defect – Impurity distribution – Configuration observation

When process integration becomes more complicated, the requirement for the separation of process integration becomes greater. With the introduction of CMP (Chemical Mechanical Polishing), almost ideal planarization can be realized. This topic is the main subject of this book.

In the next section, element isolation and multi-interconnection will be discussed as a representative example of how planarization CMP technology is used in a device process.

2.2.1 Element Isolation Structure

One of the technologies that significantly advanced Si ICs was the planar element isolation method using SiO_2. This technology was developed in 1960, but when elements were micronized and an "n" channel MOS transistor was introduced, it became insufficient. In 1970, LOCOS (Local Oxidation of Silicon) replaced it. The LOCOS method, as shown in Figure 2.5, is a method of creating a thick oxidation film in the other element isolation region by selectively adhering a Si_3N_4 film onto the active region (in which the element is formed) that is extremely difficult to oxidize at high temperatures. Since almost half of the thickness of a field oxidation film is embedded into a silicon substrate, step height can be reduced, which makes it easier to complete the next process. However, LOCOS also met with difficulties when the element dimension became less than 9.5 μm. Problems with this method included:

1. Oxidation films invade beneath the Si_3N_4 films and make the active regions narrow.
2. About 90% of boron at a level of 10^{13} cm^{-2} injected to form a channel stopper is taken into a field oxidation film.
3. Because 10 times the amount of boron is injected, boron invasion into an active region is much more pervasive.

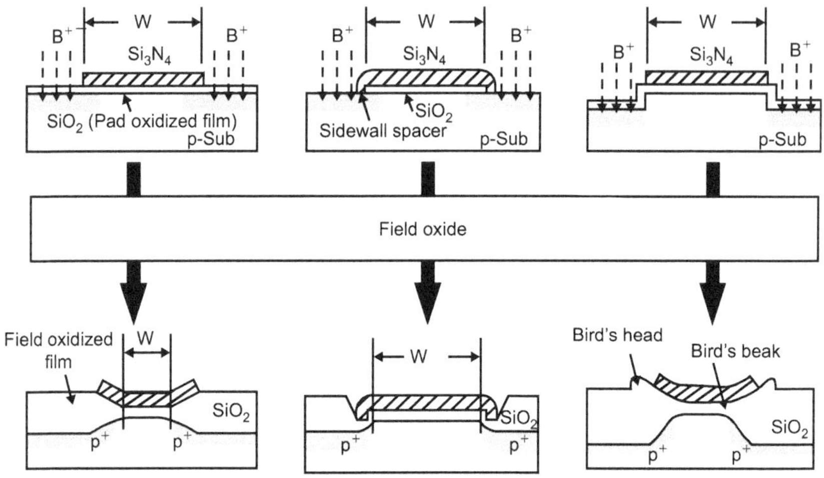

Figure 2.5 LOCOS (Local Oxidation of Silicon) method

This invasion width is about the same thickness as the field oxidation film; thus, when an active region with a width of 0.5 μm and an element isolation region with a width of 0.5 μm are placed next to each other, if a field oxidation film with a thickness of 0.5 μm is formed, the active region disappears. Transistor density of a current processor is less than one tenth of DRAM; but because a processor contains ultra high speed primary cash memory, the same problem occurs.

As a tentative improvement measure, a method for preventing the invasion of a field oxidation film by covering the sidewall with a Si_3N_4 film has been suggested. This method was tried in the 1970s on a bigger pattern, but there was no technique to uniformly embed SiO_2 in a deep wide groove; instead, polycrystalline Si was embedded, the surface of which was thinly oxidized and no impurity was added in the whole groove. This method was inefficient, however, because the existing capacity could not be reduced as expected, and it was time consuming and expensive. It was therefore applied only to ultra-high-speed bipolar ICs.

Subsequently, STI (Shallow Trench Isolation) technology in which ozone and TEOS (tetraethoxy silane) were used and SiO_2 was embedded in the groove by using plasma energy. As shown in Figure 2.6, after embedding SiO_2, a groove could be planarized by using CMP technology. Because the roughness on a pattern varies, to get ideal planarization, the extra etching on the larger parts of the embedded region must be controlled by attaching something like a block layer. With CMP the

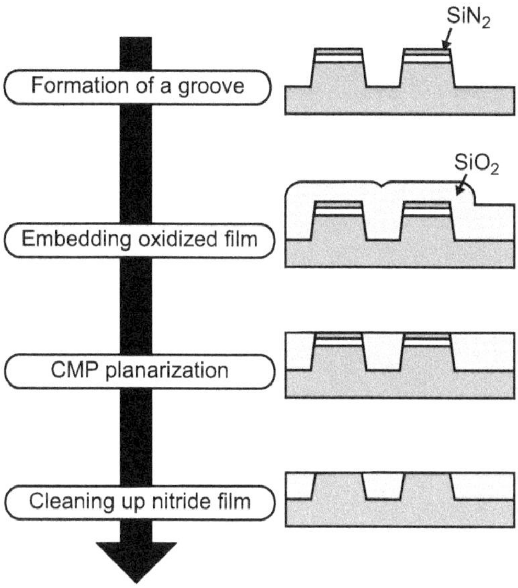

Figure 2.6 A method to prevent the invasion of a field oxidation film

required embedding could be achieved over the entire tip surface. However, there are still challenging matters to be resolved, such as the change of sub-threshold property which produces a different threshold voltage at the corners. There are difficulties in forming a suitable channel stopper on the sidewall of a groove, and these variations produce a change in transistor properties at the edge of an active region, but the structure gained is essential to produce a device smaller than 0.25 μm.

2.2.2 Multilayer Interconnections

Multilayer interconnections connect the various elements which are fabricated on a Si substrate. The design is intended to embed an electrode into n^+, p^+ and a gate electrode on a substrate through interconnection holes; then another interconnection layer is placed on top, with interlayer insulating films between. A microprocessor has more than 7 layers of interconnections. Conventional methods form an interconnection electrode by etching, as shown in Figure 2.7(a), but this can only achieve about five layers. Since the method to embed an interconnection electrode into a groove dug on an insulating film was designed, over 10 layers are possible; this is called the Damascene method (Figure 2.7(b)). Because contact holes are embedded at the same time as the interconnections are fabricated, it is called dual-Damascene (See Figure 2.8).

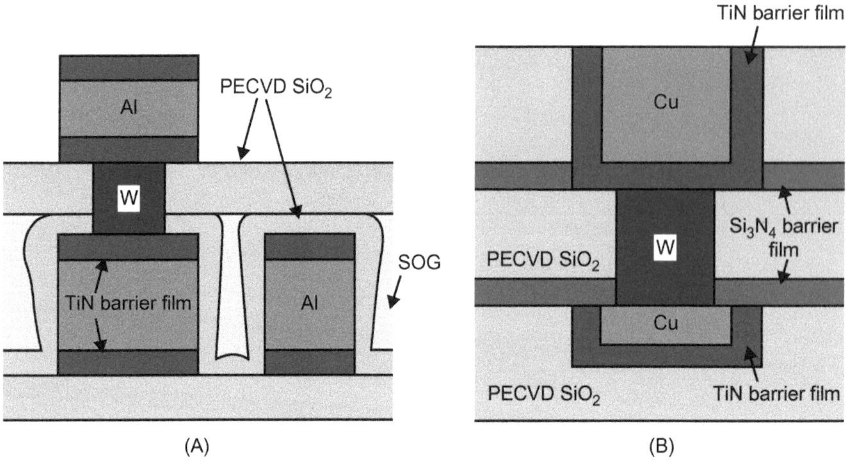

Figure 2.7 Interconnection electrode is formed by etching by (A) the conventional process (B) the Damascene process

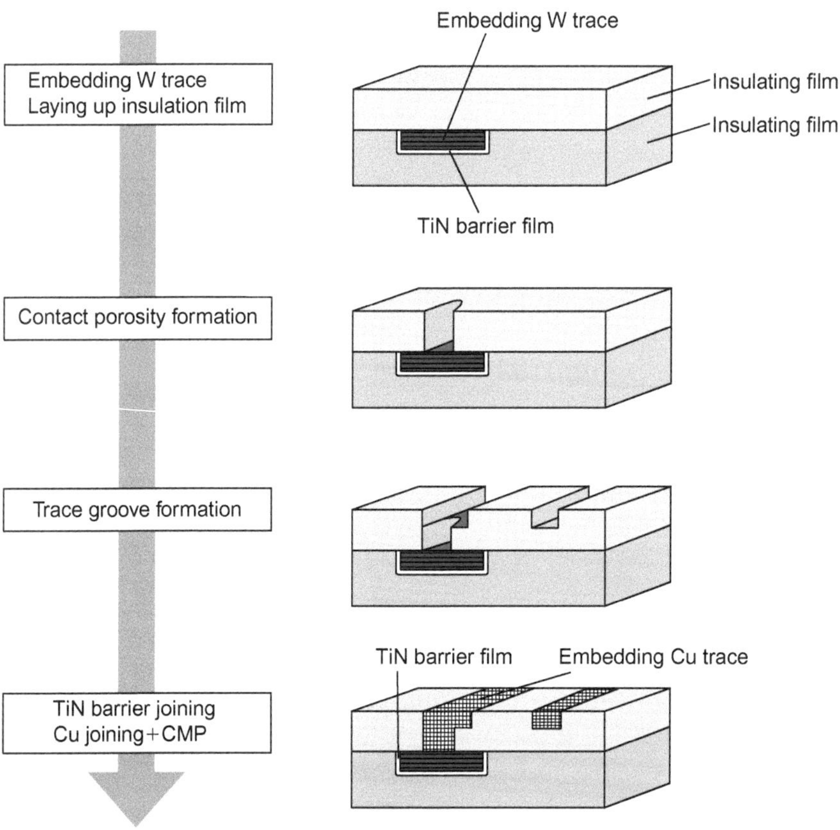

Figure 2.8 Cu wiring by dual-Damascene process

A conventional Al (aluminum) interconnection with a base and a step height suffered from the problem that the Al layer was thin at the step, and thus the mechanical strength of the insulating film attached at the top was reduced. Also, because the step part was hidden, an etching residue remained. The Damascene method can resolve these problems. Furthermore, the low resistivity and high reliability that it can produce made it popular, and allowed the possibility of etching a Cu (copper) interconnection. Most importantly, the combination of planarization and CMP can produce an almost ideal planarization structure, and embedding a Cu interconnection is now easily achievable. Without planarization CMP, today's semiconductor devices would not exist.

REFERENCES

1. T. Doi, Details of Semiconductor CMP Technology (in Japanese), Kogyo-chosakai Publishing Co., Tokyo, Japan, 2001.
2. S. Wolf, Microchip Manufacturing, Lattice Press, California, 2003.
3. T. Doi, M. Kinoshita, H. Hiyama (Eds.), A Library of CMP Planarization Technology & Application (in Japanese), Global Net Corp., 2006.
4. <http://public.itrs.net/Common/2005ITRS/Home2005.htm/>.
5. K. Nojiri, K. Tsumokuni, K. Hirobe, A. Koike, K. Ito, S. Kishino, Extended Abstract, 17th Conference on SSDM, Tokyo, 337 (1985).
6. N. Matsukawa, H. Nozawa, J. Matsunaga, S. Kohyama, IEEE Trans. on ED ED-29 (1982) 561.
7. C.L. Borst, W.N. Gill, R.J. Gutmann, Chemical-Mechanical Polishing of Low Dielectric Constant Polymers and Organosilicate Glasses, Kluwer Academic Publishers, Boston, 2002.

CHAPTER 3

The Current Situation in Ultra-Precision Technology – Silicon Single Crystals as an Example

A diverse range of functional materials, including semiconductors, dielectric, magnetic and optical materials, as well as ceramics, metals, plastic and glass are all used in electronic/optical parts, and have demanding requirements. High performance devices and parts are made by use of free ultra-precision cutting, grinding and abrasives (lapping and polishing) technologies and by machining processes which are suitable for the various types of work and parts.

CMP is an ultra-precision polishing procedure that occurs at the end of the machining process; it is a very important process step, because it directly affects the performance of the resulting device.

In order to design an efficient machining process, it is necessary to understand the pre-machining process that precedes it. In this chapter, crystal growth, final finish machining, precision cleaning and defect inspection of a substrate are summarized by using a semiconductor silicon single crystal destined for an LSI as an example.

Figure 3.1 shows schematic pictures of ultra-LSI device fabrication as a fabrication flow chart. Process steps from creating a silicon crystal, to making a wafer from the crystal, to processing and then completing the device, are shown. In the first step, a silicon wafer is produced through a process of pulling up a single silicon crystal. Next, peripheral grinding (notch produced), slicing, beveling, lapping (or grinding), etching, ultra-precision polishing (mirror polishing) and then ultra-precision cleaning occur. Precision machining technology is unrivaled in these processes, collectively called the wafer-making process. The next step is divided into two: the wafer process (called pre-process) in which the device is generally processed in a wafer state, and the assembly/test process (called post-process) in which a tip is produced and elements are assembled.

Advances in CMP/Polishing Technologies
DOI: 10.1016/B978-1-4377-7859-5.00003-X

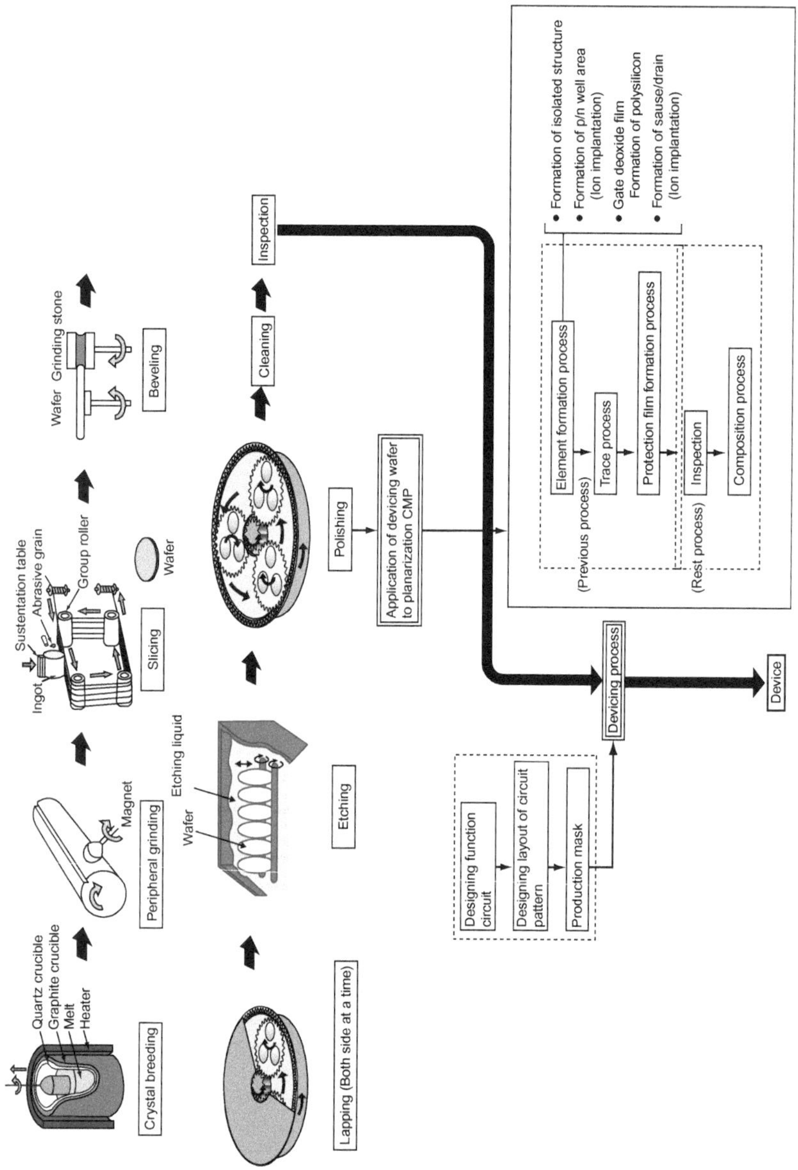

Figure 3.1 A schematic picture of ultra LSI device fabrication process

3.1 PRODUCTION OF SINGLE CRYSTAL SILICON

The engineering methods for producing a single crystal are broadly classified into two types: the Czochralski (CZ) method and the Floating Zone (FZ) method. Both involve careful solidification of molten silicon, which has been previously made from heated polycrystalline silicon. Figure 3.2 is a schematic picture of single crystal silicon production by the CZ method. In

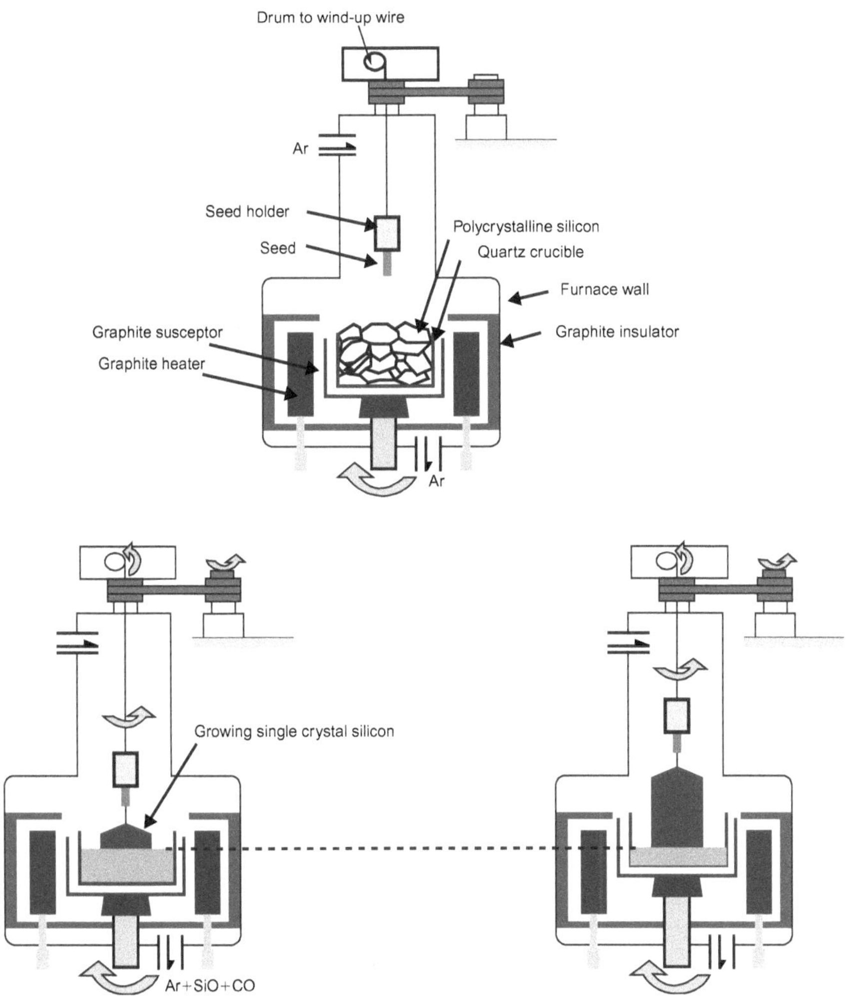

Figure 3.2 A schematic picture of single crystal silicon production with the CZ method

the first step, a quartz crucible is filled with polycrystalline silicon, and then placed in a CZ furnace, where it is melted by a graphite heater. There are different shapes of polycrystalline silicon, e.g. cylinders or lumps, which are used for different purposes. The melting point of silicon is 1420°C, so the heat is kept inside the furnace by a graphite heat insulator, and the furnace wall is cooled with cold water.

A round rod or square pillar of single crystal silicon, called a seed crystal, is prepared. This is brought down into the molten silicon and, soon after it touches the surface, the interface instantly solidifies. The molten silicon that is contacted solidifies as a single crystal. When the seed crystal is pulled up slightly, the solidified part cools, and this part becomes the seed crystal for the next crystallization step; the molten silicon right below the seed crystal solidifies as a single crystal. By continuing this action, the bottom of an original seed crystal successively solidifies. When the pulling speed increases and the temperature of the molten silicon gets higher, the crystal's diameter gets smaller. With skillful control of the pulling speed and temperature, it is possible to grow a crystal with any given diameter. Figure 3.3 shows a single crystal ingot having a diameter of 300 mm and a weight of 350 kg grown with the CZ method. Today, mass production of single crystals of such size is routine. Experiments have grown single crystals with a diameter of 450 mm successfully. It is expected that mass production of devices using 450 mm wafers will occur by 2012.

If control of the temperature fails during the growth of a single crystal ingot with {100} plane, leading to a slight decrease in temperature at the solidification interface, the {111} plane starts to grow soon after. When a quartz crucible melts by reacting to molten silicon, a part of it is taken as oxygen into a crystal from the solidified interface.

Molten silicon contains complex convection currents, which can make crystal growth unstable. Molten silicon has the physical properties of liquid metal, and so the convection currents can be controlled by applying a magnetic field outside the furnace with an ultra electrical conductive coil. This is called the MCZ (Magnetic CZ) method. When these currents are controlled, the solution in a quartz crucible is controlled (reducing the amount of oxygen dissolving in the melt), and, as a result, it is possible to lower the oxygen concentration of the resultant crystal. All crystals with a diameter of 300 mm are produced with the MCZ method.

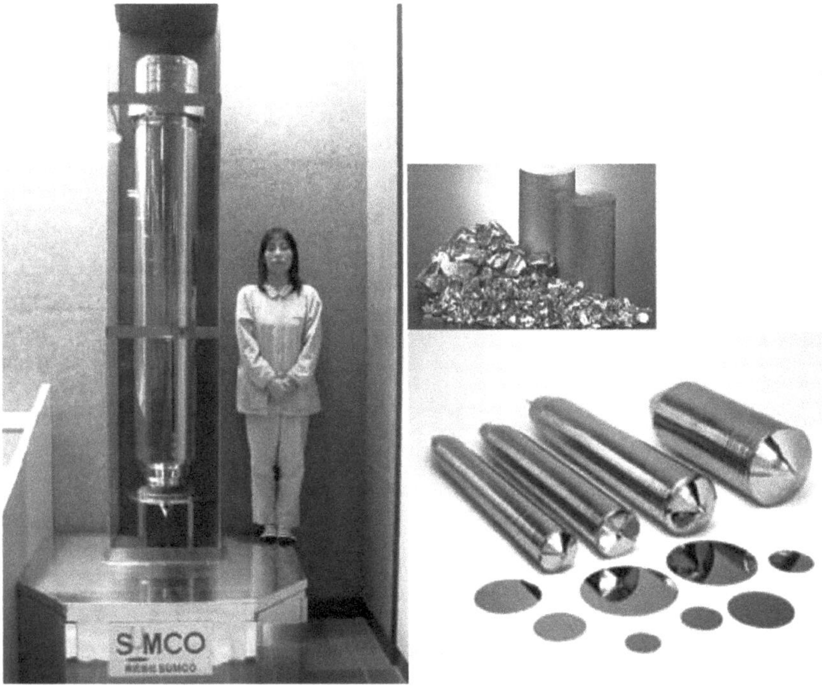

Figure 3.3 A single crystal ingot having a diameter of 300 mm and a weight of 350 kg grown with the CZ method (SUMCO)

Doping with boron or phosphorus is required to create the necessary electrical conductivity in single crystal silicon, and almost all silicon used in an industrial environment is boron doped.

3.2 SLICING: PRE- AND POST-PROCESS

The first step in wafer machining is to cut an ingot that has been crystal grown. Peripheral grinding of the ingot then occurs, followed by orientation processing (planarization or notching) and then beveling.

Peripheral grinding and orientation flattening are precision machining processes, performed to achieve a precise and consistent wafer diameter, so it can be precisely carried and positioned in further processing. Beveling (chamfering) of wafer peripheries after slicing is important to prevent breakdown. Single crystal silicon contains cleavage planes and hence small scratches or chipping on the wafer edges can cause it to break. If this occurred during transport in a clean room, the automatic machining line would all have to be halted, which would be a big problem.

3.2.1 Slicing

3.2.1.1 Introduction

Silicon solar cells are manufactured from silicon ingots by slicing them into individual wafers. Historically, the wafers have been produced with the use of inner diameter (ID) blade saws.

A shift to the use of wire saws as a cutting tool has dramatically increased the productivity of the slicing operation. This is due to a reduction in both the cutting losses (kerf losses), and in the amount of damaged silicon left after cutting. The wire slicing leaves a much smoother surface, with less saw damage to the wafers. A typical ID blade cut will consume over 250 mm of silicon, which is lost as waste, and require etching of 100 mm of saw damage on the silicon wafer. By contrast, a wire cut will consume less than 180 mm of silicon with less than 25 mm damage to be etched. This has nearly doubled the amount of wafers produced per length of ingot supplied; Figure 3.4 shows the increase in yielded wafers per inch (2.54 cm) of ingot. This technology change has dramatically reduced the cost of producing silicon solar cells.

3.2.1.2 Abrasive Wire

Over 80% of global solar cell production requires the cutting of multi- or monocrystalline silicon blocks into wafers. Multi-wire sawing is the main slicing technique used in the photovoltaic and microelectronics industry. This technology, which allows for a high throughput, small kerf loss and excellent surface quality, enables the wafers to be used without any

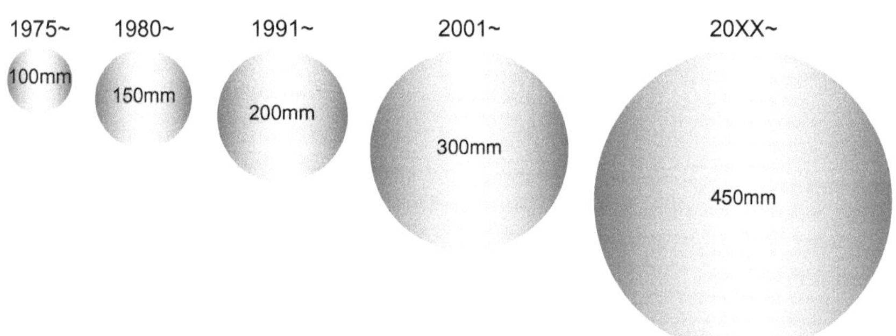

Figure 3.4 Increase in yielded wafers per inch (2.54 cm) of ingot

further machining. Currently, solar wafers of 156 mm × 156 mm can be as thin as 250 μm.

Loose-Abrasive Wire

Wire saws using the Free Abrasive Machining (FAM) technique are an emerging technology for large diameter thin crystal wafer production in the semiconductor and photovoltaic (PV) industry. FAM's advantages include the ability to produce very thin wafers with small kerf loss, and high yield and productivity.

Wire saws have been used for cutting hard materials like granite slabs and other varieties of stone since medieval times. However, the requirements of the technique in terms of wafer thickness and quality are very different in electronics and photovoltaic applications than in the traditional use of the process. An evaluation of the wire saw cutting process has shown that it is a poorly understood phenomenon, and no model exists for simulation design and control. Even so, in crystal cutting applications, it has shown potential for producing a better surface finish and thinner wafers at much higher yield than ID saws. This, coupled with the total absence of any commercial US technology in this field, makes the process worthy of detailed study. The primary objective of such a study would be to evaluate current wire saw slicing technology, and compare it with ID saw technology that is most commonly used for wafer slicing. Such a comparison is expected to shed light on why the wire saw is superior to other conventional crystal sawing methods, thus indicating the importance of concentrating efforts on the wire saw.

Comparison Between Wire Saw and ID Saw

Figures 3.5 and 3.6 show schematics of a wire saw. In the wire saw, a single strand of thin wire (175 μm in diameter) moves from a feed reel to a take-up reel. In between, the wire goes through the entrance of a storage system called the "carriage" and into a rectangular arrangement of fixed shafts with replaceable wire guides which have hundreds of grooves. The wire wraps around the wire guides creating a multiple net of parallel wires known as web. The ingot crystal is fed through this web together with abrasive slurry to produce a cut.

Current wire saws can cut up to four ingots simultaneously from four sides of the web; a vast improvement in productivity over the ID saw. Since the wire is so thin, the kerf loss in the wire saw cutting process is minimal. Additionally, the future use of large diameter crystals favors wire

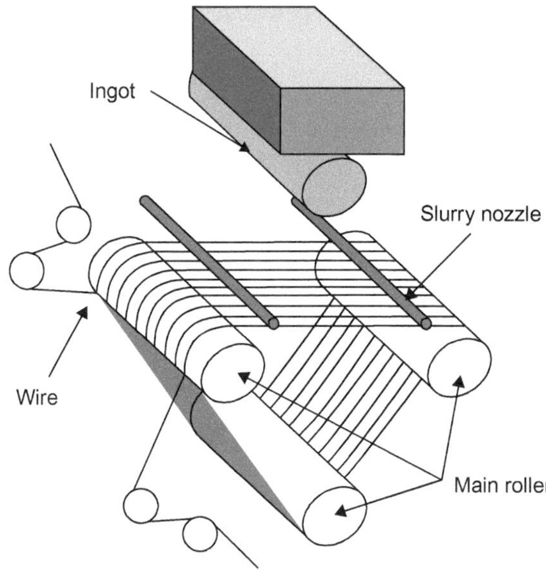

Figure 3.5 Geometric description of the wire slicing operation

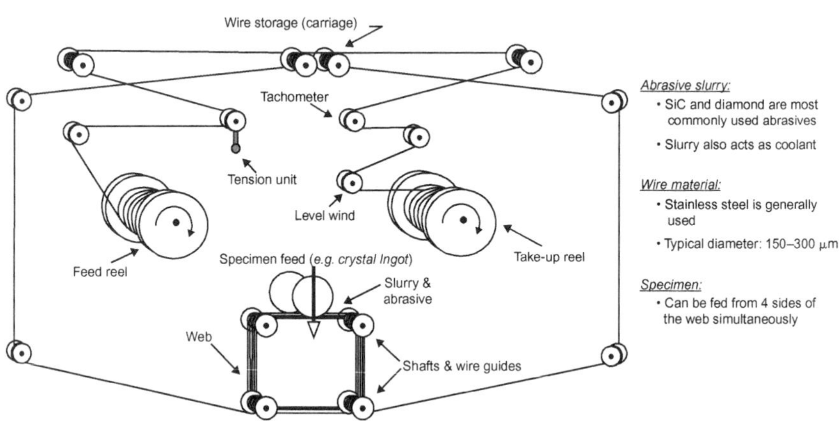

Figure 3.6 Conceptual wire saw slicing setup

sawing, because the ingot diameter capacity of the wire saw is limited only by the shaft spacing of the wire guides of the web and the travel. Table 3.1 summarizes the principal differences between the wire saw and the ID saw.

The FAM process of the wire saw produces considerably shallower damage and more uniform surfaces than the grinding process of the ID saw. The low residual stresses in wire-saw-cut crystal wafers make the

Table 3.1 Principal Differences Between the Wire Saw and the ID Saw

Property	Wire Saw	ID Saw
Method of cutting	Lapping	Grinding
Typical cut surface features	Wire marks	Chipping and fracture
Depth of damage	Uniform 10 to 15 μm	Variable 20 to 30 μm
Productivity	300 to 2000 sq.in./hr	200 to 400 sq.in/hr
Number of wafers cut per run	As high as 3200 wafers	A single wafer per cut
Kerf loss per cut	Typically 200 to 300 μm	From 300 to 500 μm
Minimum thickness of cut wafer	As low as 200 μm	300 μm
Yield of 0.025 inch thick wafers	31 wafers per inch	27 wafers per inch
Maximum diameter of cut crystal	300 mm-diameter crystals	Up to 200 mm-diameter

wire saw more amenable to cutting very thin wafers, because the constructive interference between stress fields that occurs when two cuts are made very close to each other, becomes less significant. Higher residual stresses are not desirable since they lead to breakage of thin wafers in both the cutting process and post-process handling.

In PV applications, where post processing is not required, it is of cardinal importance to have wafer surfaces with minimal damage. This may not be as important in the semiconductor industry, where the sliced substrate is lapped, edged and polished well beyond the damaged region. Nevertheless, lower residual stresses in the substrate will make it easier to handle the cut wafers without breaking them, and to produce more homogeneous wafers for the fabrication processes. The other area where minimal damage to the cut surface is important is in the slicing of optical material crystals (e.g. $KTaO_3$). Here, residual damage to the material adjacent to the cut produces unacceptable levels of birefringence in materials with high strain-optic coefficients.

Also the well-known process of obtaining a smooth surface by burnishing, which uses much lower working forces than polishing, makes the wire saw better at giving a good surface finish than the ID saw. The control of the wire saw in terms of wire stiffness and vibration and slurry concentration management can be facilitated by properly modeling the manufacturing processes. Compared to grinding and other cutting processes, the wire saw uses much milder forces during the cutting process and thus has the potential for more energetically efficient cutting. This also can be advantageous in processes in which a temperature rise in the crystal at the local contact area is not permissible due to possible changes in its microstructure.

Currently, the cost of consumables (abrasives and slurry base) for wire saw operation is still higher than that for the ID saw. The cost and ease of process control of the ID saw gives it a slight edge over the wire saw, but this can mostly be attributed to a total lack of research into the wire saw manufacturing processes. Development of systematic modeling and control strategies could improve the wire saw manufacturing process and make it more cost-effective.

Fundamentals of Free Abrasive Machining

Wire saw technology has made enormous progress from the woodcutter's saw to current machinery. This has mostly been due to practice and experience; little systematic theoretical understanding of this process exists. Currently, as industry moves towards improving the wafer manufacturing process, this understanding is becoming essential. In this way, the process can be optimized, through analytical model and control, in order to meet the stringent requirements of efficiency and productivity.

The FAM process is prevalent in the wire saw slicing in the form of three-body abrasion. This involves rolling abrasives in between a continuously moving wire and the crystal (Figure 3.7), by which process they become indented into the crystal.

This indentation allows the penetration of the hard abrasives (mostly SiC) into the comparatively soft silicon crystal. As the abrasives are carried at high speed by the wire, they remove material from the crystal in the form of chips. This chipping is a function of the force applied on the abrasive, and the hardness differential between the abrasive and the crystal. In addition to these factors, the total material removal rate is a

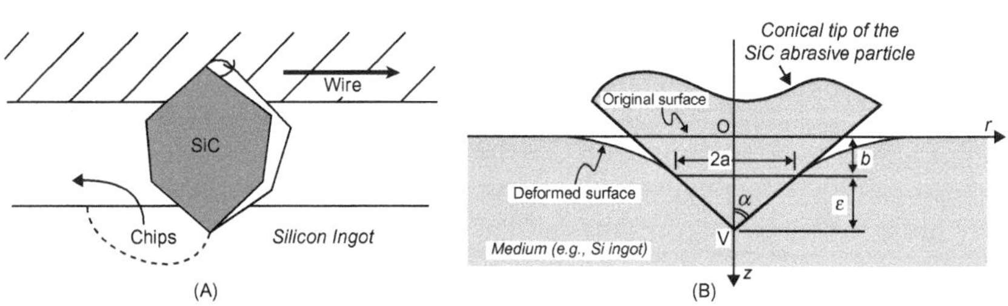

Figure 3.7 (A) Schematic of rolling-indenting model of free abrasive machining; (B) Elastoplastic modeling of an abrasive particle indenting on substrate surface

function of the concentration of abrasives in the slurry and the speed of the wire.

This chipping phenomenon is a very complex interaction involving fracture mechanics, but a model was developed as a first approximation by mapping the stresses underneath a conical abrasive indenting into an infinitely elastic work piece. From this analytical model the likely pattern of cracking was derived. This matches the well-known cracking patterns (lateral cracks, radial cracks and surface (Palmquist) cracks) obtained in indentation studies. Another industrially significant result obtained from this model is an estimation of the amount of material removed in one indenting action, which was approximated by using the Von-Misses yield stress contour in an infinitely elastic medium. Although this model was developed for a single abrasive, the results can be extended to material removal by multiple abrasives. Thus an overall material removal rate as a function of wire speed and the abrasive concentration can be obtained.

It has been observed that in the wire saw slicing process, the slurry which the wire brings (on account of surface tension) into the cutting zone does not, in its entirety, enter the cutting zone. A considerable amount of this slurry gets thrown out, without participating in the cutting action. This has been postulated to be due to elastohydrodynamic interactions between the highly viscous slurry (either oil or ethylene glycol base) and the axially moving semi-compliant wire under tension. We have developed an elastohydrodynamic model which will relate the slurry film thickness to the slurry viscosity and the wire speed. As a first approximation the slurry is considered as a Newtonian fluid and the situation is modeled as a slider bearing. To analyze this situation, the customary Reynold's equation from lubrication theory needs to be solved, along with the elasticity equation for a wire under tension, with the crystal being approximated as rigid.

The so obtained "working zone slurry flow pattern" can be helpful in deciding the number of abrasives in action in the cutting zone. The velocity profile within the laminar flow region of the slurry between the wire and the crystal can also explain the rolling of abrasives so that they come in contact with the moving wire. Also the hydrodynamic film thickness in conjunction with the size of the abrasive will govern whether the cutting regime is in the polishing zone (very small removal rate) or the machining zone. This can shed some insight on the abrasive grit size needed to cause a cutting or polishing action, as a function of the wire speed and the slurry viscosity.

Fixed-Abrasive Wire

A multi-wire saw with a loose-abrasive wire has become an established technology for slicing silicon ingots in mass production. In wafer slicing with a loose-abrasive wire, a slurry comprising a lubricant (generally glycol-based chemicals or non water-soluble cutting fluid) and a green silicon carbide (GC) abrasive is used. This generates a large amount of industrial waste, including waste slurry and wires. A fixed-abrasive wire, in which diamond grit is attached to a bare wire, has been developed as an alternative. An advantage of this latter method is that it is possible to slice wafers by supplying a water-soluble coolant (grinding fluid) without the need for any slurry. Moreover, the fixed-abrasive wire saw has a higher slicing speed and the spent grinding fluid can be recycled. Although a fixed-abrasive wire with diamond grit attached by electroplating has been successfully developed, it has high production costs. In order to reduce these costs, we have recently developed a new technique where the diamond grit is fixed onto the bare wire by resin bonding instead of electroplating.

Watanabe et al. investigated the properties of polycrystalline silicon wafers sliced with a fixed-abrasive wire with resin-bonded diamond grit (Figure 3.8). They fabricated solar cells using these wafers and studied their electrical characteristics. From these experiments, they showed that wafers sliced with a fixed-abrasive wire with resin-bonded diamond grit have a saw-damage layer that is more than a factor of two thinner than those of wafers sliced with a loose-abrasive wire. They also argued that fixed-abrasive wires with resin-bonded diamond grit are suited for the fabrication of solar cells, particularly thin silicon cells where wafer thicknesses approach or become less than 150 mm.

This group also developed a fixed-abrasive wire with resin bonding for slicing single crystal and polycrystalline silicon ingots, and investigated the properties of the resulting sliced polycrystalline wafers. Carrier lifetime measurements indicated that the saw-damage-layer thicknesses of the wafers that were sliced with fixed-abrasive wires were more than a factor of two (from 10 to 4 mm) smaller than those of the wafers sliced with the loose-abrasive wires, when the slicing speed of the fixed-abrasive wires was approximately 2.5-fold higher than that of the loose-abrasive wire.

A difference in the depth of saw-damage layers was also observed in cross-sectional TEM images. In addition to the difference in the slicing method, the presence of the resin bond of the fixed-abrasive wire can contribute to the observed reduction in the thickness of the damage layer.

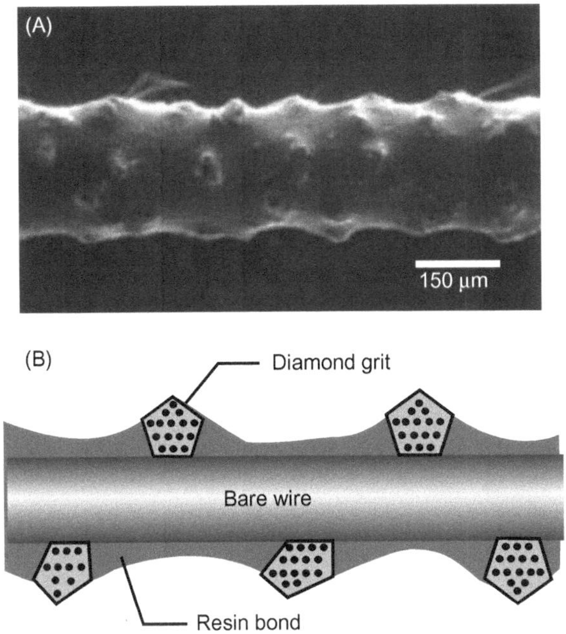

Figure 3.8 Photograph (A) and schematic cross-section (B) of the fixed-abrasive wire

Furthermore, the saw-damage etching depth of 4–7 mm was acceptable for the fabrication of high-efficiency solar cells when the wafers were sliced with the fixed-abrasive wire. A conversion efficiency of 15.3% was demonstrated using non-textured wafers with an etching depth of only 7 mm. The saw-damage layers on both sides of the wafers must be removed prior to cell fabrication. By replacing the loose-abrasive wire with a fixed-abrasive wire, one can reduce the thicknesses of the damaged layer from 20 to 8 mm. This reduction will be significant when the thickness of the crystalline silicon solar cells approaches or becomes less than 150 mm in the future. Therefore, wafer slicing using the fixed-abrasive wire is promising as a next generation slicing technique for solar cell fabrication, not only because of its high slicing speed but also because of its advantage in the fabrication of thin cells.

Saw Damage on the Surface of Silicon Wafer

It is well known that mechanical damage induces stacking faults on the surface of silicon crystals during oxidation. Metal impurities tend to adhere to these defects during the heat treatment of the silicon. This phenomenon is called gettering, which is useful for improving process yield in the production of semiconductor devices. In order to induce the formation of these defects, silicon slurry is blasted on to the back side of the silicon wafers during their production.

The pressure employed in this blasting process is in the range of 0.3–1.7 kgf/cm^2. As a result, a small amount of damage is created. This damage is less than that created by scratching the silicon surface with a fine sandpaper. However, due to the silica blasting process, stacking faults 20 mm long are formed during the oxidation treatment at 1150°C for 2 h. These oxidation-induced stacking faults (OSF) on the back side of the silicon wafer can getter any metal impurities located on the top of a silicon wafer of 800 mm thickness.

During the slicing of silicon ingots into wafers with wire saws, mechanical damage is left on the silicon surface. This saw damage is more pronounced than the damage produced by silicon slurry blasting. Consequently, the saw damage is expected to induce micro-cracks and or a strained layer in the silicon crystal lattice. It would be interesting to determine the possibility of utilizing the defects created by saw damage as gettering sources in the silicon crystal.

Cost reduction is a primary concern in the production of silicon solar cells. Ingot preparation and the slicing process constitute more than 60% of the total production costs. Lowering wafer production costs would reduce the production cost of each solar cell. This could be achieved by using cheaper raw materials which contain higher concentrations of metal impurities, but such impurities are known to reduce the conversion efficiency of solar cells. Therefore, the use of gettering technology is needed in order to keep the conversion efficiency at a level that is commercially viable.

A conversion efficiency of around 15% is acceptable. In solar cell fabrication, the gettering of metal impurities with aluminum and phosphorus has been investigated. Compared to these two techniques, extrinsic gettering with defects on both sides of the silicon wafers seems much simpler and more cost-effective.

In summary, recent studies show:

- Saw damage induces dislocation loops on the silicon surface during the oxidation process.
- Small amounts of mechanical damage produced by light scratching with sandpaper or silica wet blasting induces stacking faults during oxidation of the silicon crystal. Severe mechanical damage produced

by heavy scratching with sandpaper or by the wire sawing process induces dislocation loops in the silicon crystal during oxidation treatment.
- There must be some critical value of mechanical damage above which dislocation loops are generated during oxidation. Stacking faults are generated if the mechanical damage is below this value.
- In the silicon crystal, it is considered that OSFs are generated due to the silicon self-interstitial collapse mechanism.

3.2.1.3 Electrical Discharge Machining (EDM)

Monocrystalline silicon is now the most commonly used substrate material for the production of microelectronic components worldwide. Conventional inner diameter (ID) saws and wire saws are the two main types of equipment used for slicing silicon ingots. They use contact force to slice the ingots. This can easily result in harmful surface micro-cracks or other damage to the wafer of a mechanical abrasive nature. Chonan et al. analyzed the stress caused by silicon-wafer slicing, and found that it arose from the tension of wire, spinning, and loading from the work piece. The vibration and deflection of a silicon-wafer slicer has been further studied and simulated but the improvement in performance is limited. Additionally, an ID saw has a lower throughput when it is used for slicing large ingots.

Wire saw technology is now replacing ID saw slicing technology, but the process has problems relating to dirty work environments, low efficiency and instability from the wire system. Thinner wafers are more profitable because of the high material cost. At the same time, wafer size becomes larger to accommodate more chips. These factors also make the slicing process an important topic to be re-examined and reinvestigated.

Improvements can either be achieved by adapting existing methods, or introducing brand new procedures, or indeed by using modifications of other existing technologies. Wire electrical discharge machining (WEDM) technology is a potential new candidate for silicon slicing, which can be compared with existing procedures to see if it can eventually substitute for the ID saw, or even compete with wire saws in some manufacturing orders (Figure 3.9).

In 1995 Luo et al. used rapid-wire WEDM to slice heavy-doped silicon ingots. Thin wafers with little geometrical error were produced with this process. High contact resistance on the ingot surface is solved by nickel-plating. The surface roughness, cutting efficiency and micro-structures were

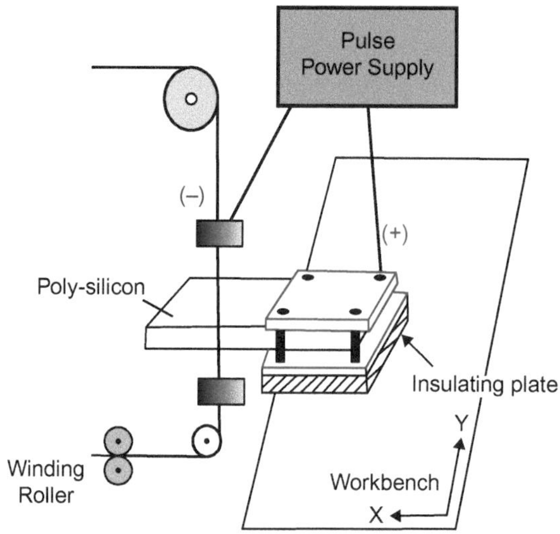

Figure 3.9 Schematic diagram of WEDM

observed under different energy intensities, but the effects of this procedure were not well demonstrated in the study.

WEDM is now a well-developed technology. Surface roughness and stability problems are, in general, solved for most surface slicing rates. Water immersed and water flushing are two main types of machine used worldwide, and they have been proved to have better manufacturing efficiency and quality than those of the rapid-wire type. If WEDM technology is to be utilized in the silicon slicing process, both these kinds of machines should be selected to conduct the relevant experiments. The former types use deionized water as the dielectric fluid and their wires are not used repeatedly, while the rapid-wire type uses kerosene as the dielectric fluid and the wire is used repeatedly. Since they have different power source characteristics and hardware configurations, different problems and phenomena may be encountered in their use, hence different procedures will need to be adopted for manufacturing processes. Sliced wafers will need to be produced under different operating conditions, such as current on time, current off time, water pressure and wire tension. The results will then be compared with the wire saw and ID saw for kerf ratio, average cutting speed and surface roughness.

The thickness of defects which need to be polished can be predicted from a SEM photograph of the cross-section of the sliced wafer. The kerf is related to the wire diameter and the discharge energy level. It will have a depth of 0.289 mm or so if a 0.25 mm copper wire is used with large current on time, reducing to approx. 0.22 mm with a 0.2 mm molybdenum wire. The total material waste rate can be estimated from the wafer thickness, kerf and the predicted thickness of any surface defects. On average, this is less than 28%. Due to some fundamental differences between these three machining methods, not all the conditions are assumed to be even. At least, the non-contact slicing scheme does reduce the geometric errors of the wafer. If the total batch production rate is to compete with that of a wire saw system, WEDM will need a multi-wire system of three or five parallel wires, with separate discharge power control.

It has been verified that doped ingots can be sliced by the WEDM method. If the discharge parameters match the material characteristics, wire breakage can be avoided and the slicing efficiency is acceptable if the throughput is small. The surface roughness is reasonable and geometric errors are quite satisfactory. Generally speaking, this method can be a substitute for current wire or ID saws in certain situations, but to date it is not suitable for pure silicon ingots.

Yu et al. studied the influences of WEDM parameters on machining poly-silicon and concluded that:

- Increasing pulse-on time and open circuit voltage has the greatest influence on cutting speed. Increasing both also increases energy supply and the material removal rate. However, such increases raise the peak current, further increasing machining groove width and worsening surface roughness.
- Raising the wire speed and flushing rate improves the cutting speed, machining groove width and surface roughness. However, as the flushing rate reaches a certain level, its improving effect on the machining rate disappears.
- Strengthening the tension significantly improves the machining groove width because wire electrode vibration decreases.
- WEDM gives a particular geometry to poly-silicon material.

3.2.1.4 Wire Electrolytic-Spark Hybrid Machining

Research over the last decade has merely verified that low resistance silicon ingots can be sliced by the WEDM method. The unique advantages of a long cutting edge (maximal cutting edge > 500 mm), a small cutting kerf (minimal kerf < 0.05 mm), a small cutting taper and a homogeneous surface for producing thinner and larger scale wafers have not been sufficiently explored, so this research and development has not made the transition from the laboratory into production processes in the photovoltaic and microelectronics industries.

Water immersed and water flushing types are the two main categories of low speed (LS)-WEDM used worldwide. They have been proven to have better manufacturing efficiency and quality, but a relatively high price compared with high speed (HS)-WEDM. Of these alternatives, the former uses deionized water as the dielectric fluid and the wire electrode is not used repeatedly, but the latter uses water, emulsifier and oil as the dielectric fluid and the wire electrode is used repeatedly. The discharge process and material removal mechanisms are quite different, especially when slicing silicon ingots.

Some improvements have been made to the HS-WEDM machine. Firstly, the total thickness variation (TTV) and profile bending rate, defined as warp, are extremely important in wafer slicing. To control these factors, the wire tension controller and wire guide are mounted on the wire winding system, which avoids the cyclical stripes caused by the reversing motion of the traveling wire on the wafer surface, as shown in Figure 3.10.

The dielectric fluid normally used in HS-WEDM is a mixture of water, emulsifier and oil. These produce debris, gas, heat and oil in the spark gap, which can be degraded and react to form a colloidal substance, which is adhesive and aggregates in the gap. When this occurs, the debris evacuation capacity is decreased, and fresh dielectric fluid cannot reach the gap, which seriously influences the machining rate, surface quality and wire breaking. To overcome this problem, a hybrid electrolyte with

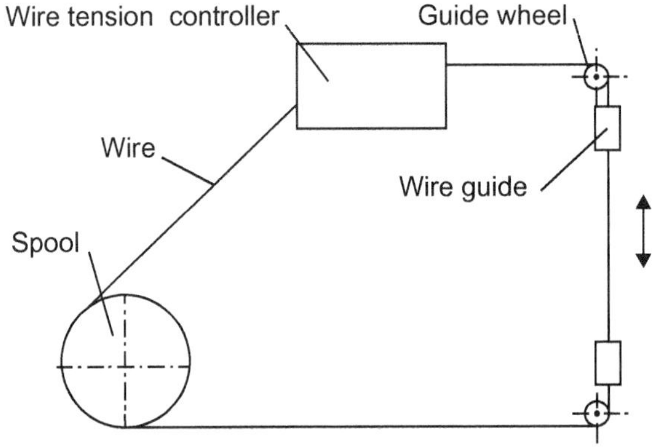

Figure 3.10 Improved wire winding system

superb detergent capability has been developed for electrolytic-spark machining (Liu, 2006). In comparison with wire electrolytic-spark hybrid machining, the same silicon material can be sliced by normal WEDM (using deionized water), HS-WEDM (using water, emulsifier and oil) and multi-wire saw methods, respectively.

Wire electrolytic-spark hybrid machining, which combines electric discharge and anodic etching into a single process, has been investigated experimentally. Some improvements, such as a new wire winding system, hybrid electrolyte and high efficiency pulse generator, have been adopted in HS-WEDM machines. Based on the experimental results discussed above, the following conclusions can be drawn:

- Due to the relatively high conductivity of the hybrid electrolyte, chemical etching occurs simultaneously during the electrical discharge process.
- The cutting rate in wire electrolytic-spark hybrid machining is currently greater than 600 mm^2/min, and wire consumption is quite low.
- Microscopic examination shows that many micron- and submicron-sized holes with high aspect ratio and smooth internal profile are located in the craters, which may be introduced by high temperature electrolytic reaction. Micro-cracks are evidently diminished and surface texture is quite even giving a dark color. Meanwhile, wire material residue is not observed by energy dispersive spectroscopy (EDS).
- On the basis of non-contact electrolytic-spark erosion, the wafer thickness and kerf could be reduced with further research, meaning the technique is favorable for slicing large-scale, thin wafers.

It has been demonstrated that wire electrolytic-spark hybrid machining has the capability to slice and texture silicon wafers simultaneously, especially the low resistance silicon used in the photovoltaic industry. This is a novel approach to cost reduction.

3.2.2 Beveling

Because single crystal silicon is very hard and brittle, it breaks along a given crystal axis direction; hence chipping occurs easily on the wafer edges during processing of the device. Broken silicon is a cause of low device quality and poor yield. The beveling process is the chamfering of the wafer periphery in order to avoid chipping (Figure 3.11), and to prevent abnormal growth of silicon at the wafer periphery (crown) by epitaxial growth. Beveled profile geometries are generally classified into two categories: Metal Oxide Semiconductor (MOS) and Bipolar Device (BIP) (Figure 3.12).

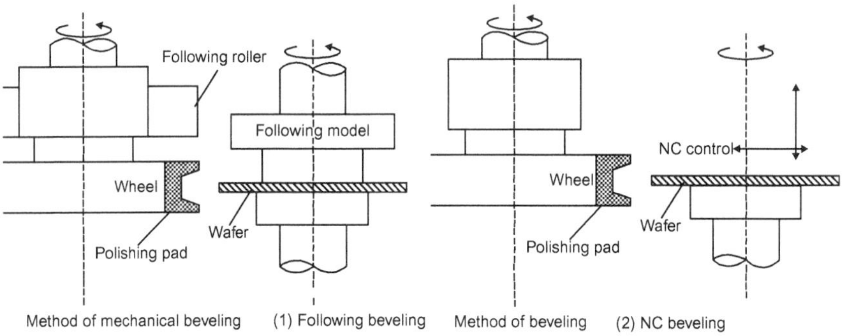

Figure 3.11 Method of mechanical beveling

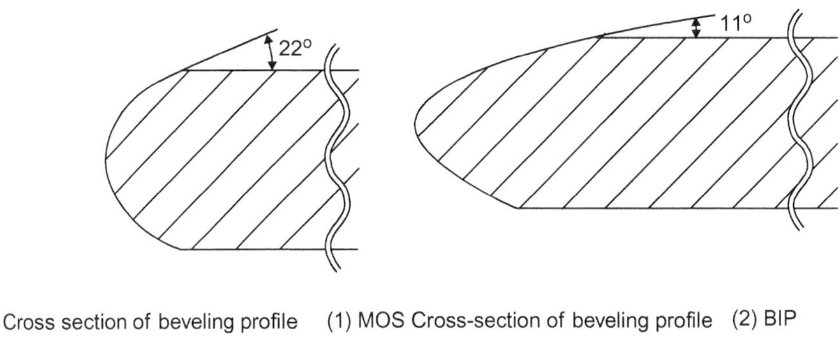

Figure 3.12 Cross-section of beveling profile; (1) MOS and (2) BIP

The surface geometries required for MOS and BIP are different. MOS is used to prevent chipping or breakdown during device processing, while BIP is used to avoid crown in epitaxial growth. Both are chamfering processes, usually using a diamond wheel (#800–#1500). Leaving the layers with varied pressure created by machining with a diamond wheel as they are causes problems, so they are removed by an etching process.

In recent years, when dealing with a device with higher integration, the demand to reduce dust from a chamfered part during device processing has increased, so not only is the geometry of a beveled surface important, but also reduction of chamfered roughness. Simple chamfering was done in the past, but now mirror chamfering, shown in Figure 3.13, is performed. As devices becomes highly integrated, and design rooms become smaller, the effect particles have on device yield increases. The silicon wafer device process requires dust prevention measures, which mirror chamfering provides. Currently, however, such processes are used exclusively for highly integrated wafers that have a required surface finish of 200 nm.

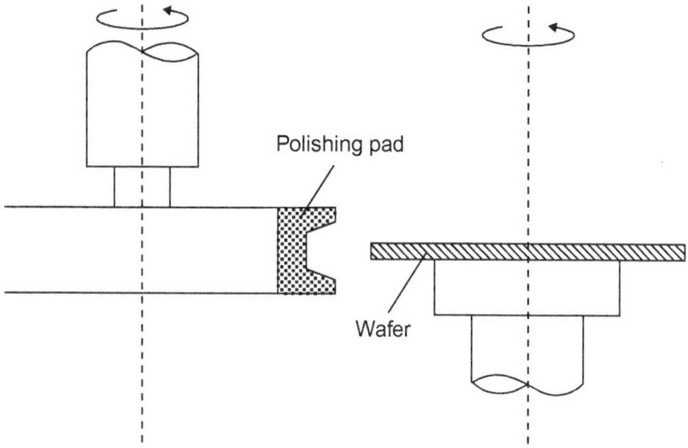

Figure 3.13 Edge mirror polishing

3.3 LAPPING OF SILICON

3.3.1 Introduction

Lapping can be defined as an abrasive machining process used to obtain materials with high dimensional accuracy and excellent surface finish. Processes where the surface characteristics get the maximum attention are called ultra-precision machining operations. They alter surface attributes such as roughness, waviness, flatness, roundness etc. Typical examples of such processes are grinding, honing or lapping. Lapping has never been used to the same extent as grinding or other precision machining operations, probably because machinists prefer to use fixed-abrasive tools and cutters rather than abrasive slurry. They view lapping as a messy and inexact process. To some, lapping is more of an art than a machining method. However, modern lapping machines can produce exceptionally flat surfaces with a flatness of less than one light band (helium light, wavelength 23.2 millionths of an inch).

The process can also be defined as a work process in which the work piece and tool slide over one another on a loosely applied medium (lapping compound) and are subjected to a rotation that continually changes direction. Though lapping tends to decrease original surface roughness, its main purpose is to remove material and modify shape. It is used less for finish and more for form accuracy, such as flatness in the case of flat objects, or sphericity in the case of balls.

The characteristics of the lapping process are as follows:

- Material removal rate is low as compared to other abrasive processes, like grinding, due to low cutting speeds and shallow penetration of fine abrasive grains into the work surface.
- Lapping does not generate excess heat, hence it can be considered as a cool process. This means that there is no thermal damage to the work piece.
- Three basic forms can be machined by lapping, namely flat, cylindrical and spherical, although sometimes it can be used for shapes such as screw threads.
- Since relatively low forces are exerted on the work piece, lapping can be used quite effectively for fragile and brittle materials.
- Excellent accuracy can be obtained for flat and cylindrical surfaces.
- In addition to producing flat surfaces, lapping also produces functionally needed surface characteristics such as random lay pattern, adaptability to microscopic examination at very high magnifications etc.

3.3.1.1 Brief History

The beginning of the lapping process can be traced to the Stone Age. According to A.W. Stahli, prehistoric man first used lapping to make tools for hunting and farming. A primitive lapping machine and schematic of the lapping process are illustrated in Figures 3.14 and 3.15, respectively.

A primitive lapping machine consisted of a lap rotated by a weighted stick. Sand was put on the stone lap through a hole. The sketch in Figure 3.14 was taken from a model at the German Museum, Munich and is based on archaeological findings. Lapping was used for polishing and finishing optical lenses in the late sixteenth century, such as were used in Galileo's telescope. With the emergence of precision optical instruments in the late nineteenth century, such as binoculars and range finders, powered lapping machines were used extensively in optical shops throughout Europe and the USA. Since there was no significant requirement for producing quantities of metallic components with excellent finishes and tolerances, lapping was not used in the metal working industry. The situation started to change around World War One. The introduction of the mechanical shaft seal gave the greatest impetus to lapping technology. These are used in pumps, compressors and turbines to prevent leakage in the rotating components. Such seals need excellent accuracy, hence manufacturing them by hand lapping was an arduous task. This created a

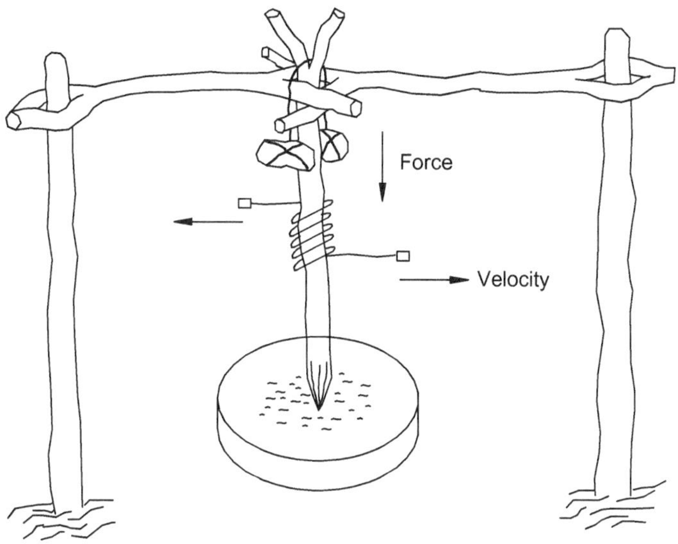

Figure 3.14 Primitive lapping machine

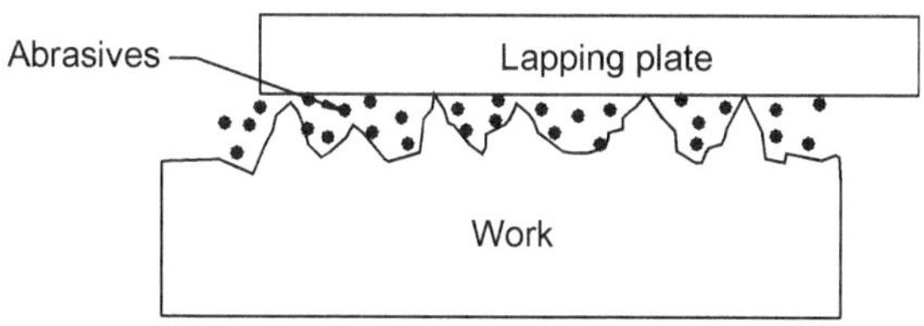

Figure 3.15 Schematic of lapping process

necessity for motorized lapping machines for delivering fast and consistent results.

3.3.2 Lapping Mechanism

The fundamental principle of lapping consists of transferring the flatness of the lapping plate to the work piece. This is achieved when abrasive grains in a liquid vehicle called a slurry are guided across a work piece by means of a rotating tool or plate. The tool rotates in the same direction as the work piece, and the slurry is fed into the rotating plate to form an abrasive film between the plate and work piece or group of work pieces. The work pieces are held in position by control rings. As the plate and rings rotate, force is applied by using a suitable weight. The work pieces are then abraded to the required thickness.

The abrasion mechanism that takes place in ductile or brittle machining can be explained as follows. Suppose that we need to obtain a flat and smooth surface on a work piece. Consider that we have an exactly flat surface that can be used as the lapping plate. The flatness of the lapping plate has to be transferred to the work piece. The hardness of the work piece is greater than that of the lapping plate, and the hardness of the abrasive grains are greater than that of the work piece. Let us consider a single grain between the lapping plate and the work piece. When we apply a force to the work piece, a type of mutual hardness tester, as in a hardness test, is produced, with the grain as the indenter. The work piece and the lapping plate act as the test pieces. Due to the applied force, the grain penetrates to a certain depth in both the work piece and the lapping plate. Now, if we slide the work piece relative to the lapping plate, three abrasive cutting mechanisms can occur simultaneously; that is rolling, sliding and embedding (charged plate). These mechanisms are shown in Figures 3.16–3.18.

As the work piece moves relative to the lapping plate, the abrasive grain is free to roll between them. If the penetration is deep enough, the advancing edge of the grain will dislodge a chip from the work piece or the lapping plate – or both – by shearing. There will also be a pit with a raised edge left behind as a result of plastic deformation. This can be explained by shear theory, in which, when the load of the indenter is large enough, the shear stress is bigger than the sum of shear strength and internal friction force, and the surrounding material gets sheared off, and a conical hole is formed. This explains material removal by rolling.

If the grains are rectangular, with a width larger than their thickness, the probability of sliding increases. In a sliding motion, the wear process is a two-body abrasion for a short period of time. In abrasion by sliding there would be only one indentation on the lapping plate surface, and

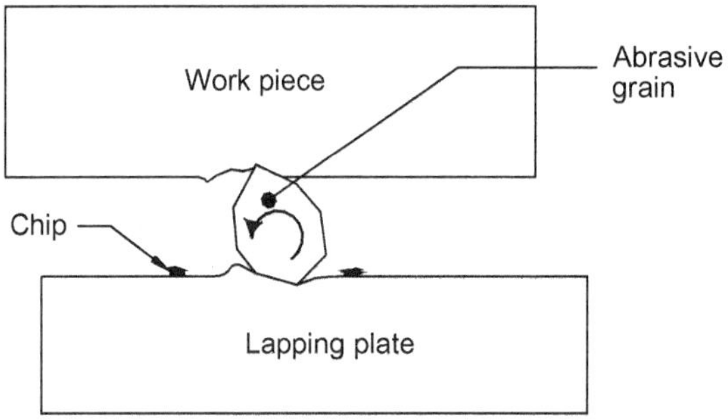

Figure 3.16 Abrasion by rolling

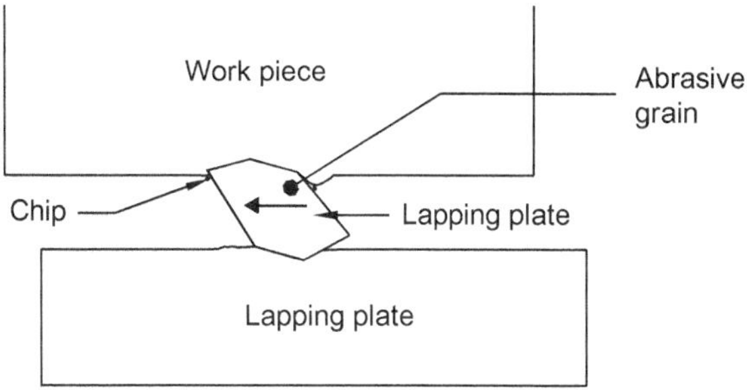

Figure 3.17 Abrasion by sliding

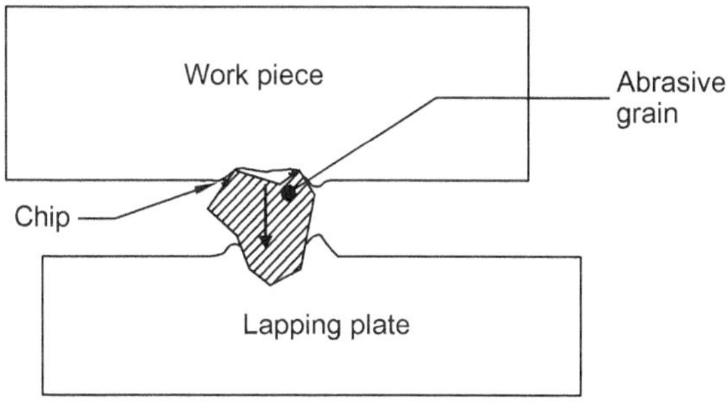

Figure 3.18 Abrasion by embedding

there would be a groove on the work piece surface. Rolling and sliding are both three-body abrasions, in which the material removal mechanism and the movement of the particle depends upon the particle morphology, particle size, applied load, surface hardness and the hardness ratio of the two mating surfaces.

Since the hardness of the grain is higher than that of the plate and the work piece, some grains get embedded in the plate. These embedded grains remove material from the work piece by two-body abrasion. In two-body abrasion hard particles get embedded into a softer surface and plow out material, yielding distinctive marks. These are continuous grooves, which may or may not result in chip formation depending upon the effective rake angle and depth of penetration.

Experiments have shown that in lapping, rolling comprises 45% of the action, sliding 15% and the balance of 40% is embedding.

3.3.3 Mathematical Model

The lapping process appears to be very simple but its mechanism is actually very complicated. It is necessary to find an equation that defines lapping conditions and stock removal rate. The complexity of the lapping mechanism makes this difficult.

3.3.3.1 Ductile and Brittle Regime

The mechanism of lapping can be classified into two separate theories depending upon the hardness of the work piece. The ductile regime is where material is removed by plastic deformation, as is commonly observed in metallic materials. The brittle regime is the machining state where the material removal takes place by the formation and propagation of cracks.

Moore and King reported that the dominance of either state depends on the material hardness (H_w) and the critical indentation depth (d_{cri}). The latter can be defined as the minimum depth required to cause fracture. Lawn showed that:

$$d_{cri} = \left(\frac{E_w}{H_w}\right)\left(\frac{K_t}{H_w}\right)^2 \tag{3.1}$$

where E_w = elastic modulus and K_t = material fracture toughness.

According to Williams and Hyncica there is a "critical size ratio" for the characteristic particle size to film thickness (h_f) which divides it into the two mechanisms. The longest diagonal of the abrasive cross-section D_1 is used as the characteristic particle size. It was seen that when this ratio is low, the worn surface was similar to that produced by three-body abrasion. This was because the film thickness was high, and so the abrasive particles could easily roll, slide or get embedded into the lapping plate. When the ratio was high, i.e. when the film thickness was low, the abrasive grains underwent more embedding than rolling and sliding. Hence the surface produced was similar to that produced due to two-body abrasion.

$$\frac{D_1}{h_f} = \sec\beta \tag{3.2}$$

Through experiments, the critical size ratio was found to be 2.

Another important factor in the cutting mechanism of lapping is the ratio of the work piece H_w and H_p. Assuming that the abrasive does not fracture, the indentation depths on the work piece and the plate will depend on their individual hardness; the hardness ratio can be stated as:

$$R_H = H_w/H_p \tag{3.3}$$

3.3.3.2 Calculation of Material Removal Rate

According to the two criteria discussed above, the machining mechanisms in lapping can be classified into four states, namely:

1. Two-body ductile machining.
2. Two-body brittle machining.
3. Three-body ductile machining.
4. Three-body brittle machining.

Several studies have been conducted into these mechanisms, in order to find an equation for the material removal rate. Change in the process characteristics with time was excluded from these studies.

The material removal rate in the case of two-body abrasion for a ductile material is greatly influenced by the size of the plastic deformation zones. The equation can be written as follows:

$$\frac{d(Q_W, 3, \mathrm{d})}{d_t} = \frac{2\pi}{3} d_w R_t^2 \frac{2V_a}{\pi D_m} \tag{3.4}$$

where:

Q_w = volume wear in mm^3

W = width of the groove

d_w = depth of the abrasive penetration

In the three body ductile mode, the abrasives roll on the surface making indentations which are not deep enough to cause fracture. Due to indentation, some of the material gets detached from the surface of the work piece.

Crack area is the key element for calculating material removal rate under a brittle regime. The crack dimensions are shown in Figure 3.19 below, where:

a = radius of indentation mark

b = radius of plastic zone

c = lateral crack length

h = depth of lateral crack = b (approximately)

Ψ = half indenter tip angle

The length of the crack can be calculated as follows;

$$c = K_1(cot\psi)^{5/12} \frac{E_w^{3/8}}{K_c^{1/2} H_w^{1/2}} P_i^{5/8} \tag{3.5}$$

where K is a constant that depends on the number of radial and lateral cracks.

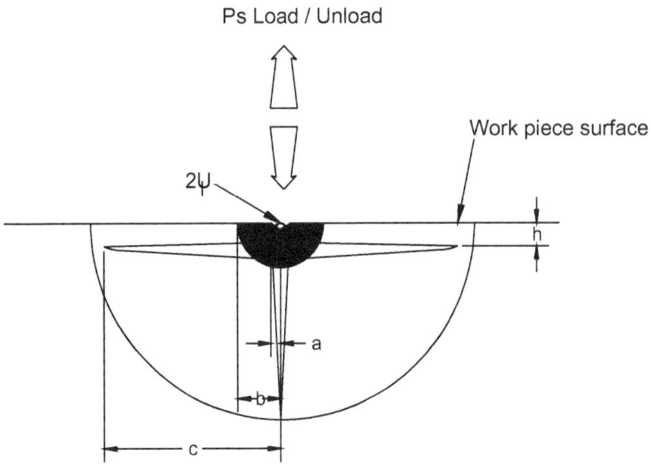

Figure 3.19 Median and lateral cracks in an indenter

For two-body abrasion in the brittle regime, the material removal rate is related to the groove length and, due to fracture, the depth and length of the lateral crack.

In reality the material removal rate does not take place by a single mechanism, but by a combination of all of them. This is caused by the non-uniform depth of penetration by the abrasive particles due to their size distribution. The total number of abrasive particles associated with the particular distribution is determined by the slurry abrasive concentration. To be more precise, when calculating the material removal rate, the abrasive particle size distribution and slurry concentration also have to be considered.

3.3.4 Kinematics

The relationship between lapping conditions and material removal rate (MRR) is obtained by making the following assumptions:

1. The load on each work piece is evenly distributed, contact between lap and work pieces is uniform and all work pieces are finished uniformly.
2. The material removal rate per unit lapping distance is the same, and is independent of the work piece velocity when lapping.
3. The material removal rate is assumed equal without reference to lapping condition.
4. The effect of pressure on the material removal rate is independent of work piece size.

However when dilute compound is used, abrasion by the pointed edges of the abrasive would be rapid and the MRR would differ for small and large work pieces.

The time necessary to finish a definite dimension is calculated under the assumptions mentioned above and these new conditions:

1. The abrasive to vehicle ratio is 1:1 by weight. The lapping vehicle used is grease.
2. All loads are supported by the abrasive and no metallic contact between lap and work piece occurs.
3. The work piece is quenched steel and its Vickers hardness is 600 to 800.
4. The MRR with lapping pressure of 2 kg/cm^2 is taken as standard. The effect of pressure on MRR is calculated on the assumption that the MRR is proportional to pressure.
5. White fused alumina is used as the abrasive.

3.3.4.1 Sliding Velocity

The sliding velocity on each point of a work piece is different. The MRR is dependent on, and sometimes proportional to, the sliding velocity of the work piece on a lap. Hence it is necessary to calculate this velocity. The motion of the work piece in most cases is divided into three components as shown in Figure 3.20. These components are:

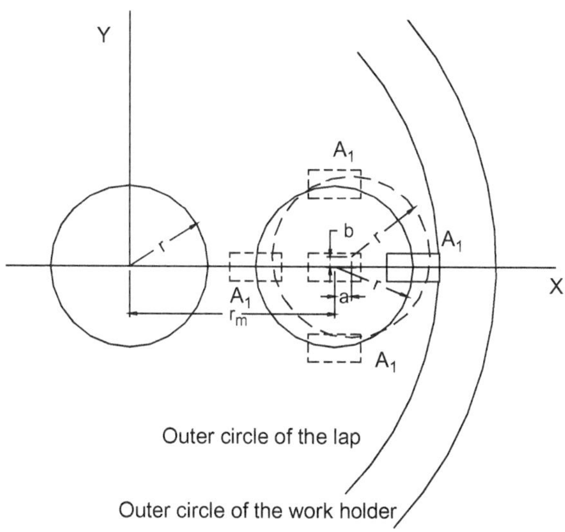

Figure 3.20 Rotation and revolution of the work holder

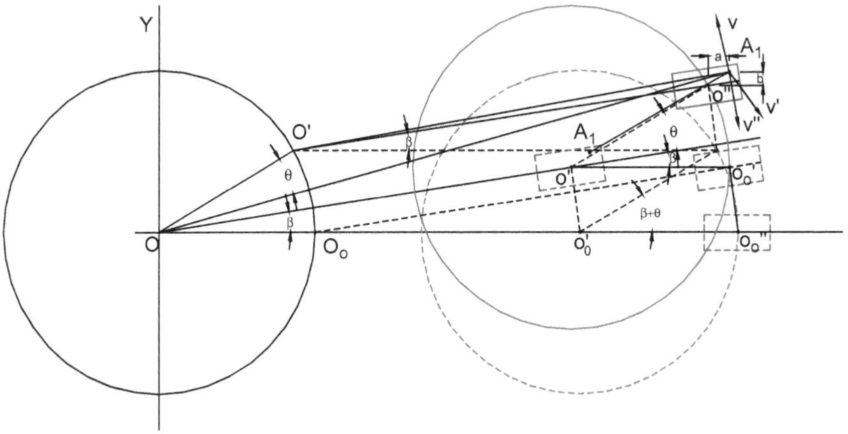

Figure 3.21 Sliding velocity on each point of work piece relative to the lap

- Rotation of a lap. O is the center of the rotation and the speed of rotation is N rpm.
- Revolution of a work holder around an axis of a lap. The speed of revolution is n rpm.
- Rotation of the work holder around the axis of the holder itself. The speed of rotation is N′ rpm.

Anticlockwise rotation and revolution is taken as positive. The revolution of the work holder can be explained by Figure 3.21. A revolution in this context means that if the work holder on a circle O whose radius is r, without changing its relative position in reference to XOY coordinates, which are fixed in space. The work piece revolves in circles having the same radius r, keeping the axes O″x and O″y parallel to OX and OY respectively, where O″x and O″y are the coordinates fixed to the work piece.

Rotation of the work holder around its axis of the holder indicates its rotation about the coordinates XOY and is not the rotation about the axis OO, which is the axis of the work holder itself.

As shown in Figure 3.21, the center of the lap is O and the coordinate system XOY is fixed in space and is independent of movements of either lap or work pieces. A voluntary position of a work piece is indicated. When the work piece rotates about the axis of the work holder by an angle β3, the position of the work piece is O1″.

The velocity of an arbitrary point A is calculated as the vector sum of the following three velocity components:

1. The sliding velocity due to rotation of the lap. This component is perpendicular to OA, and in the same direction to the rotation of the lap.

$$v = 2\pi \cdot OA \cdot N = 2\pi \mathrm{RN} \tag{3.6}$$

2. The sliding velocity due to revolution of the work holder. This component is perpendicular to A′A and in the opposite direction to the revolution.

$$v' = 2\pi \cdot A'A \cdot n = 2\pi \mathrm{rn} \tag{3.7}$$

3. The sliding velocity due to rotation of the work holder. This component is perpendicular to CA and in the reverse direction to the rotation.

$$v'' = 2\pi \cdot CA \cdot N' = 2\pi R'N' \tag{3.8}$$

The sum of the three components and its X, Y components are:

$$\begin{aligned} V &= (V_x^2 + V_y^2) \\ &= 2\pi[(r_m + a^2 + b^2)(N - N')^2 + 2r\{(r_m + a)\ cos\ \theta + b\ sin\ \theta\} \\ &\quad \times (N - N')(N - n) + r^2(N - n)^2]^{1/2} \end{aligned} \tag{3.9}$$

It is found that V is independent of the angle β. The sliding velocity on each point of a work piece is equal when:

$$N = N' \tag{3.10}$$

and the equal sliding velocity is:

$$V_{N=N'} = 2\pi V - n \tag{3.11}$$

In the other conditions, the sliding velocity on each point of work piece is different and dependent of the work piece relative to the lap.

3.3.4.2 Equation for Streaks on a Work Piece

Streaks are striations on the lap or work piece produced during the process of lapping. When a work piece does not move, the streaks on it form circles. An example is indicated as A in Figure 3.22. When the lap rotates by an angle α, the center of the work piece rotates by an angle (n/N) around O′ and (N′/N) around C as shown in the figure. As a result of these rotations, A, C and O″ change their positions and are displaced to A_o, C′ and O_o' respectively. Therefore, when a coordinate system $xO_o'y$ is fixed on a work piece, an equation for streaks under this coordinate system can be calculated by the following procedure.

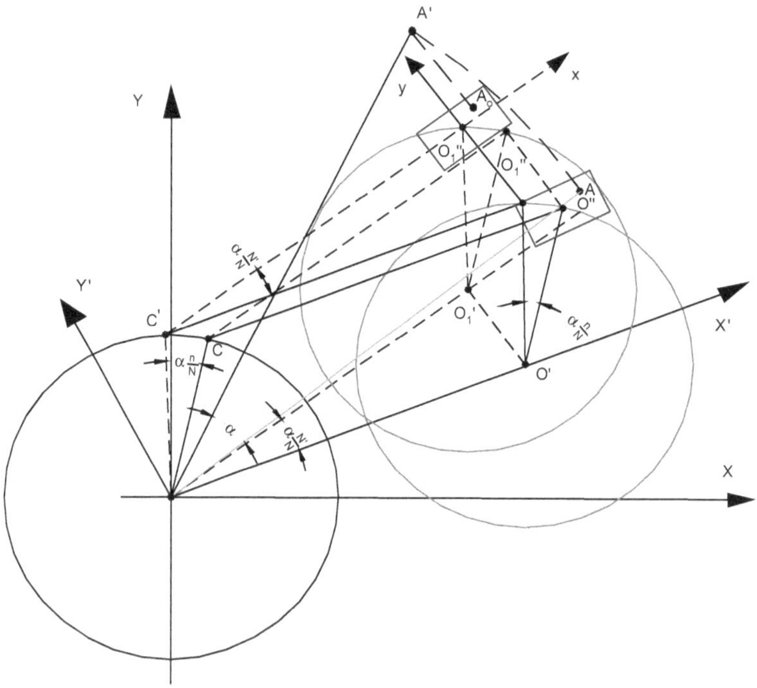

Figure 3.22 Streaks on a work piece

1. Calculate the coordinates of A′ about X′OY′ system.
2. Move the origin to O_o'.
3. Rotate the coordinate axes by α (N′/N)

   The x and y coordinates are obtained as follows:

$$x = r_m + r\cos\theta + a\cos\left\{\alpha\left(1 - \frac{N'}{N}\right)\right\}$$
$$-(r\sin\theta + b)\sin\left\{\alpha\left(1 - \frac{N'}{N}\right)\right\} - r\cos\left(\theta + a\frac{n}{N} + a\frac{N'}{N}\right) - r_m$$
$$y = r_m + r\cos\theta + a\sin\left\{\alpha\left(1 - \frac{N'}{N}\right)\right\}$$
$$-(r\sin\theta + b)\cos\left\{\alpha\left(1 - \frac{N'}{N}\right)\right\} - r\sin\left(\theta + a\frac{n}{N} + a\frac{N'}{N}\right) \tag{3.12}$$

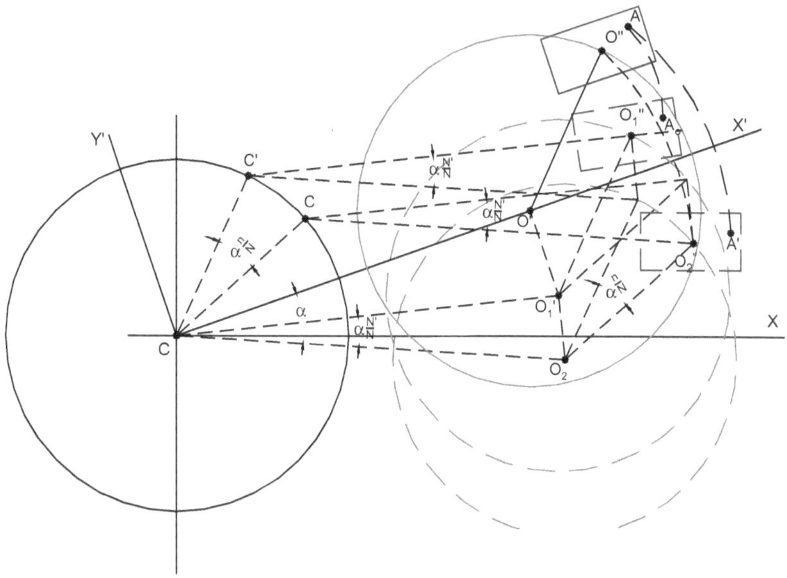

Figure 3.23 Streaks on a lap piece

3.3.4.3 Equations for Streaks on a Lap

When the work piece does not move, the streaks on the lap can be represented by the circle AA′ as shown in Figure 3.23. Point A′ moves to A_o, when the lap rotates by an angle α.

$$X' = (r_m + a)\cos\left\{\alpha\left(1 - \frac{N'}{N}\right)\right\} + b\sin\left\{\alpha\left(1 - \frac{N'}{N}\right)\right\} + r\cos\left\{\theta - \alpha\left(1 - \frac{n}{N}\right)\right\} \tag{3.13}$$

Hence the equations for AA_o give the equation of streaks on the lap. The equations are calculated in reference to a coordinate system X′OY′ fixed to the lap.

$$Y' = -(r_m + a)\sin\left\{\alpha\left(1 - \frac{N'}{N}\right)\right\} + b\sin\left\{\alpha\left(1 - \frac{N'}{N}\right)\right\} + r\sin\left\{\theta - \alpha\left(1 - \frac{n}{N}\right)\right\} \tag{3.14}$$

The kinematics of lapping machines can be calculated by using equations (3.9), (3.12) and (3.14).

3.3.4.4 Effects of Different Parameters on Material Removal Rate

Effect of Lapping Pressure on the MRR

It was originally believed that the material removal rate (Q) was proportional to the pressure on work pieces. Dr. Okoshi proved that this was wrong and showed that this rate increased with increasing pressure in the low pressure range, decreased with increasing pressure in the medium pressure range and finally increased with increasing pressure (Figure 3.24). Dr. Okoshi used lapping compounds which contained small amounts of abrasive (1% by weight).

Dr. K. Sato, Dr. H. Tsuwa and Matsunaga did further research in this field. They found out that the stock removal rate was practically constant under ordinary conditions, namely if:

- Large abrasive (over 5 μm) is used.
- Abrasive to vehicle ratio is between 3:7 and 5:5.
- Compound is continuously fed between the lap and work piece during lapping operation.

When the compound was fed discontinuously, or when the abrasive to vehicle ratio was very low, the maximum and minimum points appeared on the stock removal rate/pressure curve as shown in the graph.

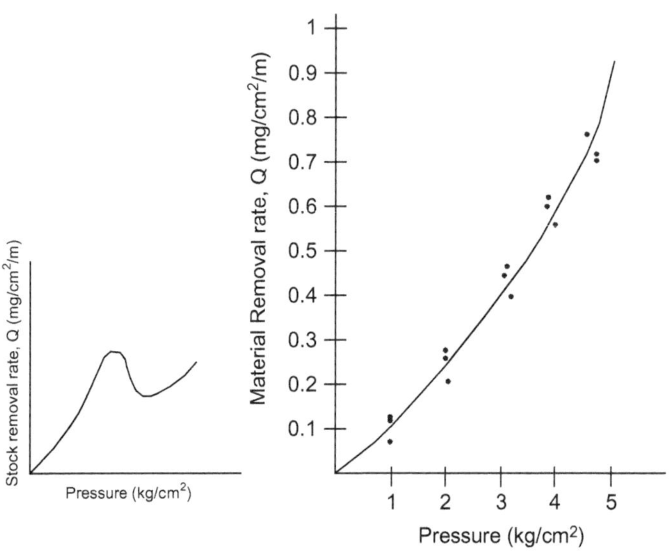

Figure 3.24 Effect of lapping pressure on the MRR

These experiments showed that the stock removal rate was nearly proportional to pressure under ordinary conditions. However, the curve is not straight.

Effect of Abrasive Size on the MRR

There are two theories which differ regarding the relationship between abrasive size and stock removal rate. Experiments conducted by Dr. Tsuwa and Dr. Shigemitsu showed that Q is not significantly affected by using coarser abrasives. This theory is supported by experiments conducted by Dr. Matsunaga.

The other theory states that the stock removal rate increases significantly with an increase in abrasive grain size. Dr. Sato showed that the stock removal rate decreased greatly with decreasing grain size.

Dr. Matsunaga obtained some interesting results when he conducted the lapping process by feeding the compound continuously. The abrasive used was white alumina, the vehicle used was grease, the abrasive to grease ratio used was 1:1 and a pressure of 2 kg/cm^2 was applied. The graph obtained is shown in Figure 3.25.

The graph shows that the stock removal rate increases with the increase in abrasive grain size. However, the graph does not pass through the origin, but through a specific value on the Y-axis. Nearly the same material removal rate was obtained when abrasive grains of about 3000 mesh size were used. The stock removal rate was also nearly the same when chromium oxide was used.

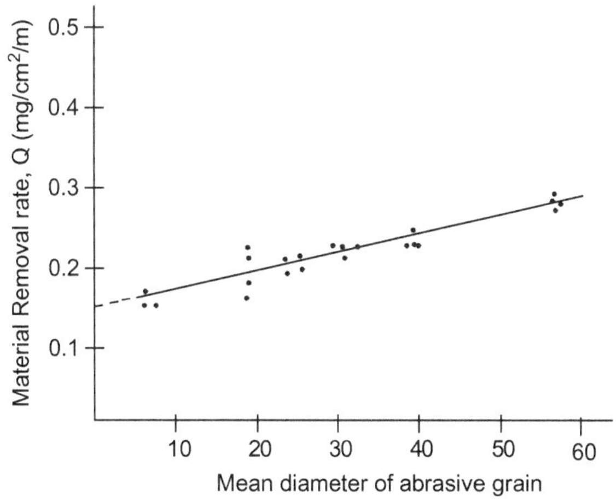

Figure 3.25 Effect of abrasive size on the MRR

3.3.5 Types of Lapping (Focus: Double-Sided Lapping)

The lapping process can be carried out in three different ways. The focus of my research is on double-sided lapping. Lapping can be one sided or double sided, depending upon the type of component machined, and cylindrical lapping is used for cylindrical surfaces. Lapping with bonded abrasives is a recent development in which the abrasives are first embedded in the lapping plate. This technique differs from conventional lapping in its principal characteristics, and closely resembles a grinding process.

The different types of lapping processes can be classified as:

- Single-sided lapping – the most common process used for lapping components requiring a single side flat surface.
- Double-sided lapping – used for lapping components in which consideration is given to parallelism, flatness and precise control of form.
- Cylindrical lapping – used for lapping of cylindrical components like shafts, bearing needles, rollers etc.
- Lapping with bonded abrasives – a recent development in lapping which can utilize the principle of ELID.

3.3.5.1 Double-Sided Lapping

Although work pieces can be lapped satisfactorily on a conventional single-sided lapping machine, the process becomes time consuming and tedious if both sides of the work piece need to be lapped. Lapping both sides of a work piece on a single-sided lapping machine needs the work piece to be turned and requires skilled workmanship. This is highly undesirable in mass production processes due to the time factor involved. Hence double-sided lapping machines were developed.

3.3.5.2 Mechanism

In double-sided lapping, the lapping abrading mechanism is applied to both sides of the work piece simultaneously. The work pieces are held in position by steel or plastic carriers, and sandwiched between two rotating plates. Usually, the plates rotate in opposite directions. The slurry is fed through holes in the top plate to the interface. Lapping takes place at the top and bottom faces of the work piece (Figure 3.26).

3.3.5.3 Machine Set-up

A typical machine set-up for double-sided lapping is shown in Figure 3.27. The work pieces are held in a carrier, which is kept flat on

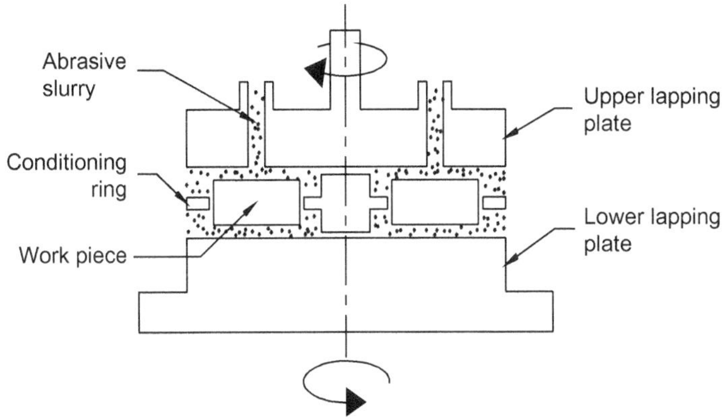

Figure 3.26 Schematic mechanism of double-sided lapping

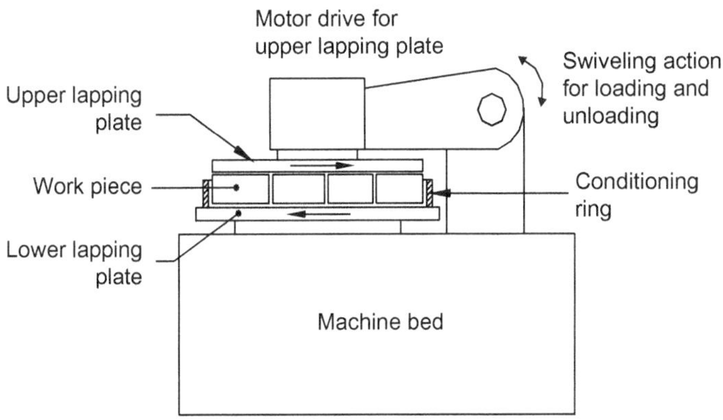

Figure 3.27 Schematic of double-sided lapping machine

the lower wheel. A center inner pin ring driving device rotates the carriers in an epicyclic motion. A stationary outer ring confines the epicyclic motion of the carrier. The carriers are always thinner than the work pieces and hold the surfaces of the work piece to be machined against the lap and abrasive in an ever-changing pattern. The lower plate is mounted at a fixed level while the upper plate can be lowered or raised by a hydraulic or pneumatic system. The lapping plates rotate in the opposite or the same direction. The speed of the cycle is automatically controlled by a time delay relay to ensure a gentle starting speed. With variable speed drive mechanisms, several different lapping speeds can be produced (Figure 3.28).

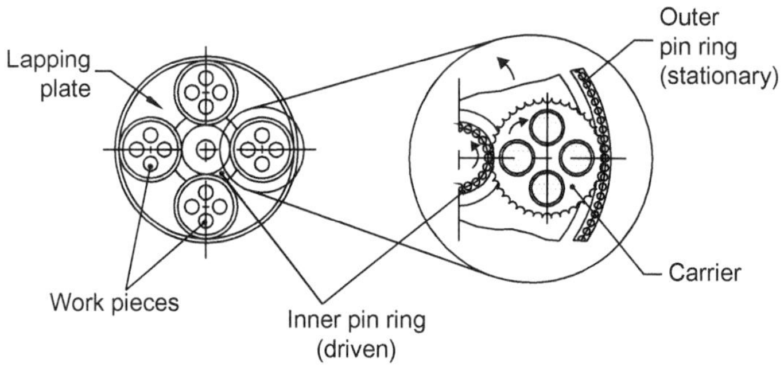

Figure 3.28 Relative motions of the carrier, lapping plate and the inner ring

3.3.5.4 Process Cycle

A typical process cycle for double-sided lapping follows this sequence:

- The top lapping plate is raised to the loading position and the work pieces are mounted in the carriers.
- The top lapping wheel is lowered gently to rest lightly on the top of the work pieces.
- The lower lapping plate, the upper lapping plate and the carriers start rotating slowly under automatic or manual control.

Abrasive is fed automatically through the top lapping plate during the cycle to provide the necessary abrasive action on both sides of the complete load of the work piece.

The pressure on the top wheel is adjusted such that it is initially light. As the high spots are machined away, the pressure increases gradually until the optimum pressure for the job is reached. The vertical travel of the top lapping plate is shown in Figure 3.29. This cycle continues until the size desired for the work piece is obtained.

At this point the top lapping wheel is lifted, and the finished work pieces are unloaded.

3.3.5.5 Advantages of Double-Sided Lapping

Double-sided flat lapping offers the following advantages:

1. Two sides of a work piece can be lapped in the time required to machine one side.
2. A large number of work pieces can be lapped simultaneously.
3. Magnetic as well as non-magnetic parts can be lapped.
4. It can machine any kind of stable material from plastic to diamond.

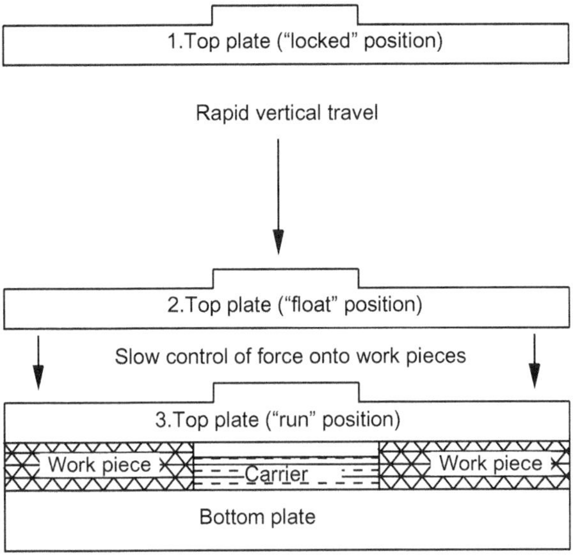

Figure 3.29 Vertical travel of top lapping plate

5. Double-sided lapping is the best available method to obtain close tolerances for flatness, parallelism and size.
6. The stock is removed from both sides of the work piece simultaneously, which helps to relieve internal stresses, thus achieving flatness.
7. No clamping force is required to hold the work pieces in position and simple holder design eliminates stresses in the work pieces, thus improving flatness, parallelism and size.
8. Heat generation is minimal; hence, the work piece is exposed to minimal stress and thermal damage.
9. The cutting rate is uniform and repeatable. No dulling of the abrasive takes place, as in the case of a single point turning tool or milling cutter, because fresh sharp abrasive particles are gradually fed to the lapping area during the processing cycle.
10. The operating costs are low because of less handling, higher efficiency, greater repeatability and the feasibility of combining machining operations.

3.3.5.6 Parameters of Double-Sided Lapping

The following process parameters are important for the double-sided lapping process:

- Variable pressure during lapping cycle.

To lap away taper, bow and size variation it is necessary to accurately control the lapping pressure. When a load of work pieces are placed in the machine and the top lapping plate is lowered to contact the top of the work pieces, the top of the lapping plate is actually contacting only a few high points. In the past, dead weight top wheels were used which produced excessive initial pressure, but not enough pressure during the process cycle to remove stock. This reduced the efficiency of the machine. Today's lapping machines can start the cycle with an appropriate amount of pressure to wear out the initial high points. Once the high points are removed, the pressure is automatically increased to an optimum and maintained at that level for sufficient time to complete the lapping. When the work pieces are machined almost to the finish size, a means to automatically reduce the pressure should be incorporated in order to provide a light cutting action and to allow sufficient time to remove any imperfections caused by higher pressure. Consistent and dependable results are obtained on today's double-sided lapping machines by controlling the pressure throughout the cycle.

- Abrasive slurry flow rate.

For high accuracy in size, flatness, surface finish, parallelism and uniform production, a continuous and even flow of abrasive slurry must be maintained. The abrasive slurry must be consistent in its concentration and the flow rate must be uniform in order to maintain the desired uniform cutting action.

3.3.5.7 Lapping with Bonded Abrasives

This kind of lapping lacks one of the principal characteristics of conventional lapping. It replaces the loose abrasives with a fixed abrasive lapping plate, and replaces the slurry with a water-based fluid.

The mechanism of lapping with bonded abrasives is similar to abrasion by embedding in single-sided lapping. The work piece is pressed against a rotating lapping plate, which is embedded with abrasives. The grains remove material by the mechanism of two-body abrasion. The schematic mechanism of lapping with bonded abrasives is shown in Figure 3.30. In their general appearance and function the lapping machines that use this principle closely resemble grinding machines. The reason why this process can be considered as a lapping rather than a grinding process is explained in Table 3.2.

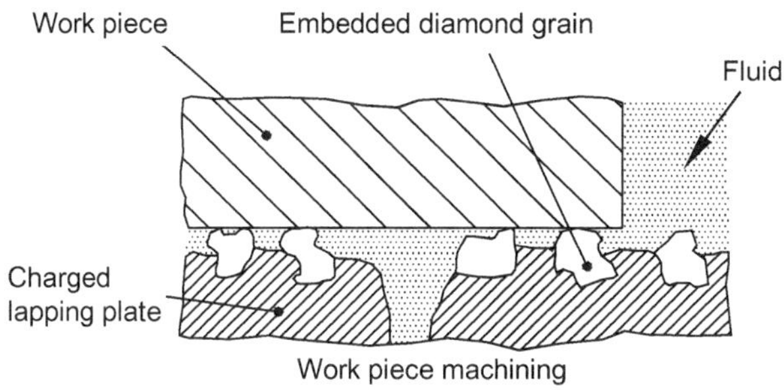

Figure 3.30 Lapping due to embedded abrasive grains

Table 3.2 Comparison of Surface Grinding with Bonded Abrasive Lapping

Aspect	Surface Grinding	Bonded Abrasive Lapping
Objectives	Produce a work piece having a surface of specific geometric form and size whose roughness upper limits are usually specified by a single value.	Improve a number of different characteristics of the surface of a work piece whose form and size have already been developed to the final or near final condition.
Material removal	Material removal rate should be high to remove the excess material within the shortest operational time and within the specified limits.	Material removal rate is low and is limited to those areas of the surface which interfere with the required surface texture.
Abrasives	Uses grinding wheels with grit sizes, which permit the largest chip cross-section without leaving furrows on the work piece surface, which exceed the specified roughness limits.	Operates with abrasives whose grit sizes are fine enough to avoid penetration into the work piece surface to a depth detrimental to any of the required surface characteristics.
Cutting speeds	It is a high speed cutting speed The speed is limited only by the capacities of equipment and tools.	It is a low speed operation. High speeds can spoil the surface texture as a lot of heat is generated. Speeds in the order of one-fifth, or often less, of grinding speeds are used.
Fixed infeed versus controlled pressure	Grinding machines usually operate with positive infeed set in advance at one or two fixed rates.	Lapping applies a specific pressure resulting in an infeed rate which adjusts itself to the conditions of the work piece surface.

3.3.5.8 Process Capabilities for Lapping

Accuracy and straightness to <0.125 μm are obtained in this process in selected parts. Surface finishes of the order of 0.025 μm have been obtained on hardened ceramics by setting optimum lapping parameters (Table 3.3). This finish is unique to this form of lapping. There is a negligible difference in the measurements of surface finish between the finish given by lapping processes and other advanced finishing processes.

3.3.6 Abrasives and Lubricants

There are a variety of abrasives that can be used for lapping, either in natural or artificial crystalline forms. Natural abrasives are diamond,

Table 3.3 Surface Roughness Results for Lapping

Process	Work Piece	Ra nm	Rq nm	Rmax nm	Reference
Lapping with sintered cast iron plate with CBN abrasives	Tungsten carbide			100	Hagiuda (1981)
Lapping with tin lapping plate and diamond abrasives	Poly Mn-Zn ferrite	0.9			Touge (1996)
Lapping with tin lapping plate with various fine particles	Single crystal sapphire Single crystal Mn-Zn Poly Mn-Zn			<1 <1 5	Namba (1978)
Lapping wheel with low bonding strength	Single crystal Si BK7, lime glass, AlN	 <10	2–3 2–3	20	Tani (1986) Kawata (1993)
Synthetic fabric-faced lap with diamond paste	Electroless Ni		<1		Parks (1997)
Lapping with Mylar film + foamed SiC substrate	Black filter glass		1		Parks (1997)
Lapping with fluorocarbon foam polisher with SiO_2 powder	Fused silica			0.3	Kasai (1990)

corundum, garnet, emery, quartz and other softer materials found in the earth. Artificial abrasives include diamond, Borazon™ CBN, silicon carbide, aluminum oxide, boron carbide and various aluminas, which are divided into two groups, fused and unfused.

3.3.6.1 Types of Abrasives

(1) Diamond

Diamond can be found in the earth in its natural form, or can be man-made. It is the hardest and the sharpest abrasive known. Because of its hardness, it is normally used to work with very hard materials such as tungsten carbide and ceramics. Soft materials should not be lapped with diamond abrasives because of its embedding tendency. A lapping plate charged with diamond abrasives cuts fast and produces a fine finish. Recent developments in man-made abrasives have made synthetic fine diamond powders increasingly available for industrial applications. Their use is limited due to their high cost.

Two types of diamond abrasives are commonly used: polycrystalline and monocrystalline diamond.

The features and advantages of polycrystalline diamond include the following:

- Higher cutting rates.
- Very uniform surface finish.
- More uniform particle size distribution.
- Higher removal rates (self sharpening abrasives).
- Harder/tougher particles.
- Blocky shaped.
- Extremely rough surface (more cutting points).
- Surface area 300% greater than monocrystalline diamond.

(2) Cubic Boron Nitride (CBN)

Cubic boron nitride is commonly known as Borazon™ CBN, and is a man-made synthetic abrasive. CBN is made from cubic boron nitride grains bonded by ceramic material. It is as hard as diamond on the Mohs scale, making it well suited for ferrous materials in lapping operations, as it will not carbonize when interacting with iron (Fe), as diamond abrasives may. Typically, CBN is used for lapping 52100 bearing steel, cast iron, die steel, tool steel, stellite, super alloys and in some cases ceramic materials.

(3) Norbide Abrasive

Norbide abrasives are fused abrasives with high grain strength. The grain has a hexagonal structure and is not easily friable. It is lower in hardness than CBN. This abrasive is not used extensively, and is reserved for unusual or special applications.

(4) Silicon Carbide

Silicon carbide is a hard fused crystalline abrasive. When used for lapping either high or low tensile strength materials, a fast cutting action is achieved with low crystal breakdown. Materials lapped with silicon carbide have a frosty or gray finish. It is best suited for rough lapping operations such as forged or hardened gears, valves, tool room work and general maintenance where polishing is not essential.

(5) Aluminum Oxide

Aluminum oxide is also a fused crystalline abrasive with a hardness just under that of silicon carbide. It has a very hard crystal structure, which is slowly dulled, and its hardness increases gradually resulting in its fracture. It is normally used for rough lapping operations and never for fine finishes or precision lapping.

(6) Fused Aluminas

Another form of fused alumina oxide is the 38 white aluminum oxide abrasive. It is white in color and has very friable crystals. During lapping operations, the pressure on the friable crystals keeps breaking them down into smaller particles which perform the finishing operations. A very low root mean square (rms) surface finish and good polish is achieved with fused alumina.

(7) Corundum

Corundum is the second hardest natural abrasive found in the earth. It has a softer crystalline structure than silicon carbide and aluminum oxide and hence its hardness is lower than either. It breaks down readily and is important for lapping a great variety of medium-hard metals with the hardness range Rockwell C 35–45. It gives a medium polish or reflective finish.

(8) Garnet

Garnet is a natural abrasive found in the earth. It has a blocky crystalline structure, hence it does not embed readily into the lapped parts. The biggest advantage of using garnet abrasives is that they can be safely and effectively used for lapping cast iron gears, brass or bronze running seals and instrument gears where such non-embedding qualities are desired. It gives a medium polishing quality.

(9) Unfused Alumina (hydrated-calcined)

A wide variety of unfused aluminas are available – gamma, alpha, hydrated and calcined. Hydrated alumina is relatively soft, and is used for polishing. Calcined aluminas are produced by heat treatment and the degree of the calcination determines the characteristics of the product. Mild, medium and high degrees of calcination correspond to soft, medium and hard characteristics of the abrasives respectively. They are, unlike blocky crystals, composed of flat or platy crystals with a thickness of about one-sixth of their diameter. Because of their platy shape, unfused alumina allows more even pressure to be distributed over larger surface areas. This contrasts with fused alumina, which has a small surface area. The disk-shaped particles work by a shaving action, as opposed to the rolling and gouging action of ordinary abrasives, and hence are not likely to scratch the work piece being lapped. The calcined types are recommended for lapping and polishing harder materials, with hardness range Rockwell C 45–63.

(10) Micro-Graded Flours

Micro-graded flours, also called optical finishing flours, are fused oxides, and are known as flours because of their fine grit size, which normally ranges from 500 to 1000. These abrasives were developed to meet the requirements of the glass and optical industries. They have a blocky crystalline structure, which is very effective when used for lapping soft materials. Pressure exerted on these abrasives will cause them to enter the material, and cut, rather than fracture. The softer the material, the lower the pressure needed. These abrasives are used with special lapping machines where the abrasive is mixed with water, glycerine and various oils and is fed from an agitated tank onto the lapping area producing a matt finish. As the abrasives are not friable, their use for precision lapping is not justified. They are normally used for rough lapping.

*Courtesy of Norton Company, Worcester, Mass

(11) Linde Powders

These powders are up to 99.98% alumina. They are used for polishing hard materials in the range Rockwell C 45–63. For extreme finishing operations, Linde powders are excellent for lapping steel balls or ball bearings on which they produce a very high polish or luster. They can also be used with pitch laps for producing mirror finishes.

(12) Other Abrasive Materials

See Tables 3.4 and 3.5.

- Red rouge (Ferric oxide): Jeweler's rouge for polishing soft materials.
- Green rouge (chromium oxide): For polishing hard materials such as chrome plate, stainless steel etc.
- Natural emery: It is one of the oldest abrasives but has little place in modern lapping due to its impurities.
- Cerium oxide: Best suited for glass polishing and used in the optical industry.
- Titanium oxide and other abrasives are used only for special applications.

Table 3.4 Hardness of Various Abrasives

Mohs Scale of Hardness

No.	Abrasive Type	Hardness
1	Diamond	10.0
2	Cubic Boron Nitride (CBN)	9.9
3	Norbide, boron carbide	9.7
4	Crystolon, silicon carbide	9.5
5	Alundum, aluminum oxide	9.0
6	38 White aluminum oxide	9.0
7	Linde "A" alpha alumina	9.0
8	Linde "C" alpha alumina	9.0
9	Linde "B" gamma alumina	8.0
10	Corundum	9.0
11	Calcinated alumina	8.5 to 9.0
12	Green rouge, chromium oxide	8.5
13	Cerium oxide	8.0
14	Garnet	8.0 to 9.0
15	Quartz	7.0
16	Red rouge, ferric oxide	6.5
17	Aluminas (hydrates)	5.0 to 7.0

*Courtesy http://www.mfginfo.com/mfg/lapvalve/lapractical.htm

Table 3.5 Average Particle Size of Abrasive Grain

Size	In Inches	Average Microns
100	0.0068	173
120	0.0056	142
150	0.0048	122
180	0.0034	86
220	0.0026	66
240	0.00248	63
280	0.00175	44
320	0.00128	32
400	0.0009	23
500	0.00065	16
600	0.00033	8
900	0.00024	6
1000		5
1200		3

*Courtesy http://www.mfginfo.com/mfg/lapvalve/lapractical.htm

3.3.7 Equipment

3.3.7.1 Lapping Plate

The accuracy and the flatness achieved in the lapping process depend entirely on the lapping plate. The surface of the lapping is finished, prior to charging, by scraping and testing with a standard surface plate, or by a method of scraping in which three plates are lapped together to secure a plane surface. All the high spots on the lapping plate have to be removed, or they must be uniform and close together.

The type and composition of the lapping plate is important, as the work piece may become either scratched or contaminated with abrasives. If the lapping plate is hard, it resists becoming embedded with abrasive particles; therefore the grains roll more than they slide. In this situation, material removal occurs by rolling, in which stress-induced micro-fractures remove the material. This may be problematic, as the abrasives tend to get embedded in the work piece and contaminate it. Alternatively, however, if the lapping plate is soft it allows the abrasives to become embedded in it, meaning that more sliding motions take place, and material removal is primarily by a plowing action. In this situation, abrasives do not become embedded in the work piece. A softer lapping plate gives a good final surface finish but at the expense of loss of planarity. To obtain both, modern lapping machines often use a hard lapping plate and very fine grit abrasives. Generally laps are made of cast iron, copper, brass, lead or ceramics. Whatever material is used, the lap should be softer than the work piece, or the work piece will become charged and cut the lap.

To produce a planar surface it is important that the distribution of abrasive slurry is uniform between the work piece and the lap. There is a tendency for abrasives to build up in some areas, resulting in raised local pressure and hence an increase in the material removal rate in those areas. Also, the leading edges of the lap will tend to engage the grains more quickly causing a barrel lap shape. This may give rise to a loss of planarity and surface finish. Consideration should therefore be given to the rotation of the lapping plate and uniform distribution of the slurry.

In roughing operations, the lapping plate is scored by narrow grooves, usually located about a half inch apart and extending both lengthwise and crosswise, forming a series of squares similar to those on a checker board. This scored surface helps to produce better and faster cutting in the roughing operation. For finishing operations, a planar lapping plate with no scores is used.

3.3.7.2 Charging of the Lapping Plate

To produce a perfectly smooth, flat surface that is free from scratches, the lapping plate needs to be charged with very fine abrasives. Abrasive slurry is gradually poured on the rotating lapping plate (Figure 3.31). After some time the lapping plate is examined for high and bright spots. If there are any visible bright spots, the charging is continued until the entire surface has a gray appearance. Once the lapping plate is charged it needs to be used without applying more abrasives until it ceases to cut. Overcharging and an excessive amount of abrasives results in rolling action between the work piece and the lapping plate, which results in

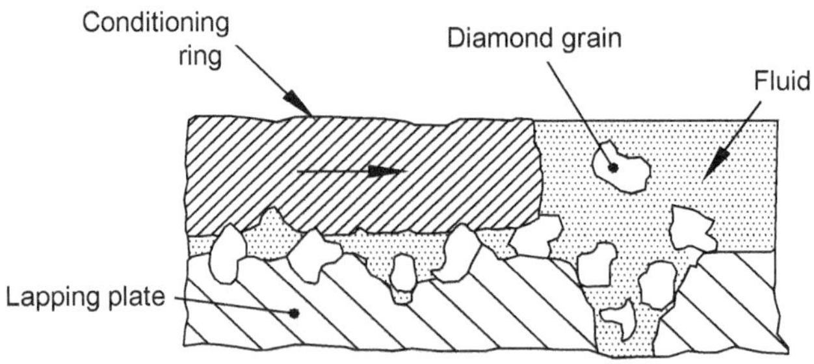

Figure 3.31 Charging of lapping plate

inaccuracy. To prevent scratching, several conditions must be observed and maintained. The room where lapping is being performed should be reasonably free from dust and other impurities.

3.3.7.3 Different Types of Lapping Plate

Plates used for lapping can be made from iron, copper, tin, lead and tin, ceramic or composite materials. A description of each kind of plate is given below.

1. Iron plate (Figure 3.32)
   The primary characteristics of plates made from iron are as follows:
   - They serve as an excellent primary/roughing lap plate, with a long service life.
   - They are often used as an alternative to cast iron plates.
   - They produce a good surface finish on most materials, especially metals and ceramics.
   - Lapping plates made from iron are generally used with coarse to medium size diamond abrasives.
2. Copper plate (Figure 3.33)
   The important features of lapping plates made from copper are as follows:
   - They are the most widely used type of lap plate.
   - They are excellent when primary and finishing lap are combined in a one-step operation.
   - They are suitable for virtually any solid material including metals, ceramics, glass and plastics.
   - This kind of plate is typically used with medium to fine diamond sizes.
   - They minimize fracturing and chipping tendencies when lapping crystal components.

Figure 3.32 Iron lapping plate

Figure 3.33 Copper plate

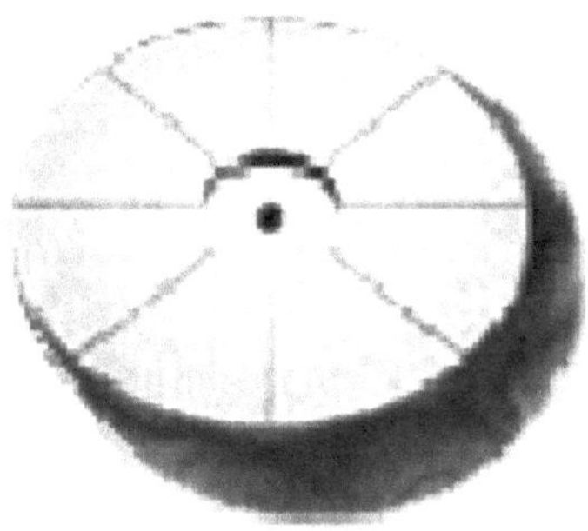

Figure 3.34 Tin plate

3. Tin plate (Figure 3.34)
   Some of the important characteristics of tin lapping plates are as follows:
   - Often used where lead-type contamination cannot be tolerated.
   - Suitable for charging extra-fine abrasives.
   - Used with fine to extra fine (micron to nanometer) diamond sizes.
4. Tin/Lead plates (Figure 3.35)
   The important characteristics of these kind of plates are:
   - They are most widely used finishing lap/polishing plate.
   - They are often used in place of polishing pads.
   - They are suitable for metal, ceramic and other materials.
   - They are typically used with fine to very fine diamonds.
   - They minimize fracturing and chipping tendencies when lapping crystal components.
5. Ceramic plate (Figure 3.36)
   - These plates are generally used to lap/polish ceramic parts and other stain-sensitive materials.

Figure 3.35 Tin/Lead plate

Figure 3.36 Ceramic plate

- They are used in applications where metallic-type contamination cannot be tolerated.
- They are an affordable, more machinable alternative to "natural" ceramic plates.
- They are very flexible – used with coarse to fine diamond sizes.

6. Composite laps (Figure 3.37)
Special composite laps are made from a unique blend of powdered metal or ceramic, combined in a resin system. Their special formulations allow composite laps to take a more uniform charge of diamond, compared to pure metal plates. The result is a more consistent and controlled performance. Composite plates are also superior in applications where lapping and polishing are combined into one step. Some of the significant features of these types of plates are:
   - These plates can produce finishes up to 0.2 μm Ra and flatness up to 1/20 wavelength (0.000001").
   - These lapping plates are ideal for finishing two or more dissimilar materials to a uniform plane of flatness.

Figure 3.37 Composite laps

3.3.7.4 Carriers Used in Double-Sided Lapping

The work pieces are held in position in the single-sided lapping process by using carrier plates. Carriers can be machined using punch and die tooling, CNC machining, EDM and laser cutting. Work holes can be custom-cut to suit any application and any geometry. The material used for manufacturing carriers depends on the type of work piece to be lapped.

- Steel – Spring steel is the strongest, most durable carrier material. Closer tolerances of thickness and flatness are obtainable with spring steel material, making it ideal for thin work.
- Lamitex – Lamitex, also known as G-10 and FR-4, is an epoxy-based material reinforced with fiberglass. It is an excellent alternative to steel, offering excellent machining capabilities, high mechanical strength, resistance to corrosion, and low water absorbing properties. It is an excellent choice for large, thick parts.
- PVC – PVC is a thermoplastic vinyl. Combining chemical inertness, high strength-to-weight ratio and economical cost, PVC is an excellent choice for corrosion-resistant applications. PVC can be sanded to achieve tight thickness tolerances and is commonly used for lapping thick parts.
- Phenolic – Phenolic is a resin binder reinforced with cotton fabric. It is very wear resistant and stronger than most plastics. It is typically used with oil-based slurries due to its water absorbing qualities. It is an excellent choice when steel or fiberglass are not permitted in a process.
- Lexan – A plastic without fiber reinforcement, this material is less strong than Lamitex but does not cause scratches in polishing applications.

3.3.7.5 Ultrasonic Cleaner

The final step in lapping is the cleaning process. This removes the dirt, contaminants and other particles deposited on the work piece during lapping. The contaminants or dirt could be grease or wax deposits, which get deposited on the surface of the work piece. Originally, the cleaning process was done by scrubbing, soaking, and steaming, but this did not give satisfactory results, so an ultrasonic cleaner is now used (Figure 3.38). These cleaners are powerful enough to remove heavy oils, grease, waxes or slurries, yet gentle enough to be able to clean delicate parts without damaging them.

Ultrasonic sound waves moving through a cleaning solution create an effect called cavitation, which is the process of rapidly forming and imploding microscopic bubbles. Cavitation can be considered as a kind of "cold boiling," as the bubbles in the water are created not by heating, but by ultrasonic sound waves. These waves create water bubbles, which explode, thus removing the contaminants (Figure 3.39). Since no heat is produced there is no thermal damage to the work piece. This deep cleaning action removes even the most stubborn dirt and contaminants.

3.3.8 Abrasive Slurry

The vehicle, also referred to as the abrasive slurry, is the liquid carrier or suspension agent in which the abrasive is delivered to the lap plate. The abrasive grains are transported to the lapping zone suspended in carriers like oil or water. To prevent rusting of machine components certain suspension agents are added. These suspension agents also prevent settling of the abrasive particles. The suspension agents do not aid the lapping

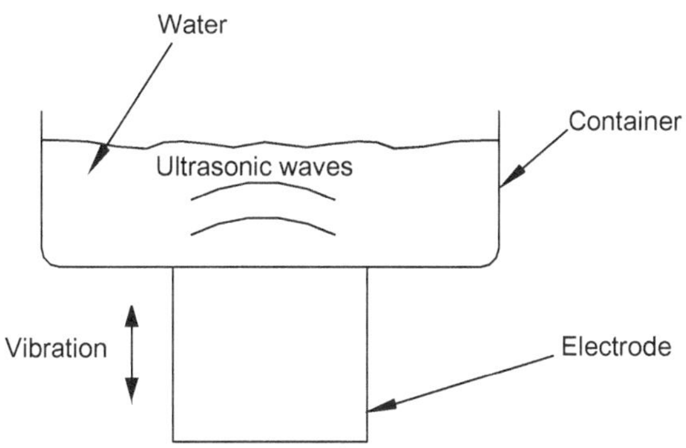

Figure 3.38 Schematic of ultrasonic cleaner

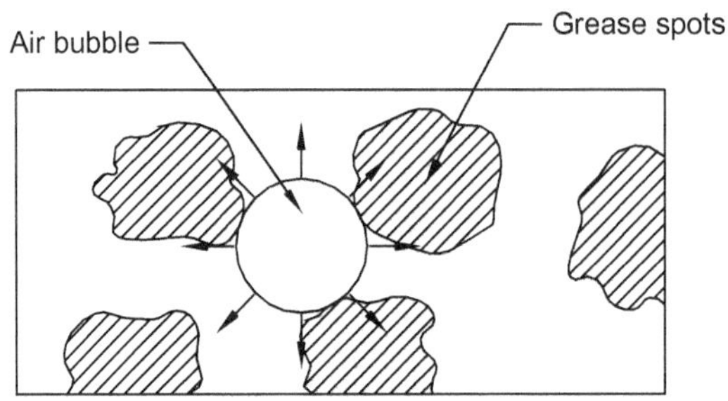

Figure 3.39 Cleaning action due to bubbles

method in any way. Liquid carriers are available in various viscosities to deal with almost any process.

The important characteristics of a liquid vehicle are its suspension, lubricity, scratch prevention, and recyclability. The most important characteristic of the vehicle is to suspend and uniformly disperse the abrasive particles at the interface between the work piece and the lap plate. During the lapping action the abrasive grains are continuously broken down into smaller pieces and the effectiveness of the grain size is gradually reduced. This means that the abrasive slurry has to be continuously replenished. If the fluid is delivered to the same position on the lap, the increased concentration of slurry in that area leads to uneven wear of the lap.

The liquid carriers can be oil or water. Water-based vehicles are commonly used due to the environmental impact of the disposal of waste oil-based carriers. In addition to this, oil is expensive and has a tendency of penetrating into the pores of some materials.

The functions performed by a good abrasive slurry or vehicle are:

1. To carry the abrasive grains and to position them to work most efficiently.
2. To lubricate the surface and carry away the abraded material from the work pieces.
3. The vehicle carries away the heat that is generated in the process of lapping.
4. The lapping fluids, in some cases, accelerate the lapping process by inducing chemical reactions.

3.3.8.1 Abrasives

The abrasives most commonly used in the double-sided lapping process for silicon wafers are corundum (Al_2O_3), silicon carbide (SiC) or specially prepared mixtures such as Al_2O_3 and ZrO_2. For single-step lapping operations, grain sizes of 12−15 μm are generally used.

Material removal during silicon lapping takes place mainly through cracking, splitting and brittle fracture of the crystalline material. This is caused by abrasive grains that roll between the surface of the wafer and the grinding/lapping wheel. During the double grinding process, the total number of coarse active grains dictates the material removal rate. The smaller grains do not make any contribution to the removal process.

In lapping experiments with SiC abrasives, higher material removal rates could be achieved by increasing the average path velocity by increasing the pressure and using larger sized grains. Also, the surface quality of the wafer is independent of the grinding/lapping pressure and path velocity.

3.3.9 Introduction to Silicon

Silicon is the material most commonly used for microchip production. Various processes are needed to transform a silicon crystal into wafers. In 1997, about 150 million silicon wafers of different sizes were manufactured, representing a worldwide revenue of $6.2 billion. Silicon is the most important building block of semiconductors and is found in many types of microelectronic applications, such as computer systems, telecommunications equipment, automobiles, electronic products, and automation and control systems in industry and also in defense systems. More than 90% of all semiconductor components are made of silicon. Silicon has a diamond-like crystal structure with covalently bonded atoms in contrast to carbon steel, which has a cubic body-centered structure with metallic bonding. Silicon has a high hardness in the domain of 1000 Vickers as compared with 200 for steel. Silicon has a very low fracture toughness; that of steel is almost 100 times higher.

Silicon wafers are used for the production of microchips. Monocrystalline silicon is presently the most commonly used substrate for the production of microelectronic components worldwide. The precise machining of silicon is critical to high quality standards. Various processes are needed to make a silicon crystal into wafers. One such process is surface lapping. This section discusses the requirements of the silicon wafer fine grinding and lapping processes.

3.3.10 Lapping of Silicon Wafers – Experiments

Semiconductor wafers are an integral part of the electronics industry and the demand for them continues to increase. They are usually made from silicon.

After semiconductor wafers are sliced into the required length, they are quite rough and have saw marks on both sides. Back lapping is the thinning of semiconductor wafers by removing material from the rear – i.e. the unpolished or unprocessed face.

Since the wafers are extremely fragile, suitable consideration needs to be given to the residual strength of the sample after back lapping. Final thickness requirements can vary between 0.01 mm and 0.1 mm. A small wafer of about 0.01 mm thickness can be more easily manipulated than a large 125 mm diameter wafer of 0.1 mm thickness. If the final thickness of the wafer to be lapped is very small it must be suitably supported, using a substrate. This substrate can be another wafer, or glass or ceramic. When high accuracy is desired the substrate's thickness to diameter ratio should be at least 1:7. The substrate can be permanently bonded to the wafer depending upon the application. Achieving a thickness tolerance of +/− 0.001 mm over the area of a small device is comparatively simple but a larger tolerance would be required for the whole wafer. Generally the surfaces of the wafers are highly parallel but the wafer profile is not flat. To achieve flatness the wafer has to be held flat since any variations in the flatness will directly affect the thickness of the wafers.

3.3.10.1 Mounting the Wafer

From the technical considerations discussed above, it is apparent that the wafer must be held flat to avoid variations in thickness. This can be done by cementing the wafer to the mounting plate by using wax or suitable adhesive. In addition to the mounting plate, a pressure plate is also required (Figure 3.40). The wax bond of the substrate to the mounting plate should have a lower softening point than that used to bond the wafer to the substrate or the substrate can be held in a vacuum jig. Permanent bonding can be done by using adhesives like epoxy resins.

3.3.10.2 Double-Sided Lapping Kinematics

The machining kinematics of the double-sided lapping procedure are fairly simple (Figure 3.41). Here, the work pieces, held by friction, revolve on a lapping plate. The three dominant processes that occur when mechanically machining flat surfaces are grinding, lapping and polishing. Grinding is a two-body abrasive process, with the grain fixed in the grinding tool. Lapping is a three-body abrasive process, mainly using free abrasive grains that roll in the space between the lapping plate and the work piece.

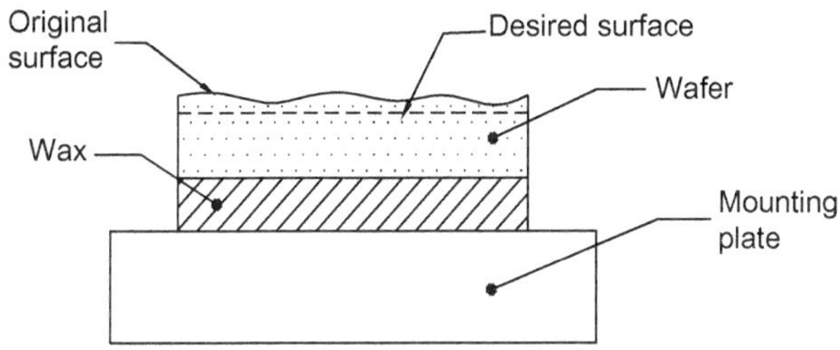

Figure 3.40 Mounting of wafer

Figure 3.41 Grinding/lapping principle

Lapping uses loose grain. In polishing, loose abrasives are temporarily fixed in a soft-polishing pad.

During lapping the work pieces are guided in a carrier system, which is driven by an inner pin ring. The carrier system rolls off the fixed outer pin ring so that the work pieces move in a cycloid motion relative to the rotating lapping wheels. Although the kinematics of the work piece motion are equivalent in fine grinding and lapping, some important differences between then may be noticed. In comparison to lapping, the effective principle during fine grinding is not bound to the spatial area but bound to the kinematically defined path of the grinding grains. Furthermore, the main effective direction is not perpendicular but parallel to the work piece surface. Material is removed during lapping by exceeding the maximum compressive strength of the base material, whilst during fine grinding the maximum shear strength is exceeded. Detailed analysis of the process shows that the mechanisms of material removal during fine grinding are similar to those encountered in conventional grinding. Depending on the parameters and the composition of the work piece, material can be removed by microcutting, microplowing, microgrooving, or due to fatigue cracking of the surface. In general, a combination of these mechanisms operate. The fine ground surface is characterized by grinding grooves, which intersect and are not oriented in any particular direction.

3.3.10.3 Lapping Process

Lapping is done by using a cast iron lapping plate and the lapping medium is medium grit (10 to 15 μm) silicon carbide powder suspended in a lapping oil. Damage penetration of the wafer can be prevented by proper control of the load. The load on the wafer and the plate speed is decreased as the final thickness is approached. After lapping, a final surface finish is obtained by polishing the semiconductor wafers.

To turn a silicon work piece into a good quality wafer, a variety of machining processes are undertaken, namely slicing, edge profiling, lapping/grinding, etching, rough polishing, final polishing and cleaning.

1. Slicing: Process to slice single crystal silicon pieces into wafers of thin disc shape.
2. Edge Profiling (chamfering): Process to chamfer the edge portion of the wafer.
3. Lapping or Grinding: Process to flatten the surface of the wafer smoothly.
4. Etching: Process to chemically remove the processing damage of the wafer without introducing further mechanical damage.
5. Rough Polishing: Process to obtain a mirror surface on the wafer.
6. Fine Polishing: Process to obtain the final mirror surface.
7. Cleaning: Process to remove the dust particles and polishing agents from the wafer surface.

Cylindrical silicon samples are sliced into thin wafers by grinding and lapping. These wafers are then lapped or ground to ensure precision thickness, flatness and parallelism. After this, the damage zone has to be removed by chemical etching. Then the wafers are polished on one side to a mirror-like finish. The individual chips are then separated by dicing after undergoing further processing steps.

In double-sided lapping or grinding, silicon wafers containing completed devices on their front side are ground on their back side, before being sliced into individual chips. Choosing different values of grit size affects the subsurface damage to the wafer. As grit size increases, the depth of subsurface cracks also increases. Sometimes, due to uneven structure of the lapping wheel, plastic deformation and brittle fracture can also occur during double-sided surface lapping.

Fine lapping of silicon wafers requires high predictability and consistency, which requires the lapping wheel to possess a self-dressing ability. There should be perfect equilibrium between the rate of wear of the abrasive grains and the rate of release of worn abrasive grains. This keeps the grinding/lapping force relatively constant.

Some other factors to keep in mind for double-sided fine lapping of silicon wafers include:

a. The lapping wheel should have a reasonable life.
b. The grinding/lapping force should be low and constant.
c. Surface and subsurface damage should be minimized.
d. The final lapped silicon wafers should be very flat.

Due to its unique requirements, fine grinding/lapping of silicon wafers presents big challenges to wheel manufacturers, lapping machine producers and engineers. To ensure the successful development of double-sided fine lapping of silicon wafers, a large amount of research work is needed.

3.3.10.4 Effects of Lapping Wheels

The slurry used to cool and lubricate the lapping wheel and the silicon surface is mixed with deionized, purified water. The slurry is usually supplied to the inner side of the wheel. Lapping wheels have significant effects on lapping performance. To ensure minimal subsurface damage to the silicon wafers, and to achieve the desired surface roughness, the grit size of wheels and abrasives should be as small as possible. The type of bond materials, the hardness of the wheel, and the concentration play significant roles in determining the self-dressing ability and the wheel wear rate. Different wheels can lap differently – some can lap hundreds of wafers, applying a constant grinding/lapping force without any dressing procedure being needed, while others lap very few before "overloading". Overloading is the phenomenon when the wheel basically stops cutting and the grinding/lapping forces increase to very high values. As a result, the wafer being lapped or grinded shows a burnt appearance.

Even the rim width of the lapping wheels can make a difference. According to research, narrower wheels demonstrate better performance; they can grind twice as many wafers as those with wider rims. However, too narrow a rim can cause the tooth segments to become very fragile and to break.

The effects of wheel type on double-sided lapping of silicon are still being researched. The proper selection of process parameters is crucial, since lapping wheels that work well under a certain parameter set may not work as well under different conditions. The lapping wheel tends to become weaker at lower wheel speed and both the nozzle position and the flow rate of the coolant affect the grinding process. Higher coolant flow rates give a better performance.

To turn a silicon work piece into flat wafers of good quality, a sequence of machining operations is required. Each operation damages the work piece slightly, and this damage has then to be removed. So, in order to develop a process flow for wafering and to choose the material removal method for each individual process, the extent and type of surface damage induced by each machining process must be known. The depth of subsurface crack on ground silicon wafers is approximately equal to half of the grit size used in the lapping wheel. The subsurface crack depth is dependent on the location of the lapped wafer.

3.3.10.5 Slurry and Its Effects

Lapping fluids (slurry) are used to cool the work piece and the lapping wheel, to transport debris away from the lapping zone, and to provide lubrication between the wheel and the work piece. Debris removal prevents the wheel from overloading and enables longer periods of lapping between dressings. Cooling and lubrication help to control lapping temperatures. Lubrication reduces lapping forces and therefore heat generation, and cooling removes heat from the wheel/work piece contact. Both coolant nozzle position and the coolant flow rate affect the grinding process. A nozzle held at an outer position allows the slurry to flow onto the area between the tooth segments and the silicon surface. An inner-positioned nozzle delivers the slurry to the base above the segment of the lapping wheel. According to tests, a delivering slurry nozzle in the inner position enables the lapping wheel to lap more wafers before being overloaded than one in the outer position, but the use of both positions together gave the best results. During any lapping process, it can be noticed that the higher the flow rate, the more wafers the wheel could lap before overloading.

Temperature control in lapping is very important, as excessive temperatures in the lapping zone may result in poor surface quality, and large temperature gradients may cause thermal warping, leading to a serious loss of dimensional and form tolerances. Hence, it is important to understand the mechanisms that govern the flow of lapping fluid in order to optimize fluid delivery.

3.3.10.6 Silicon Machining

Initially, the plate surface is wetted with slurry, in our case 0.25 μm diamond slurry. During machining, the fluid is added to the plate at a rate of 1 ml/min to maintain a fluid film of constant thickness. The work pieces are mounted on a disc-shaped work piece carrier placed inside the lower grinding wheel. Then the lapping/grinding process takes place. A work piece for surface grinding is usually held to the reciprocating worktable by a magnetic chuck. It may also be held in a vise or clamped directly to the table. Double-sided lapping is an established process in modern wafer production. Double-sided grinding/lapping is used for eliminating the saw marks. The abrasive processes of double-sided lapping influence the final geometrical wafer quality that is achievable. Subsequent etching and polishing operations are applied to remove the damaged layer and improve the surface quality, but such operations lead to a reduction in geometrical accuracy. A common feature of all lapping processes is that material removal takes place by the action of loose abrasive grains that are dispersed in a thick lapping fluid. The work piece shape is produced by the relative motion between the lapping wheel and the work piece sample coupled with the action of the loose abrasive grains between them.

For high volume production, highly powered, special purpose, double-sided lapping/grinding machines with drive systems are normally used. The main features of these lapping machines are:

a. Program and gauging controls which allow pressure-controlled operation.
b. Infinitely variable speed adjustment of the drive motors.
c. Grooved, temperature-controlled lapping wheels.

3.3.10.7 Common Lapping Wheel Problems and Solutions

1. WHEEL WILL NOT CUT: Clean with cleaning stick, then increase the amount of the work piece presented to the wheel by increasing the infeed and/or traverse rate.
2. BURN OR CHATTER: Clean with cleaning stick, then increase the traverse rate. This will result in longer wheel life, less loading or glazing, and a freer cutting wheel.

3. POOR FINISH: If the wheel is loaded or glazed, apply the same principle as for burn or chatter. If the wheel is not loaded or glazed, then decrease the infeed and/or traverse rate. A slightly loaded wheel will provide the best finish.
4. SHORT WHEEL LIFE: Decrease the infeed and/or traverse rate. Excessive pressure or traverse rate will cause the wheel to break down too rapidly.

3.3.11 Lapping Machine

Figure 3.42 is a photograph of the Melchiorre lapping machine currently in the precision micro-machining department of the University of Toledo. It clearly shows the top and the bottom lapping wheels as well as the control panel for operating the machine.

Figure 3.43 shows another overview of the machine from the front. The various controls on the machine are clearly visible in this view. The various accessories, such as the conditioning rings and the work carriers that are supplied with this machine, are just visible on the side.

The current study uses the Melchiorre machine from Melchiorre Company, Italy. It is a double-sided lapping machine, which works on the same principle as fine grinding using lapping kinematics. It has one top lapping wheel and two bottom lapping wheels, each having their own belt for speed adjustment. This machine is an SP3 model, of type 210-3P made in 1995 and it weighs 500 kg. It can handle pressures between 22 and 63 kg. Its dimensions are 900 mm × 1200 mm × 1150 mm.

Figure 3.42 Melchiorre lapping machine

Figure 3.43 Melchiorre lapping machine controls

3.3.12 Experimental Set-up

Apparatus Used

- Profilometer (Ra).
- Digital Tacometer MT137A Blue Point 1.5 V 3 A.
- Micrometer 0–1″ (0.0001″–0.001 mm) Mitutoyo.
- Diamond slurry 0.25 μm (monocrystalline).
- Copper lapping wheels abraded.
- Peristaltic pump – Used as a slurry dispenser.
- Weighing scales.
- Silicon wafer samples milled using diamond milling tool. (27 samples).
- Two 2.5 lb plates for pressure settings. (Procedure explained in a later section).

Parameter Settings

- Speeds: 50, 75, 100 rpm.
- Load: 5, 10, 20 kgf.

Procedure

1. Silicon samples.
   The silicon samples were initially milled to the required size to fit into the work piece holders. 27 samples were milled. The initial surface roughness (Ra) of each of these samples was measured using a profilometer. Two readings were taken for each sample – on the top and the bottom for accuracy and the average was taken.
2. Reactivating the Melchiorre machine.
   The parameter settings for the lapping load were 5, 10 and 20 kgf. The machine could only work in the range 22 to 63 kg pressure;

hence certain calculations had to be done to reach the needed values for pressure.

3. Pressure setting calculations.
   There is a 30 lb block placed on the top head of the machine for handling the pressure. This weight was removed and the machine top head and the pressure chamber were set in equilibrium. The top head of the machine weighed 20 lbs. Now the machine was ready to be set to the required pressure.
   A scale was made and set on the top head with distances. The calculations for the distance to place the 2.5 lb plates for the required pressures of 5, 10 and 20 kg, were as follows:

$$\text{Weight of top head } (2.5) = \text{weight plate } (d)$$

   where weight of top head = 20 lbs, weight plate = 2.5 lbs and d = distance to place weights on scale to get the required pressure.
   For a pressure of:
   5 kg: 20 (2.5) = 2.5 (d) d = 20 inches
   10 kg: 30 (2.5) = 2 (2.5) (d) d = 15 inches
   20 kg: 50 (2.5) = 2 (2.5) (d) d = 25 inches
   Hence the 2.5 lb plates were set at the above distances to obtain the respective pressures for the experiments.
4. Charging the lapping wheels.
   The slurry used was 0.25 monocrystalline diamond slurry. The peristaltic pump was attached and the slurry flow rate was set at 1.14 ml/min. The lapping wheels were lubricated and the machine was started. The wheels were run for about half an hour to set the machine into operating mode ready for experimenting.
5. Machining
   The first sets of samples were weighed and the initial Ra recorded. Their initial dimensions were also recorded using a micrometer. After the wheels were charged, the silicon samples were loaded into the work piece holder and the pressure was set at 5 kg. The bottom and top wheel speeds were set at 50 rpm. The samples were lapped for 5 minutes. and then their surface roughness was measured again. This was done for lapping times of 5, 10, 15 and 25 minutes. The final weight of the samples was also measured to determine the material removal rate. The final height was also measured. Hence, two ways were used to determine the MRR of the silicon samples that were lapped.

The above procedure was repeated for speeds of 75 and 100 rpm, and then the whole cycle was repeated for pressures of 10 and 20 kg at the three speeds.

3.3.13 Experimental Results

The surface roughness data is shown in Table 3.6 and the material removal rate data in Tables 3.7 and 3.8.

Plots

Graphs were plotted to show the effects of varying the speed and the pressure. The first set of graphs was plotted keeping speed constant and plotting surface roughness versus time at varying pressures. The second set of graphs was plotted by keeping the pressure constant and plotting surface roughness versus time at varying wheel speeds. More results from these graphs will be discussed below.

Plots with wheel speeds kept constant at different pressures are shown in Figures 3.44–3.46.

Plots with pressure kept constant at different wheel speeds are shown in Figures 3.47–3.49.

Equations Used to Calculate the Material Removal Rate

There are two ways to find out the material removal rates of the silicon work pieces that were lapped.

Equations 3.15 to 3.17:

$$\text{Initial Height} - \text{Final Height} = \Delta h(\text{change in height}) \tag{3.15}$$

$$V = \Delta h(\pi D^2/4) \tag{3.16}$$

$$V/\Delta t = MRR \tag{3.17}$$

Equations 3.16:

$$(\text{Initial Weight} - \text{Final Weight})/\rho = \text{volume} \tag{3.18}$$

where ρ = density of silicon and volume/Δt = MRR.

However, only the first set of equations was used to calculate the material removal rates due to reasons explained in the observations section below.

Table 3.6 Surface Roughness

Speed	Pressure	Time	Surface Roughness (Ra) μm		
			Ra-Piece1	Ra-Piece2	Ra-Piece3
S1	5 kg	0	0.624	0.648	0.431
		5	0.628	0.648	0.432
		10	0.439	0.65	0.427
		15	0.396	0.287	0.37
		25	0.34	0.28	0.365
		30	0.301	0.275	0.329
	10 kg	0	1.42	1.483	1.579
		5	1.096	0.832	1.003
		10	0.537	0.682	0.973
		15	0.389	0.407	0.657
		25			0.501
		30			
	20 kg	0	1.662	1.691	0.892
		5	1.463	1.38	0.563
		10	1.029	0.9	0.872
		15	0.456	0.407	0.696
		25	0.326		0.593
		30			
S2	5 kg	0	1.189	1.819	0.987
		5	0.503	0.691	0.887
		10	0.4	0.536	0.701
		15	0.359	0.422	0.431
		25	0.348	0.415	0.43
		30	0.326	0.401	0.398
	10 kg	0	1.872	1.013	0.824
		5	1.736	1.008	0.797
		10	0.729	0.452	0.709
		15	0.432	0.281	0.691
		25			0.538
		30			
	20 kg	0	1.367	1.301	1.672
		5	0.782	0.693	1.098
		10	0.639	0.601	0.732
		15	0.333	0.531	0.698
		25		0.501	
		30			
S3	5 kg	0	0.799	1.568	1.036
		5	0.701	0.748	0.529
		10	0.435	0.345	0.43

(Continued)

Table 3.6 (Continued)

Speed	Pressure	Time	Surface Roughness (Ra) μm		
			Ra-Piece1	Ra-Piece2	Ra-Piece3
		15		0.301	0.421
		25		0.298	0.401
		30		0.291	0.326
	10 kg	0	1.447	1.581	1.102
		5	0.532	0.982	1.008
		10	0.399	0.706	0.693
		15	0.206	0.431	0.512
		25			
		30			
	20 kg	0	0.756	1.108	1.211
		5	0.689	0.909	0.813
		10	0.532	0.731	0.413
		15	0.398	0.509	0.315
		25			
		30			

Observations

- Many of the work pieces did not last for the full 25 minutes at the 10 and 20 kg setting as originally planned. They either broke into pieces or disappeared into tiny fragments. This resulted in the work piece holder being lapped instead. The silicon samples were capable of lasting only 15–20 minutes. Only a few of them made it past the 20-minute mark. The reason for this may be due to too much pressure. The 5 kg range was good. The samples used were very thin and are not able to handle pressure of even 10 or 20 kg. Reduced pressures allowed the lapping time to go on for 30 minutes.
- Work piece carriers break or bend due to grease build-up or impurities in the wheel or central gear surface. A lot of care must be taken and the surface must be thoroughly cleaned before any sort of lapping procedure. The work piece carriers also ended up breaking if they were not perfectly flat. This shows that it was a very sensitive operation using really thin samples.
- Pressures of 10 and 20 kg did not really make a big difference to the lapping Ra values. Due to the samples being very thin, the surface roughness decreased very rapidly over small intervals of time, so the sample weights also decreased rapidly.

Table 3.7 MRR Data

Speed	Pressure	Time	Work Piece Height (mm)			Work Piece Weight (mg)		
			Height Piece1	Height Piece2	Height Piece3	Weight Piece1	Weight Piece2	Weight Piece3
S1	5 kg	0	0.8	0.87	0.86	185	196.6	188.5
		5	–	–	–	–	–	–
		10	–	–	–	–	–	–
		15	–	–	–	–	–	–
		25	–	–	–	–	–	–
		30	0.323	0.49	0.398	70.76	89.97	102.4
	10 kg	0	0.79	0.79	0.796	196	184.3	177.6
		5	–	–	–	–	–	–
		10	–	–	–	–	–	–
		15	–	–	–	–	–	–
		25	–	–	–	–	–	–
		30	X	X	0.319	X	X	99.76
	20 kg	0	0.793	0.793	0.79	212.2	220.2	197.1
		5	–	–	–	–	–	–
		10	–	–	–	–	–	–
		15	–	–	–	–	–	–
		25	–	–	–	–	–	–
		30	0.562	X	0.341	102.67	X	95.6
S2	5 kg	0	0.798	0.792	0.795	182.9	188.4	183.7
		5	–	–	–	–	–	–
		10	–	–	–	–	–	–
		15	–	–	–	–	–	–
		25	–	–	–	–	–	–
		30	0.328	0.373	0.298	98.67	132.96	86.8
	10 kg	0	0.792	0.793	0.789	194.5	202.4	186.1
		5	–	–	–	–	–	–
		10	–	–	–	–	–	–
		15	–	–	–	–	–	–
		25	–	–	–	–	–	–
		30	X	X	0.326	X	X	94.62
	20 kg	0	0.794	0.792	0.797	174.3	190.5	191.9
		5	–	–	–	–	–	–
		10	–	–	–	–	–	–
		15	–	–	–	–	–	–
		25	–	–	–	–	–	–
		30	X	0.239	X	X	102.45	X
S3	5 kg	0	0.795	0.795	0.788	179.1	201.3	192.5
		5	–	–	–	–	–	–

(*Continued*)

Table 3.7 (Continued)

Speed	Pressure	Time	Work Piece Height (mm)			Work Piece Weight (mg)		
			Height Piece1	Height Piece2	Height Piece3	Weight Piece1	Weight Piece2	Weight Piece3
		10	–	–	–	–	–	–
		15	–	–	–	–	–	–
		25	–	–	–	–	–	–
		30	X	0.289	0.318	X	102.45	98.217
	10 kg	0	0.795	0.795	0.802	204.9	214.9	232.5
		5	–	–	–	–	–	–
		10	–	–	–	–	–	–
		15	–	–	–	–	–	–
		25	–	–	–	–	–	–
		30	0.412	0.378	0.319	97.4	87.329	112.78
	20 kg	0	0.809	0.806	0.792	208.3	204.2	209.7
		5	–	–	–	–	–	–
		10	–	–	–	–	–	–
		15	–	–	–	–	–	–
		25	–	–	–	–	–	–
		30	0.393	0.451	0.327	90.9	98.456	102.98

Table 3.8 Material removal rates

		Pressure		
		5 kg	10 kg	20 kg
Speed	50 rpm	1.7256	1.789	1.314
	75 rpm	1.6927	1.9833	2.034
	100 rpm	1.937	1.617	1.756

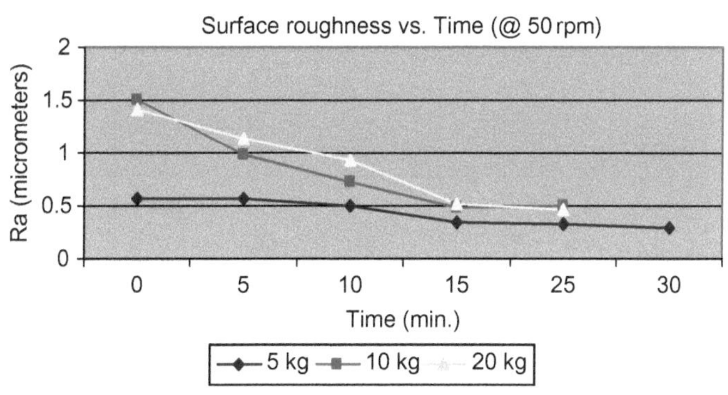

Figure 3.44 Surface roughness vs. Time (@ 50 rpm)

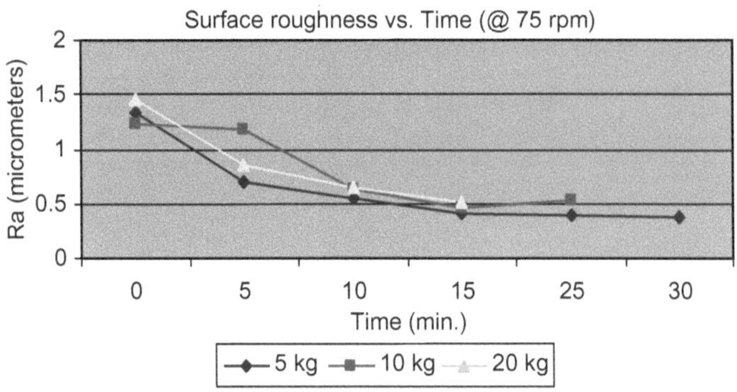

Figure 3.45 Surface roughness vs. Time (@ 75 rpm)

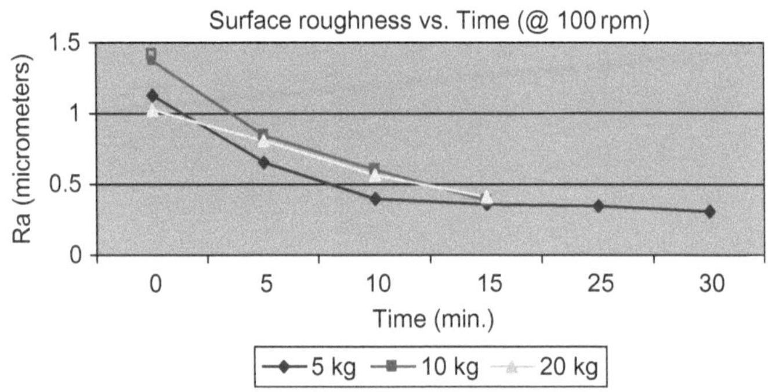

Figure 3.46 Surface roughness vs. Time (@ 100 rpm)

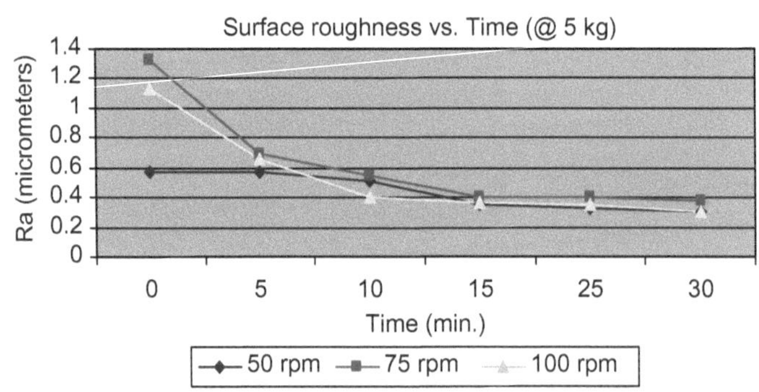

Figure 3.47 Surface roughness vs. Time (@ 5 kg)

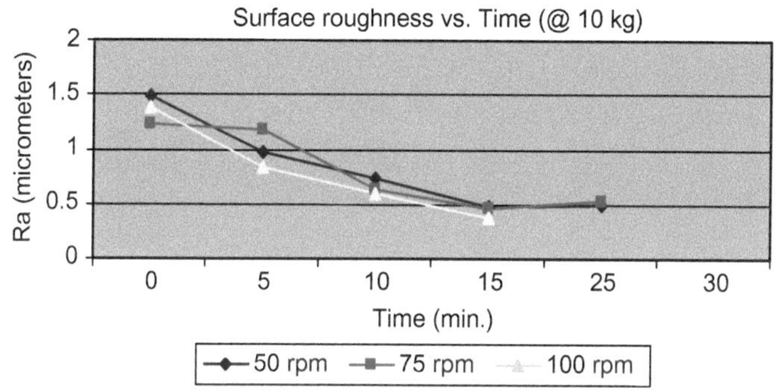

Figure 3.48 Surface roughness vs. Time (@ 10 kg)

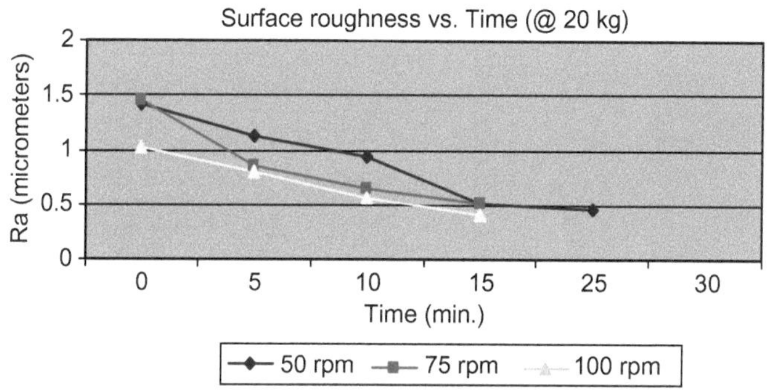

Figure 3.49 Surface roughness vs. Time (@ 20 kg)

- The surface roughness increased with time. Increasing the pressure at constant speed did not improve the quality of the surface roughness.
- Higher speeds showed a faster increase in the surface roughness values.
- A tacometer was used to calibrate the speeds of the lapping wheels.
- Due to the high viscosity of the slurry used, in some cases it was found that the final weight of the lapped pieces was greater than their initial weight. This was probably due to the work pieces absorbing a certain amount of slurry. This meant that a clear indication of the amount of material cut could not be obtained and the material removal rate (MRR) was affected. Therefore, only one of the methods was used to find the MRR.
- After the experiments, the slurry used was saved for recycling and further research.

3.3.14 Conclusions

- As the time increased, the surface roughness (Ra) decreased.
- This is the case at constant speed and constant pressure.
- At low speed (50 rpm) and low pressure (5 kg), the surface roughness did not change drastically. It only decreased by a small amount at constant pressure.
- At low speed and high pressure, the material removal rate decreased.
- The material removal rates were highest at 75 rpm.
- The material removal rates were significantly higher at faster speeds.
- The surface roughness decreased faster at higher pressures.
- The surface roughness was lowered rapidly at high speeds and pressures.

3.4 ETCHING

Etching is carried out on a wafer that has been lapped to be precisely parallel. This process chemically removes residual particles and deteriorated layers of the order of 40–50 μm thick. For wafer etching, a combination of hydrofluoric and nitric acid, which etches unselectively, is generally used because it creates less roughness and has high efficiency. The reaction is diffusion limited, and will cause deterioration in geometric precision if different etching rates occur in different regions due to local concentration variations. Etching done by rotation on its axis often brings about deterioration of parallelness by 10–20% of etched thickness.

In this section, acid and alkaline etching will be explained.

3.4.1 Acid Etching of Silicon

Generally, nitric or hydrofluoric acid, combined with acetic acid or water as a diluting agent, are used to etch silicon wafers. The etching mechanism and etching solutions have been discussed by researchers Schwartz and Klein.

The chemical reaction between silicon and the etching liquid takes place in two steps. Silicon is first oxidized by nitric acid, and then the oxide is removed by hydrofluoric acid. Kinetically, the reaction is limited either by the reaction rate or diffusion.

In a solid-liquid phase reaction, like the etching of silicon, the liquid (state A) reacts with a solid and becomes a product. Here, the rate of diffusion is Vd, and the rate of the reaction in which A becomes a product B on the solid surface is VR. When VR > Vd, diffusion of A limits the reaction, and this is called diffusion limitation. In this case, the rate of reaction relies on Vd, which is proportional to the diffusion coefficient and concentration gradient of A. This gradient is inversely proportional to the diffusion layer thickness of A, in accordance with Fick's Law. Therefore, if the liquid is agitated, making the diffusion layer smaller, the diffusion and reaction rate both increase.

When $Vd > VR$, the reaction rate is limited by the rate of conversion of $A \rightarrow B$ on the solid surface; this is called reaction limitation. Like diffusion limitation, reaction limitation is best explained as a basic A reacts on the solid surface and becomes a product (i.e. B), and in the chemical reaction of silicon, where lots of element reactions take place, a basic A is considered to be an active type and is one of the element reactions of $A \rightarrow B$.

When etching a silicon wafer, uniform surface geometry is usually created by rotating the wafer in a liquid, and hence the etching process is diffusion limited. Under conditions where the speed of the liquid differs between the wafer center and its periphery, the thickness of the diffusion layer will differ between the two and, as a result, the etching speed will vary across the wafer surface, causing deterioration in wafer geometry. This phenomenon becomes a bigger problem as the diameter of the wafer increases.

Wafers with less variation in thickness can be obtained by using a method which is limited by the chemical reaction rate. The effect that agitation of the etching liquid and the use of iced acetic acid and water as diluting agents have on the etching speed will be considered. When only iced acetic acid is used, the etching speed becomes higher as the liquid is agitated; hence the process is diffusion limited. As the amount of iced acetic acid increases, it acts as a speed reducing agent that lowers the etching rate.

In the case of water, the agitation effect of a liquid depends on the concentration C of water against nitric acid. For example, when the concentration $C < 50\%$, a liquid has an influence of agitation like in the case of iced acetic acid, and when a liquid is stirred, etching speed increases, which is diffusion limitation. When $C > 50\%$, there can be no effect of agitation of a liquid, which is reaction limitation. When both iced acetic acid and water are used as diluting agents, the rate-determining steps depend on the concentration C. A combination of HF (reducing agent), HNO_3 (oxidizing agent) and CH_3COOH (neutralizing agent) is often used for etching silicon. In this case, the reaction can be described as follows:

$$3Si + 4HNO_3 + 18HF \rightarrow 3H_2SiF_6 + 4NO + 8H_2O \qquad (3.19)$$

Table 3.9 One Example of Etching Liquid Composition and Etching Speed

No.	Nitric Acid (1)	Hydrofluoric Acid (1)	Glacial Acetic Acid (1)	Water (1)	Concentration (%)	Etching Rate (μm/min)
1	9.4 (1)	4.1 (1)	0 (1)	9.5 (1)	63	6
2	9.4	4.1	2.4	7.1	59	13
3	9.4	4.1	4.8	4.8	55	17
4	9.4	4.1	7.1	2.4	49	25
5	9.4	4.1	8.5	1	44	11
6	9.4	4.1	9.5	0	41	7

Silicon is first oxidized by HNO_3 to SiO_2 and then dissolved by HF, so the reaction progresses in two steps. The etching process should remove the distorted layers caused by machining. One example of an etching liquid composition and etching speed is shown in Table 3.9. Thickness variation is shown in Figure 3.50(a). If etching is conducted under reaction-rate limiting conditions, the reaction rate VR on the silicon surface limits the reaction; hence, it cannot be subject to the influence of varying liquid flow and so produces a wafer with uniform thickness over the entire area, as shown in Figure 3.50(b).

3.4.2 Alkaline Etching of Silicon

Etching with an alkali like KOH or NaOH is reaction limited, and the etching rate on the {100} plane is 60–100 times higher than on the {111} plane. Because of this selective etching property, etching on the {100} plane gives a very rough surface, in the shape of a reverse pyramid with depths as deep as 12 μm. This requires significant time to be spent on final polishing. However, because this etching is an elementary process reaction, geometric precision can be easily maintained after both side lapping and without taking complicated motion form made by a combination of a rotation on its axis and wafer's revolution as is the case for an acid etchant, the life of the etchant increases, thanks to the uniformity of etched thickness.

Figure 3.51 is an example of the relationship between the molar concentration of KOH/NaOH and the etching speed. Regardless of the makeup of the solution, the relationship can be represented by one line, and it is apparent that the etching rate of silicon is proportional to the concentration of OH^-.

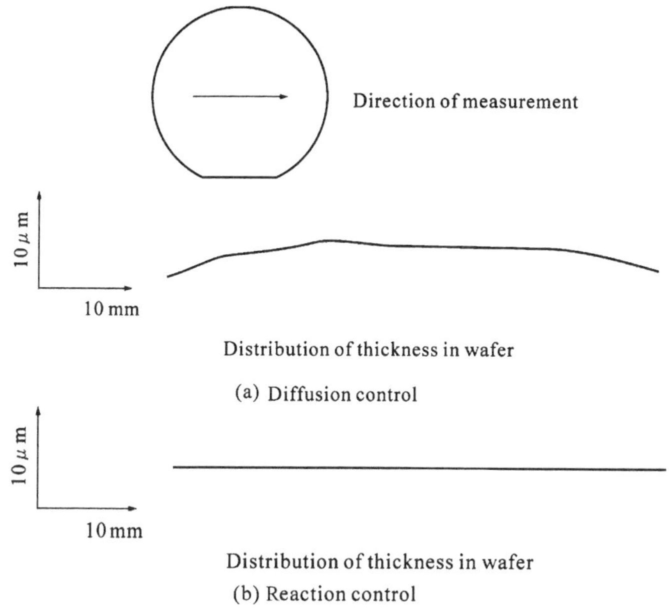

Figure 3.50 Thickness variation

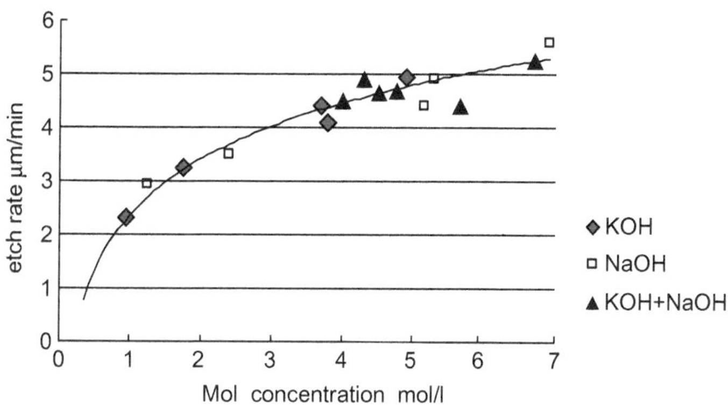

Figure 3.51 An example of the relationship between mol concentration of KOH/NaOH and etching speed

Figure 3.52 shows an example of the correlation between solution concentration and surface roughness. When the concentration is about 10 wt%, the minimum surface roughness Rz is 1.8–2 μm. When $NaNO_2W$ is added as a surfactant, an Rz of 1 μm can be achieved in either KOH or NaOH. In this case, the etched pit is very small, dense and shallow.

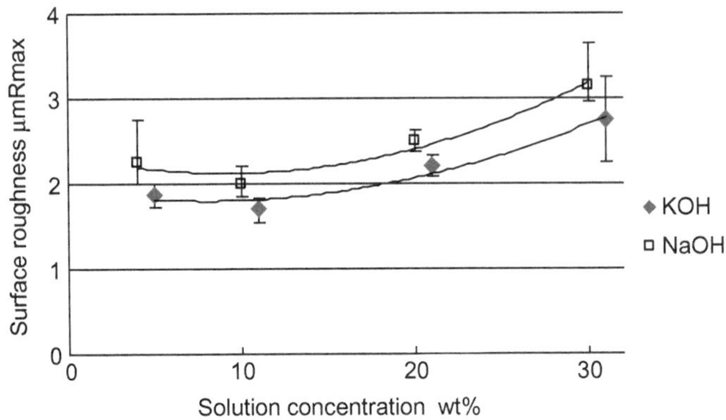

Figure 3.52 An example of the correlation between solution concentration and surface roughness

3.5 ULTRA-PRECISION POLISHING/CMP OF SILICON WAFERS

3.5.1 Overview

A wafer surface that has been chemically etched needs to be mirror finished, so it is flat and without distortion (ultra-precision polishing/CMP). Polishing/CMP of a silicon wafer is carried out by applying a fixed load to a wafer held against a pad made of synthetic leather and supplied with a slurry. The slurry commonly consists of fumed or colloidal silica in an alkaline solution. In the process, a soft silica hydrate film is formed on the wafer surface with an alkaline solution, and as machining progresses, this film is removed by the particles in the slurry.

Various methods for polishing/CMP silicon wafers have been proposed, such as combination polishing that polishes both sides to a mirror finish simultaneously, single-sided polishing in which one wafer is held onto a plate by a vacuum, and wax-free polishing in which a wafer is held by a backing pad and template without using a bonding agent. At present, the wax mount batch one-sideed polishing system is quite popular, in which a number of wafers are attached to a glass or ceramic plate, and one side is polished at a time. Figure 3.53 shows a schematic of this process. As can be seen in the figure, polishing is conducted by rotating a table on which a pad is placed and applying a load to a plate through a top ring.

The wax mount polishing process is generally done by following a flowchart, as shown in Figure 3.54. When foreign substances lie between a wafer and the plate, dimples occur, and these have to be fully cleaned.

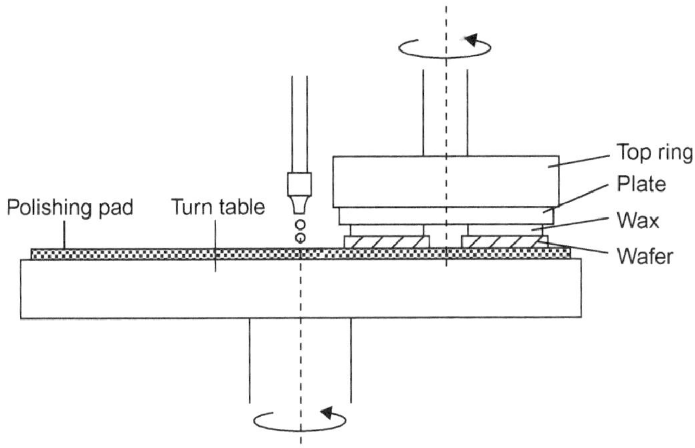

Figure 3.53 Schematic drawing of wax-mount batch type single-sided polishing method

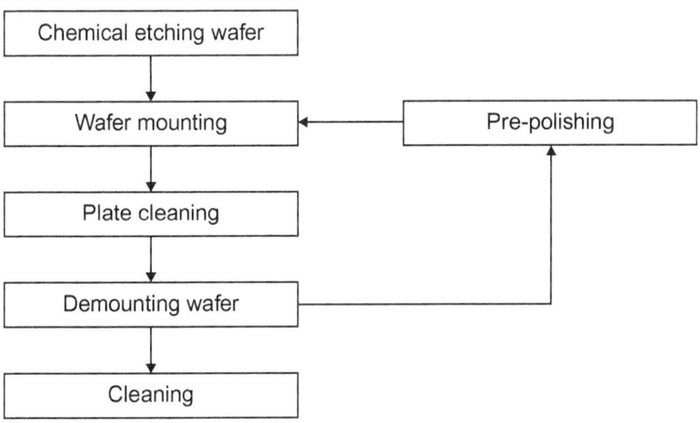

Figure 3.54 The wax-mount mirror polishing process flow

In addition, if the wafer thickness is not uniform, the wafer flatness deteriorates, so the thickness must be consistent within several μm. Three steps of polishing are necessary. The first step aims to get flatness, and the second and the third steps aim to improve the surface roughness and remove polishing scars.

During primary polishing, a hard pad (velour-type) made of a sheet of foamed urethane or polyester non-woven fabric in which urethane resin is impregnated, is used. A machining accelerating agent is added to the slurry to improve efficiency. In the second step, a suede-type pad made of urethane resin foamed on a base of a non-woven fabric (synthetic leather)

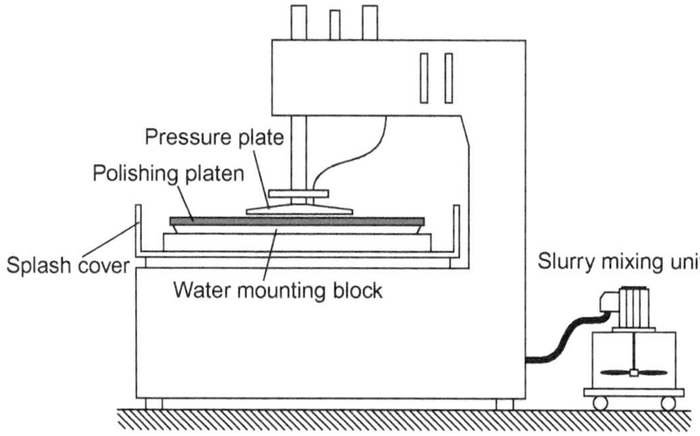

Figure 3.55 Single-sided polishing machine

is used, and a slurry with an additive to control cloudiness, called haze, observed on the wafer surface under strong light is used. Figure 3.55 shows an example of one pad for the primary and secondary polishing phases. After polishing the wafer, it is detached from the plate and sent to the cleaning process.

3.5.2 Wafer Quality and Factors Affecting It

Since a device will be fabricated on a silicon wafer after polishing/CMP, the quality and precision of a wafer will have a direct effect on the device quality. Table 3.10 shows the factors which affect the quality of the mirror-polished wafer surface. As integration increases, so do requirements for flatness. Furthermore, wafer surface roughness, particularly at the atomic level (known as micro-roughness) and residual machining distortion will have an effect on the reliability of the oxidized film.

3.5.2.1 Wafer Flatness

Some hardware factors affect wafer flatness, such as choice of polishing machine, precision of the plate that holds a wafer, unstable rotation and heat deformation of the table. Equipment manufacturers are trying to develop a highly precise machine with less instability and vibration by developing materials with low thermal expansion properties and by utilizing air bearings. Other factors affecting wafer precision are polishing conditions, the pad and the slurry. The pad plays a very important role. Figure 3.56 is an example of evaluation methods for the mechanical

Table 3.10 Quality Characteristics of Mirror-Polished Wafer

Item	Factor
Flatness (TTV, LTV)	Polisher, Polishing condition, Polishing slurry, Pad, Plate
Polishing scratch	Polisher, Polishing condition, Plate, Environment
Dimple & Peal	Polishing slurry, Pad, Plate, Environment, Wax
Micro-roughness	Polishing condition, Polishing slurry, Pad
Ripple Haze	Polishing condition, Polishing slurry, Pad
Residual (OSF)	Polishing condition, Polishing slurry, Pad, Environment

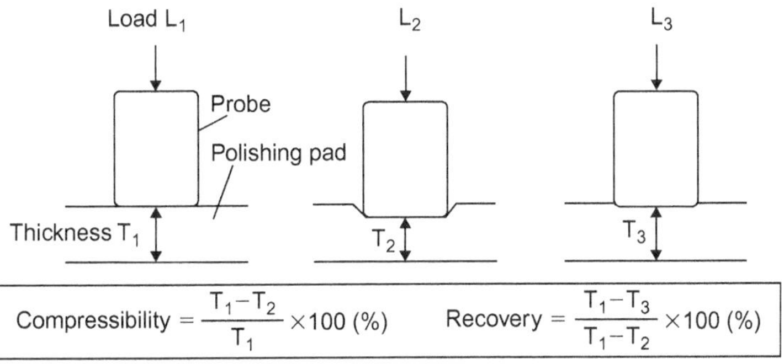

Figure 3.56 The evaluation method of a polishing pad

energy of a pad. The compression rate is equivalent to the pad's hardness. Furthermore, a smaller pad buries less of the wafer in it. The recovery rate of a pad shows the quality of the polishing fabric, which relates to creeping property. These differences in the mechanical energy properties of pads greatly influence the flatness of the resulting wafer. Figure 3.57 shows the flatness (surface roughness) of wafers polished by pads made of non-woven fabrics, at different compression rates. If the compression rate of a pad is high, the dull edge produced on the wafer periphery deteriorates the flatness. Therefore, in order to produce a wafer with high flatness, a pad with a lower compression rate should be used. Such a pad may cause scratches, however, or degrade the surface roughness, so it is important to consider the properties of different pads before choosing one.

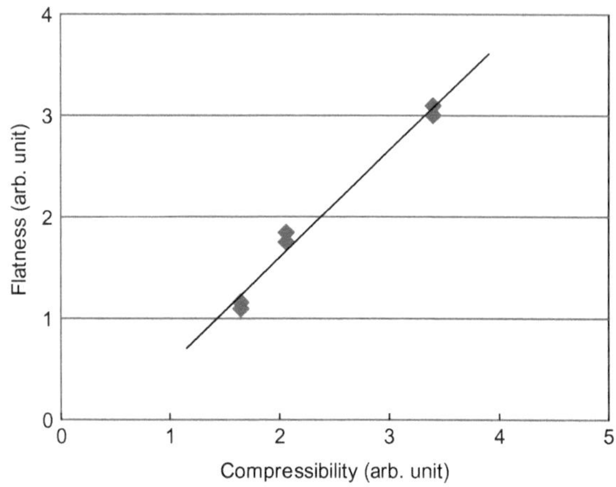

Figure 3.57 The relationship between wafer flatness and compressibility of a polishing pad

3.5.2.2 Wafer Surface Roughness (Micro-Roughness)

There is no clear definition of the micro-roughness of a silicon wafer, but surface roughness below 10 nm in P-V (peak-to-valley) value is called micro-roughness. Figure 3.58 shows types of measuring equipment for wafer surface roughness and their analytical ability in horizontal and vertical directions. There are also two evaluation criteria for mirror-polished wafer surface roughness: ripple and haze. Ripple is quantitatively evaluated with a phase shift interferometer and refers to the wave region of several to several hundred microns. In order to control ripple, it is important to control the pad surface roughness (surface geometry).

Essentially, haze is still inspected visually under strong light; however, it is now possible to evaluate wafer surface geometry at the atomic level thanks to the development of a scan-type probe microscope. Haze is considered to be microscopic surface roughness in the size range of several to tens of nanometers. The roughness cycle of the haze component is of the same order of magnitude as the colloidal silica used as an abrasive; so, if it can be presumed that haze is largely affected by the size of the colloidal silica, it is possible to improve haze by using small particles. However, small particle colloidal silica is very unstable and can easily agglomerate. Developing a stable dispersion of small particles is a current development challenge.

Alkalis in abrasives also affect the roughness of the haze component. For example, if the pH exceeds 10, a haze defect is created suddenly. This

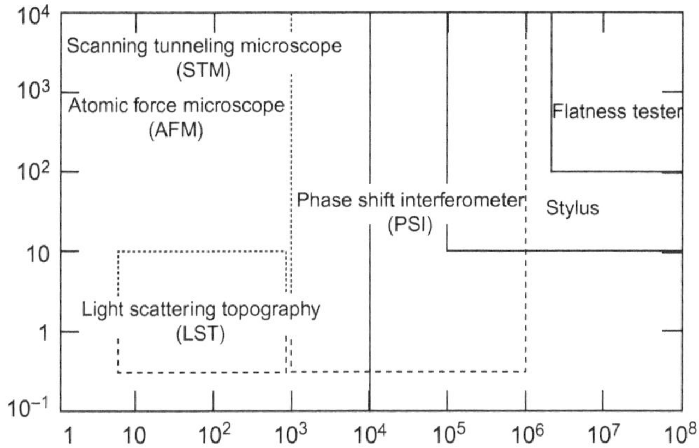

Figure 3.58 The resolution of various roughness evaluation equipment

happens because an etching reaction with the alkaline component of the slurry on the wafer surface becomes faster than the mechanical removal of the material, so the balance between the two is lost. Haze can still be seen when the pH is low, but this is attributed to the agglomeration of colloidal silica. As stated above, roughness of the haze component is determined by the balance between particle size, the chemical etching action as a result of the alkaline component, and the mechanical removal action on a silica hydrate film with particles.

3.5.2.3 Machining Distortion Caused by Polishing/CMP

(1) Oxidation Induced Stacking Fault (OSF)

Inspection of the residual machining distortion on a mirror-polished wafer is carried out by observing the surface OSF with selective etching after thermally oxidizing a polished wafer. Residual machining distortion that can cause an OSF is produced by many large alien substances that can be derived from the machining environment, like the slurry and the pad. In current mirror polishing processes, this problem is minimal, thanks to the improved working environment and cleanliness of the pad and slurry.

(2) Subsurface Scratch

A scratch-like defect can sometimes be observed when a wafer is selectively etched after it is mirror polished without the need for a heat oxidation treatment. This is called a subsurface scratch, and is a small (shallow) residual machining distortion that does not become an OSF-forming core. This type of scratch is produced by small foreign substances that get trapped in the machining environment on the slurry and pad. These residual distortions can be annealed out by a heat treatment at low temperatures (in the hundreds °C) in a short time, but they are considered not to affect device quality deleteriously in current device processes.

(3) Machining Distortion Caused by Mirror Polishing

In general, CMP is considered to be a machining method that does not produce machining distortion. However, the small amounts of distortion it causes can now be detected, thanks to developments like the thermal expansion displacement measuring method, as well as the surface life measuring technology brought by hydrogen termination with HF cleaning on the wafer surface. Figure 3.59 shows the principles of the photo-acoustic displacement measuring method. When an irradiating laser beam strikes the wafer surface, the light is absorbed and converted into heat. This heat diffuses in the silicon bulk, but if there is distortion on the surface, it prevents heat from dispersing, resulting in thermal expansion of the wafer surface. When this photo-acoustic displacement is measured by a heterodyne interferometer, the presence of very shallow residual distortions can be confirmed. However, multi-stepped alkaline cleaning is carried out in the ordinary cleaning of a silicon wafer, and thus machining distortion caused by CMP is considered to be removed completely in the current cleaning process.

3.6 PRECISION CLEANING (WET CLEANING)

3.6.1 In the Beginning

Cleaning technology for semiconductor processes was based on RCA cleaning, developed by engineers like Kern of RCA Co. in the US 40 years ago. This method uses two types of cleaning agents; SC-1 (or APM) ($NH_4OH/H_2O_2/H_2O$) and SC-2 ($HCl/H_2O_2/H_2O$). Various improvements have since been added. The standards required for wafer cleaning become more demanding each year. During micronization, wafers with higher aspects and larger diameters need to be considered, and also cleaning costs, as well as environmental impact, need to be reduced.

Dry cleaning is being considered as devices are becoming micronized; however, particles are not always removed during dry cleaning, hence wet cleaning is still an important technology. As semiconductor device

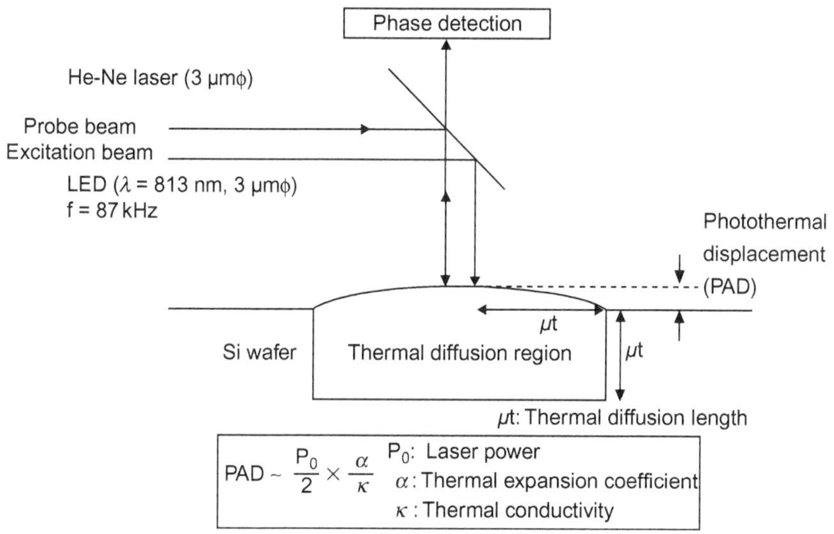

Figure 3.59 A schematic drawing of the photothermal displacement measurement

integration progresses, the requirement for cleanliness during the fabrication process becomes much higher. When micronization of a pattern progresses and the chip surface area gets larger, very small contaminants, even in extremely small amounts, have an effect on device properties. In general, particles on a wafer should be controlled below one tenth of a design rule set for particle size, and, in the case of impurity metal, on a wafer it is known that the electrical properties are affected when the impurity is over 10^{10} atoms/cm^2. Therefore, wafer cleaning is essential, requiring a prompt development of a cleaning technology for a more thorough cleaning effect.

3.6.2 The Cleaning Process and Its Details

In order to establish an effective cleaning process it is important to select the most suitable cleaning method for substances to be removed and to combine such processes efficiently. Table 3.11 shows one example of an efficient cleaning process based on RCA cleaning. Organic matter is removed in the first step. If organic matter adheres to the wafer surface, its wettability is poor, and the necessary cleaning cannot be achieved. Next, particles originating from the pre-process are removed. Both organic matter and particles can be removed at the same time by using SC-1 (APM) cleaning liquid. The next step is to remove natural oxide

Table 3.11 Effective Cleaning Sequence

Step	Materials	Cleaning Solution
1	Organic impurity	SPM, APM
2	Particle	APM
3	Native oxide	DHF, BHF
4	Metallic impurity	HPM, SPM
5	Dry	Spin-dry, Vapor-dry

SPM : concentrated sulfuric acid + hydrogen peroxide (3:1 ~ 4:1) (100 ~ 130°C)
APM : annmonia water + hydrogen peroxide + aq (1:1:5) (~80°C)
DHF: diluted hydrofluoric acid
BHF: buffered hydrofluoric acid
HPM: hydrochloric acid + hydrogen peroxide + aq

films. This is done because full removal of metal impurities is not possible if there are films on the surface. After this step, metal impurities are removed. After all these steps, the wafer cleaning process is complete, and the drying process begins.

The steps mentioned above represent the basic cleaning process. Recently, however, a trial was conducted to reduce the number of steps by using a multi-functional cleaning liquid. For example, adding a chelating agent to the APM cleaning liquid, enables it to remove metal impurities. Table 3.12 summarizes the substances to be removed, and suitable cleaning methods.

In this section, residual harmful substances on a silicon wafer are discussed, including particles, metal impurities and organic matter after CMP.

3.6.2.1 Particles

Particles that appear in a clean room are considered to be wafer contaminants because they are shed from a human body (i.e. skin, dandruff), materials, equipment and reacted products. The overall composition of contaminants is divided into organic matter, inorganic matter (silicon, oxide, etc.) and metals (SUS, alumina, etc.). In order to efficiently remove these particle contaminants, an alkali-based cleaning liquid (e.g. ammonia or choline) is best. In this case, particles on a wafer are efficiently removed by the lift-off effect of the particle based on the silicon etching, and by controlling re-adherence of particles on the wafer surface as Zeta potential of the silicon surface and particles are the same polarity in alkaline solution.

Table 3.12 (A) Adequate Cleaning Methods for Each Material

Material	Cleaning Methods	
Particle	Chemical cleaning	APM ($NH_4OH/H_2O_2/H_2O$)
	Mechanical cleaning	Ultrasonic Brush scrubbing Jet scrubbing
Metallic impurity	Wet cleaning	HPM ($HCl/H_2O_2/H_2O$) SPM ($H_2SO_4/H_2O_2/H_2O$) DHF (HF/H_2O) FPM ($HF/H_2O_2/H_2O$)
	Dry cleaning	Cl_2 + UV ($\lambda < 400$ nm)
Organic impurity	Wet cleaning	SPM, APM Ozone-water
	Dry cleaning	UV/O_3, O_2-plasma
Native oxide	Wet cleaning	DHF, BHF ($HF/NH_4OH/H_2O$)
	Dry cleaning	HF-vapor

Table 3.12 (B) Problems with the Conventional Cleaning Solutions

Cleaning Solution	Problems
APM	• Short lifetime of solution • Deposition of metallic impurities in solution • Increase of surface micro-roughness
DHF, BHF	• Metal plating (Cu, Au, Pt) • Particle adhesion • Increase of surface micro-roughness
HPM, SPM	• High temperature and high concentration of solution

Jet or brush scrubbing or sonication cleaning removes particles physically by mechanical action, which is effective for the removal of particularly large particles. However, during such physical cleaning, it has to be noted that insulation breaks down due to static electricity, creating scars and other physical damage. During sonication, a frequency region of several 10 kHz is used for the effective removal of much smaller particles. By combining physical cleaning with alkaline-based wet cleaning, particles are removed more efficiently.

3.6.2.2 Metal Impurities

Metal impurities on a wafer are generally removed by an acid solution, like HPM and SPM cleaning liquids. DHF is used to remove the metal impurities in natural oxide films. It has been reported that metal impurities are also effectively removed by chlorine radicals generated by irradiating Cl_2 gas with UV in dry cleaning.

3.6.2.3 Organic Matter

H_2SO_4/H_2O_2 (SPM) or APM cleaning liquid is used to remove organic matter. These cleaning liquids are highly acidic; this acidity oxidizes the organic matter on the wafer surface. More recently, cleaning liquids made of ultrapure water to which a very small amount of ozone has been added have been used, since they have been proved to show the same removal ability against organic matter as SPM and APM. Dry cleaning, like oxidized plasma or UV/ozone, has been used for removing organic matter, but as much drier processes are used, much more effective removal methods will be required.

3.7 INSPECTION OF CRYSTAL SUBSTRATE

If a substrate has been manufactured well without defects, an inspection process should not be necessary. However, in order to keep the yield as high as possible, inspection is indispensable at almost every stage of the aforementioned processes. Inspection is aimed mainly at screening the defects found at each step of the process. For example, in silicon single crystal growth, factors to be inspected include perfection in crystal structure, crystal defects and impurities; in slicing, warp, waviness, and higher order shape of the entire wafer are defect possibilities; surface roughness at chamfered parts are possible defects of the beveling process; in lapping, warp, waviness, and damaged layers are possible; in etching, surface defects, like etched pittings, are possible; in ultra-precision polishing, possible defects include higher order shape, waviness, edge roll-off, micro-scratches, nanotopography, micro-roughness, haze, slurry residue and wafer strain; and, finally, in ultra-precision cleaning, particles, metal impurities and residual organic matter are all possible defects. In addition to these inspections, electrical properties, like resistivity, are tested.

During the inspection that occurs after the polishing process, a general optical microscope is used. Optical interferometry and confocal laser microscopy are used for measuring wafer geometry, and light scattering is used for detecting particles and micro-scratches. These methods are combined with the SEM (Scanning Electron Microscope) and AFM (Atomic Force Microscope) to conduct a more detailed evaluation. Details of these measuring methods will be described in Chapter 4 when surface defects and inspection tools are discussed. Analysis and assessment of crystal defects and impurities are merely summarized at this point.

3.7.1 Inspection of Crystal Defects

The yield and cost of integrated circuits depend on the purity of the single crystal ingot, the number of crystal defects and the diameter of a wafer. How well defects are controlled becomes a critical factor in managing the performance and productivity of electronic devices. A rough classification by type of crystal defects follows.

Dislocation: Dislocation is a type of defect where some of the atoms are misaligned due to a slip in the silicon crystal lattice. This is referred to as a linear defect. When stress occurs due to thermal fluctuations during the CZ method, dislocations take place, which causes a leakage of electric current in an LSI. There are two types of dislocation, edge dislocation (Figure 3.60) and screw dislocation (Figure 3.61), according to the slip direction. Often a point defect, such as a vacancy, exists near a dislocation.

OSF (Oxidation-induced Stacking Fault): OSF is a defect caused by oxidation and heat treatment. It is a type of plane defect in which the discontinuity is caused by inserting or removing a close-packed {111} atomic plane. This originates from crystal growth rate, temperature gradient, and oxygen concentration in the crystal, all of which form a core to cause a ring-shaped OSF.

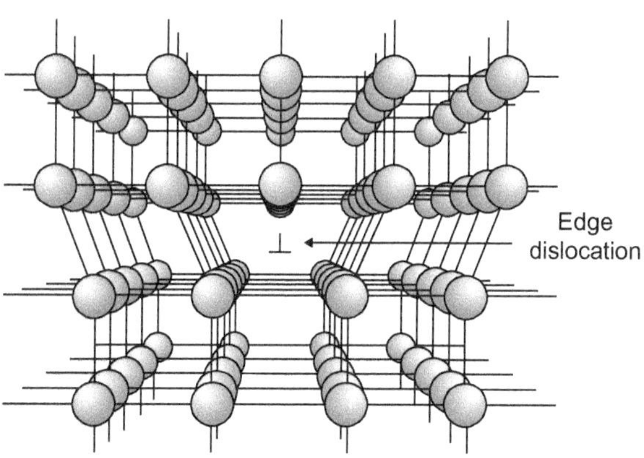

Figure 3.60 Edge dislocation

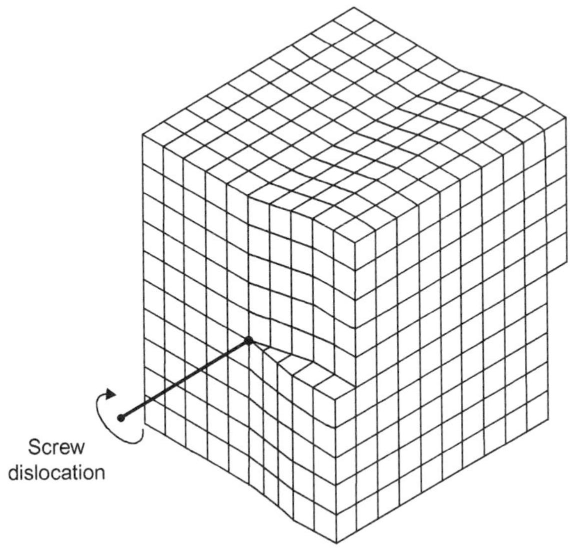

Figure 3.61 Screw dislocation

BMD (Bulk Micro Defect): BMDs are formed as SiO_2 by separation of interstitial oxygen in the thermal process. This is closely related to the oxygen concentration and is necessary for gettering technique during the device process.

COP (Crystal Originated Particle): COP is a cavity defect in the form of a regular octahedron about 0.1 μm in size in a single crystal, and its inside is coated with SiO_2. This is originated by crystal growth rate, temperature gradient, and thermal history in the cooling process of the crystal. It is smaller than an OSF and causes problems in very fine LSI patterns.

Among these defects, dislocation and stacking faults have significant influence on device properties. There are various methods to closely examine these crystal defects, which are classified by inspection probe type.

3.7.1.1 Electron Beam Probe

The TEM (Transmission Electron Microscope) is the most typical and direct observation method in use. In this technique, a sample that has been thinned to below 100 nm by FIB (Focused Ion Beam) or argon sputtering is irradiated by a parallel electron beam. This forms an image when focused by an electron lens (Figure 3.62). Resolution is determined by the aberration of an objective lens and the wavelength of the electron beam (energy of electron beam = acceleration voltage), and within

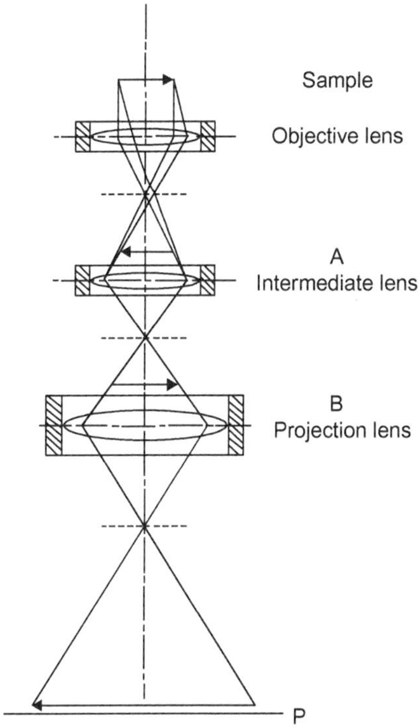

Figure 3.62 Principle of Transmission Electron Microscope (TEM): Microscopic appearance

$0.1 \sim 0.2$ nm of theoretical resolution has been obtained. With this method each one of the atoms can be observed, making it a mainstream tool for observing microstructure. This method is particularly useful for detecting dislocation and stacking faults, as it is possible to observe a damaged layer directly from observed atomic arrangements (Figure 3.63). This method can also be used to observe electron beam diffraction patterns in order to evaluate crystal structure. In the case of defect inspection, if the weak beam observation method, in which the condition is altered slightly, is used, the contrast change due to distortion can be reduced.

Other methods that use electron beam diffraction are CBED (Convergent Beam Electron Diffraction) and CL (Cathode Luminescence), which is incorporated in the SEM to detect light emitted when an electron beam is irradiated. Because the CL method can cope with light transition from the bottom of the conduction band to the top of the valence band, it does not reflect element information but, rather, properties of crystal (crystal defects, impurity, carrier concentration, stress), and so it is used to uncover

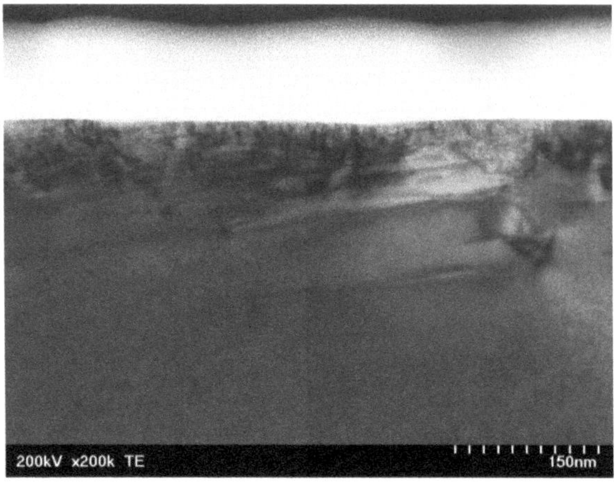

Figure 3.63 Example of layer damaged by SiC polishing observed by STEM

silicon defects, dislocation, and impurities, all of which cannot be observed by a general secondary electron image of a SEM.

PAS (Positron Annihilation Spectroscopy) is a method to observe cavity defects in a crystal. It captures energy change in gamma rays that are emitted when a positron annihilates. This method is capable of highly sensitive and non-destructive observation, and the distribution of defect depth from the sample surface to a few micrometers in depth can be observed.

3.7.1.2 X-ray Probe

A method called XRT (X-Ray diffraction Topography) irradiates a sample with X-rays focused into a sheet beam. The whole sample is scanned and the X-ray diffraction pattern is recorded. By capturing a map of the diffracted X-ray intensity, the inside of a sample can be easily observed, and inspections for dislocation in crystal orientation, change of thickness, and lattice distortion around micro-defects are possible.

XRT has been used to observe micro-defects, dislocation (slip dislocation, threading dislocation), interaction between twin boundary and a vacancy or interstitial atoms, and irradiation defects in ionic crystal and silicon single crystal radiated by various ions from an accelerator.

3.7.1.3 Laser Probe

PL (Photoluminescence Spectroscopy) uses a laser beam to capture light generated from a substance as it falls from the excited state to ground state when irradiated by a laser beam. By measuring the luminescence spectrum, it is possible to observe material imperfections and impurities.

Silicon is transparent to infrared light of 1–10 μm in wavelength. Infrared Light Scattering Tomography utilizes this property. An infrared laser is the light source, and the method captures light scattered from a defect inside a crystal with an infrared image pickup tube. If a crystal has low carrier concentration, the method can easily obtain a defect image and is particularly sensitive to micro-defects. To get a defect image, a sample is scanned with the laser and scattered light from the path of the laser is measured.

Raman Spectroscopy is a method used to evaluate crystal stress and strain. The method normally measures the shift in wavelengths of scattered light (Raman Scattering) compared to incident light in order to identify a sample substrate. In the case of a silicon single crystal, however, the Raman spectrum shifts to the high frequency side under compressive stress and to the low frequency side under tension stress, hence stress can be measured by observing the amount of the shift at the spectrum peak. One significant feature of these laser beam probing methods is that there is no need for a complex vacuum environment.

3.7.2 Impurity Inspection

Various surface analytical methods are used to observe impurities, but one method cannot cover everything, so it is useful to select a method that is suitable for a specific purpose. If necessary, several methods are used simultaneously and a cross-check is often performed. Table 3.13 contains a list of various surface analysis techniques; their names, excitation sources, detected species, detection range of elements, depth resolution and information available. The methods are classified by excitation source: ion, electron and X-ray.

3.7.2.1 Ion Excitation Source

Among the methods that use ions as the excitation source, SIMS (Secondary Ion Mass Spectrometry) is the most sensitive. It measures a number of atoms in a solid body as ionized particles (secondary ions) by utilizing a sputtering phenomenon when irradiating with the primary ion beam (Figure 3.64). With SIMS it is possible to analyze for all the elements in the periodic table, including hydrogen, at a detection limit in the ppb range (Table 3.14). By scanning the surface with a narrowly focused primary ion beam, it is possible to get a two-dimensional

Table 3.13 Various Surface Analysis Techniques for Inspecting Impurities

Name	Excitation Source	Detected Species	Detection Range of Elements	Depth Resolution	Information Available
SIMS (secondary ion mass spectrometry)	ion	secondary ion	H - U	1–30 nm	elementary composition, chemical state
TOF-SIMS (time-of-flight SIMS)	ion	secondary ion	H - U, macromolecule, cluster ion	single monolayer	elementary composition, chemical state
TEM (transmission electron microscopy)	electron	transmitted electron, X-ray	Na - U	N/A	ultra-high resolution observation, elementary composition
SEM (scanning electron microscope), EDX (energy dispersive X-ray spectroscopy), EPMA (electron probe micro analyzer)	electron	back scattered electron, X-ray	B - U	1–5 μm	elementary composition
AES (Auger electron spectroscopy)	electron	Auger electron	Li - U	1–10 nm	elementary composition (chemical state)
ESCA/XPS (electron spectroscopy for chemical analysis/X-ray photoelectron spectroscopy)	X-ray	photoelectron	Li - U	1–10 nm	elementary composition, chemical state

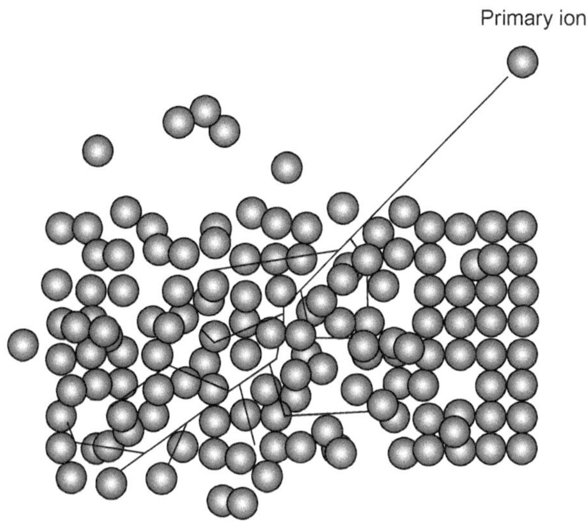

Figure 3.64 Principle image of SIMS

Table 3.14 Detection Sensitivity in SIMS

Matrix	Element	Detection Limit (at/cm^3)
Si 100 at.% = 5×10^{22} at/cm^3	H	7×10^{16}
	C	3×10^{16}
	N	5×10^{14}
	O	6×10^{16}
	B	1×10^{13}
	F	5×10^{15}
	Al	1×10^{14}
	P	1×10^{14}
	Cr	2×10^{13}
	Fe	5×10^{14}
	Ni	3×10^{14}
	Cu	8×10^{14}
	As	5×10^{13}
	Ag	5×10^{14}
	Pb	1×10^{14}

distribution of elements. Furthermore, analysis along the depth of the sample is also possible by grinding the surface with ion etching.

Since the efficiency for detecting secondary ions (secondary ion emission efficiency) largely depends on the chemical species of primary ion, the irradiation conditions and type of element, it is necessary to select the primary ion to suit the elemental analysis under consideration. In the case of an element with low electronegativity, a positive secondary ion is easily emitted by using an oxygen ion beam (O_2^+, O^-) as the primary ion, while in the case of an element with high electronegativity, a negative secondary ion is easily emitted by using cesium (Cs^+) as the primary ion. Since the emission efficiency of the secondary ion changes by matrices, to determine quantities, one must analyze a standard specimen which has a similar composition and known concentration under the same conditions as the target inspection to produce a relative sensitivity index.

3.7.2.2 Electron Excitation Source

EPMA (Electron Probe Micro Analyzer) uses an electron beam excitation source. Its representatives are EDX, EDS (Energy Dispersive X-ray Spectroscopy) and AES (Auger Electron Spectroscopy). They observe a specimen by SEM and analyze surface elements from emitted signals by irradiating a small area with an electron beam. During such irradiation, most of the energy of the incident electron is converted to heat. The electron also collides repeatedly with the atoms in a specimen, thus exciting secondary electrons and emitting electromagnetic waves of various energies (X-ray, ultraviolet ray, visible light and infrared light) (Figure 3.65). Because the detected signal is not necessarily an X-ray, this technique is called an Electron Probe Micro Analyzer. If an X-ray is emitted, the elemental analysis that uses this characteristic X-ray is called EDX; and, when an electron (Auger electron) is emitted, its kinetic

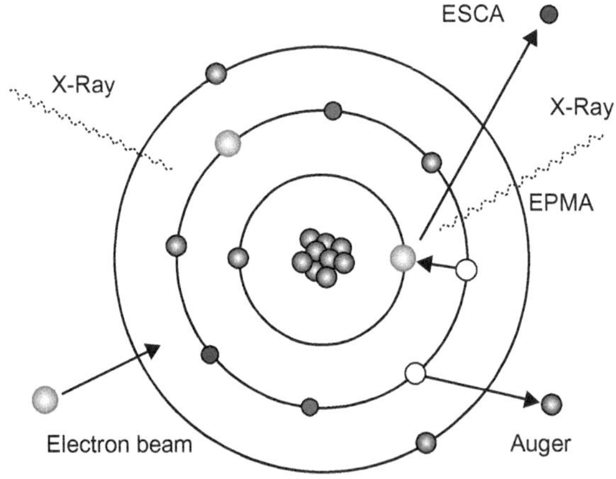

Figure 3.65 Photoelectric emission process

energy is measured and analyzed, and the technique is called AES. It is a matter of probability whether an X-ray or an Auger electron is emitted.

Since there is definite relationship between atomic number and the wavelength of the characteristic X-ray, or the kinetic energy of the Auger electron (Moseley's Rule), it is possible to carry out qualitative element analysis at the point irradiated by the incident electron. Quantitative analysis is also possible by measuring the intensity of the emitted radiation. Relatively heavy elements are detected by EDX, while relatively light elements are detected by AES. EDX can give information to a depth of 1 μm from the surface, whereas AES can only penetrate to several nm. AES makes it possible to observe a region of 10–20 nm in spatial resolution and to detect elements up to 0.1% in atomic concentration.

3.7.2.3 X-ray Excitation Source

XPS (X-ray Photoelectron Spectroscopy) is a typical analysis method using X-rays as an excitation source. It can determine binding energy by subtracting the kinetic energy of an emitted photoelectron from the constant excitation energy. This method is also known as ESCA (Electron Spectroscopy for Chemical Analysis), as it is photoelectron spectroscopy used for chemical analysis. Using monochromatic X-rays as the excitation source gives low background noise, and improves resolution. In general, a higher energy resolution is available with ESCA than with AES. Since photoelectron intensity is almost proportional to the number of atoms in a substance, quantitative analysis is possible. When observing the specimen surface while sputtering by irradiating argon ions, it is possible to analyze it in the depth direction (depth profile). Additionally, the chemical bonding state of the atoms and molecules can be analyzed by observing the chemical shift.

3.7.2.4 Infrared Light Absorption

FT-IR (Fourier Transform Infrared Spectroscopy) is the most widely used analytical technique used for solid organic matter. When infrared light irradiates a substance, specific wavelengths of light are selectively absorbed. The infrared light absorption spectrum is specific for a given substance; hence, the chemical composition of unknown substances can be found. In the technique, light emitted from an infrared light source enters an interferometer and passes through a specimen as an interference wave (interferogram). During this process, light of specific frequencies are absorbed by the vibration of specific atoms or atomic groups in the specimen.

The signal captured by the detector is Fourier-transformed to obtain the infrared spectrum of the specimen. The substance is identified by comparing the spectrum with a database of spectra of known substances. This method is complementary to the aforementioned laser Raman spectroscopy analysis, as the two methods compensate for each other's weak points.

REFERENCES

1. T. Doi, Details of Semiconductor CMP Technology (in Japanese), Kogyo-chosakai Publishing Co., Tokyo, Japan, 2001.
2. W. Zulehner, D. Huber, Czochralski-grown Silicon "Crystals 8: Silicon, Chemical Etching, Springer-Verlag, 1982, pp. 1−143.
3. W.G. Pfann, Principles of Zone-melting, Trans. Am. Inst. Min. Metall. Engg. 194 (1952) 747−753.
4. K. Hoshi, et al., CZ silicon crystal grown in tranverse magnetic field, Ext. Abstr., Electrochem. Soc. 157th Meeting (1980) 811−813.
5. P.O. Hahn, The 300 mm silicon wafer − a cost and technology challenge, Microelectronic Eng. 56 (2001) 3−13.
6. T. Doi, M. Kinoshita, H. Hiyama (Eds.), A library of CMP Planarization Technology & Application (in Japanese), Global Net Corp., 2006.
7. M.S. Kulkarni, Comprehensive upstream silicon processing for semiconductor and solar applications, The 5th International Symposium on Advanced Science and Technology of Silicon Materials (JSPS Si Symposium), (Nov. 10−14, 2008), USA.
8. T.L. Jester, Crystalline silicon manufacturing progress, Prog. Photovolt: Res. Appl. 10 (2002) 99−106.
9. I. Kao (PI), V. Prasad, J. Li, M. Bhagavat, Wafer Slicing and Wire Saw manufacturing Technology, Department of Mechanical Engineering, SUNY Stony Brook, NY 11794-2300.
10. N. Watanabe, Y. Kondo, D. Ide, T. Matsuki, H. Takato, I. Sakata, Characterization of polycrystalline silicon wafers for solar cells sliced with novel fixed-abrasive wire, Prog. Photovolt: Res. Appl. 18 (2010) 485−490.
11. J.M. Kim, Y.K. Kim, Saw-damage-induced structural defects on the surface of silicon crystals, J. Electrochem. Soc. 152(3) (2005) G189−G192.
12. I. Kao (PI), M. Bhagavat, V. Prasad, Integrated Modeling of Wiresaw in Wafer Slicing Department of Mechanical Engineering, SUNY Stony Brook, NY 11794-2300. J. Talbot, M. Chandra, K. Gupta GT Equipment Technologies, Inc., Nashua, NH 03063.
13. W.Y. Peng, Y.S. Liao, Study of electrical discharge machining technology for slicing silicon ingots, J. Mater. Process. Tech. 140 (2003) 274−279.
14. B.-H. Yu, H.-K. Lee, Y.-X. Lin, S.-J. Qin, F.-Y. Huang, B.-H. Yan, Study of Wire Electrical Discharge Machining for Poly-silicon, Asian Symposium for Precision Engineering and Nanotechnology (2009).
15. W. Wang, Z.D. Liu, Z.J. Tian, Y.H. Huang, Z.X. Liu, High efficiency slicing of low resistance silicon ingot by wire electrolytic-spark hybrid machining, J. Mater. Process. Tech. 209 (2009) 3149−3155.
16. Y. Uno, A. Okada, Y. Okamato, T. Hirano, Wire EDM Slicing of Monocrystalline Silicon Ingot, Academic Press, 1998.
17. Z.J. Pei, A. Strasbaugh, Fine Grinding of Silicon Wafers, Department of Industrial and Manufacturing Systems, Kansas State Univ., Manhattan, KS, 2000, pp. 659−672.

18. Z.J. Pei, S.R. Billingsley, S. Muira, Grinding Induced Subsurface Cracks in Silicon Wafers, MEMC Electronic Materials, Inc., St. Peters, MO, 1998, 1103–1116.
19. H.H. Gatzen, J.C. Maetzig, Nanogrinding, Institute for Microtechnology, Hanover, Germany, 1997, pp. 134–139.
20. P. Hryniewicz, A.Z. Szeri, S. Jahanmir, Coolant Flow in Surface Grinding with Non-Porous Wheels, Dept. of Mechanical Engineering, Univ. of Delaware, Newark, DE, 1998, pp. 2347–2367.
21. H.K. Tonshoff, W.V. Schmieden, I. Inasaki, W. Konig, G. Spur, Abrasive machining of silicon, Annals CIRP, 39, 1990, 621–635.
22. N. Dennis, J.L. Lucas Machinery Company, Inc., Remanufacturing double-disc grinders, Abrasives Magazine Aug–Sep. (1994) 26–27.
23. Improving the Bonds of Wheels, Abrasives Magazine Oct–Nov. (1995) 5–9.
24. I. Marinescu, Manufacturing engineering class handouts. Lapping, Micro-Grinding with Grinding Wheels.
25. Internet Source, Fundamentals of Machine Tools. <http://155.217.58.58/cgi-bin/atdl.dll/tc/9-524/toc.htm/>
26. Internet Source, Michael Weinig, Inc. <http://www.weinigusa.com/index.htm/>
27. J.L.R. Miller, Lapping and polishing technology, AES Magazine 30 (4) 9–13.
28. Masahisa, M., Fundamental Studies on Lapping, Report of the Institute of Industrial Science and University of Tokyo.
29. Internet Source, <http://www.peter-wolters.com/tools/lapping.htm/>
30. Internet Source, <http://www.lapmaster.com/>
31. Internet Source, <http://www.peter-wolters.com/tools/lapping.htm/>
32. Internet Source, <http://www.ukabrasives.com/lapping-compounds.htm/>
33. B. Schwartz, H. Robbins, Chemical etching of silicon, J. Electrochem. Soc. 123 (12) (1976) 1903–1909.
34. D.L. Klein, D.J. D'Stefan, Controlled Etching of Silicon in the HF-HNO_3 System, J. Electrochem. Soc. 109(1) (1962) 37–42.
35. T. Hamaguchi, Chemical etching of silicon, J. JSPE (in Japanese) (1984).
36. K. Koyabu, J. Watanabe, T. Karaki, J. Suzuki, Trial making of Stencil type mask for electron beam, Spring meeting of JSPE (in Japanese) 1978 pp. 505.
37. J. Suzuki, Chemical etching of Si substrates, Fall meeting of JSPE (in Japanese), 1979, p. 85.
38. I.D. Marinescu, E. Uhlman, T.K. Doi, Handbook of Lapping and Polishing, CRC Press (Taylor & Francis Group), NY, 2006.
39. M. Miyashita, et al., J. Electrochem, Soc. 139(8) (1992) 2133–2142.
40. J. Masumura, et al., Surface Science Technology Series 3, Realize Inc., 1996.
41. T. Karaki, Mechano-chemical polishing and generation of OSF, precision machining technology of crystal materials for electronics (in Japanese), Sci. Forum (1985) 625–629.
42. S. Sumie, et al., Mat. Rec. Symp. Proc. 236 (1992) 241–246.
43. W. Kern, D.A. Puotinen, Cleaning solutions based on hydrogen peroxide for use in silicon semiconductor technology, RCA Rev. 31 (1970) 187–206.
44. W. Kern, The evolution of silicon wafer cleaning technology, J. Electrochem. Soc. 137 (1990) 1887–1892.
45. F. Shimura, Semiconductor Silicon Crystal Engineering (in Japanese), Maruzen, 1993.
46. M. Takiyama, et al., Proc. 195th Workshop on Ultra Clean Technology 94 (1992).
47. T. Itoh, et al., Semiconductor World (in Japanese) 120 (1989).
48. Internet source, <http://www.azonano.com/>
49. Internet source, <http://www.tsc-web.jp/map/>
50. Internet source, <http://www.toray-research.co.jp/index.html/>

CHAPTER 4

Applications of Ultra-Precision CMP in Device Processing

4.1 OVERVIEW OF THE SIGNIFICANCE OF, AND TRENDS IN, PLANARIZATION CMP

Planarization CMP (Chemical Mechanical Polishing) technology applies ultra-precision polishing to device processing in an effort to make the profile of each interconnection layer of an ultra-LSI device almost perfectly planar. It began as a new polishing method that combined chemical and mechanical action in order to efficiently create a flat, mirror surface with no distortions in crystallography. This improved the conventional mechanical polishing method in use when semiconductor crystals (germanium, silicon, etc.) were first used in 1947. This was the first step forward for CMP, and the technology has made significant progress since then, contributing to the fabrication of high performance semiconductor devices. The technology can be said to have the relationship of "a pair of wheels to a car."

Ultra-precision processing technology, particularly CMP technology, can achieve planarity at the atomic level for bare silicon wafers. This achievement can be applied to the planarization of LSI devices, meaning that a pattern can be formed on the steps, which used to be a major lithographical challenge. Not only has flattening/planarization of inter-layer insulating films become possible with planarization CMP technology, but so has STI (shallow trench isolation), the embedding interconnection method and the Damascene interconnection method, the latter two being indispensable for the highly efficient micron-sized devices envisaged in the future.

The Damascene process has now become firmly established as an indispensible technology for producing copper (Cu) multilayer interconnections, and it also has influence over exposure technology,

Advances in CMP/Polishing Technologies
DOI: 10.1016/B978-1-4377-7859-5.00004-1

transistor design, and interconnection design. In the field of exposure, micronization has been accelerated by applying planarization CMP. In transistor design, STI has been developed, which relies on CMP. Multilayer interconnections can achieve their design performance by application of CMP technology, which has solved many difficult problems and allowed significant progress in technologies that support the development of semiconductor devices.

Ultra-LSI devices now contain multilayer interconnections to speed up device signals and actions, and Cu, which is electrically resistant, is used as the interconnecting metal. These devices use "low-k, super low-k" materials. These are insulating materials with a small dielectric constant (k), which reduce the parasitic capacity that occurs due to the ultra-micronization of the interconnection site. It was 250 nm node lock device Damascene interconnection when Cu interconnection was introduced. This extended to 180 nm node lock devices, or memory devices or system LSI, and it was after 130 nm node, when a Cu interconnection was actualized in a full scale and mass production began. SoC (System on Chip) mainly applied to logic devices, is driving LSI technology. Non-volatile memories (flash memories) has grown along with the advancement of cellular phones and digital cameras. Storing both memory and logic devices on one chip is receiving much attention at present. Cu/low-k materials are key concepts for these developments.

The most recent semiconductor devices aim for ultra-micronization, high-speed action signal, low consumption of electric power and low cost, hence wafer diameter has increased, in order to achieve improved productivity and lower costs. Currently, a 300 mm (12″) diameter wafer is the most common. Table 4.1 shows the progress made in each CMP technical field, including process/equipment technology, cleaning technology, measurement/assessment technology and device fabrication. At present, for a 40 nm generation device, Cu/low-k ($k = 2.3 - 2.7$), 10–12 interconnected layers and <20 nm in dissing/erosion are going to be applied increasingly to 300 mm wafers. Edge exclusion of 2 mm and throughput of more than 50 pcs/h in terms of ILD and interconnections can be achieved.

The ITRS (International Technology Roadmap for Semiconductors) is revised every two years, and is used as an indication of the state of development in international semiconductor technology. It shows estimated values for various technical items related to devices, indicates technical challenges, and proposes solutions for these problems, aiming

Table 4.1 The Achievement Level of Each CMP Technical Field, Including Process/Equipment Technology, Cleaning Technology, Measurement/Assessment Technology and Device Fabrication

Field	The Achievement Level for Fiscal Year 2010
Processing/equipment technology	ϕ 300 mm are the mainstream of mass production. The mass production trial starts for ϕ 450 mm. Edge exclusion (EE): 3→2 mm, setting removal rate variation <2.5% (1σ). Throughput >50/h (interlayer interconnection) Special note: Practical implementation of low-k and dual/Damascene Cu-CMP process, low polishing pressure due to Cu/low k, direct low k−CMP, application to TSV, etc. Cu−CMP: Polish uniformity WIWNU <~2%, WTWNU <~2% (1σ). Dishing <30~10 nm (20 μm isolated pattern). Erosion <20 nm (L/S: 20 μm/0.25 μm). Cu removal rate >1000 nm, barrier polishing: Cu/barrier film 2 step polishing (consider about the nanotopography problem, dishing erosion <50 nm)
Cleaning/washing technology	In the case of Cu-CMP: ϕ 300 mm: Metal impurities concentration <1×10^{10} atoms/cm^2. ϕ 300 mm: Particle size (include scratch). Surface 80/wafer (>0.225 μm). Back side 200/wafer(>0.16 μm).
Measurement/ evaluation technology	Practical realization of APC (Advanced Process Control). Integrated management of in-process and in-line monitor, interlocking movement with other process, CIM−ization. Accuracy of in-process monitor <1% (practical value on the oxidation film). Particle detection performance >0.16 μm (practical value on the oxidation film). In-process monitor (mainly used for EPD), in-line monitor (feed back the information about uniformity, planarity, deflection of wafer to the process).
Device−ization technology	Apply Cu-CMP to hp 32-45 nm generation device. Number of wiring layers 10−12 layer, wiring width: 0.05~0.8 μm, permittivity of interlayer dielectric: k = 2.3~2.7 (dishing) + (erosion) < 30 nm (Metal 1), wiring height <5~10%.

for performance improvement by scaling. According to the latest report, in the short-term, the main topics of interest until the 35 nm node can be achieved – estimated to be in 2013 – are (1) the introduction of new materials, like low-k insulating films and high-k gate insulating films (2) machining and controlling of 3-dimensional geometry and (3) improving reliability. On the other hand, after the 22 nm node is achieved, estimated to be in 2016, technical ways forward cannot yet be seen. This is called the Red Brick Wall (a large, technical obstruction), and the possibility of developing materials other than Cu/low-k needs to be discussed and reviewed.

Table 4.2 shows the progress and future estimation of the main items related to LSI devices and CMP, extracted from the ITRS. It is necessary to use Cu as a resistant stake for interconnections and low dielectric constant (low-k) materials for interlayer insulating films. Among these, porous low-k materials ($k = 1.8-2.5$) are considered indispensible for ensuring the reliability of 32 nm node multilayer interconnections. In general, a low-k material is extremely low in mechanical strength. Additionally, its adherence to CVD films (cap films), like SiO_2 and SiC, is poor, resulting in the occurrence of delamination during Cu polishing. In order to prevent this fatal phenomenon, strong low-k materials are being developed. At the same time, in order to reduce the mechanical stress on these materials, low pressure conditions are being discussed for CMP. When studying this explanation of CMP technology, these technical challenges and future prospects for the technology should be kept in mind.

Firstly, technology must be developed that can cope with the lower mechanical strength of the low-k materials used for interlayer insulating films. One measure to resolve this is to find ways of using CMP at pressures low enough to prevent the delamination of low-k materials. In the future, new processing methods, like ECMP (Electro Chemical Mechanical Polishing), which uses electrolysis, are needed. The decreasing mechanical strength of low-k materials may cause stress concentrations at the interconnection edge, so it is necessary to prepare for the creation of new defects.

The reduction of interconnection capacity due to micronization lowers the interconnection film thickness in the lower layers of the interconnection, such as the metal 1 layer (M1), which means film thickness precision is of the utmost importance, and further improvement is needed in the precision of film thickness that can be achieved

Table 4.2 The Progress and Future Estimation of Main Items Related to LSI Devices and CMP (Extracted from ITRS)

Production and Shipment Year	2001	2002	2003	2004	2005	2006	2007	2010	2013	2016
DRAM 1/2 pitch (nm)	130	115	100	90	80	70	65	45	32	22
MPU ASIC 1/2 pitch (nm)	150	130	107	90	80	70	65	50	35	25
Number of metal layers	8	8	8	9	10				11	
Edge exclusion (mm)	3				2					
Local interconnection Cu erosion (nm) @ 50% density 500 μm square array	28	24	20	18	16	14	13	5	4	3
Medium interconnection Cu erosion (nm) @ 50% density 500 μm square array	36	30	27	23	20	18	18	12	9	7
Global interconnection Cu dishing + erosion (nm) @ 80% density 15 μm width wire	67	57	50	48	40	35	32	24	17	13
Barrier/cladding thickness (nm)	16	14	12	10	9	8	7	5	3.5	2.5
Specific inductive of bulk insulating film (k)	<2.7	<2.7	<2.7	<2.4	<2.4	<2.4	<2.1	<1.9	<1.7	<1.6

Reference: MPU Interconnect Technology Requirement et al. (ITRS).

with CMP. At the same time, as the micronization of interconnection progresses, the profile surface area will shrink yet further, which causes electrons to get closer to a mean free path and a big problem of specific resistance to increase due to phenomenon of electronic dispersion at the border and interface occurs. Therefore, certain measures, such as raising the reflectivity. It is essential to minimize the surface roughness of the trench of Cu interconnections.

As micronization reduces the interconnection space and the interconnection profile surface area, reliability becomes a problem, in terms of time-dependent dielectric breakdown, electromigration, and stress-induced voiding. These factors are related to elements like cleanliness, adherence, and interface properties at the interface of Cu-cap insulating films that are planarized by CMP; therefore, the quality of insulating films and the metal surface must be improved, and it is also necessary to review Cu alloy and metal cap technology, as well as planarization CMP technology.

High expectations are placed on planarization CMP for achieving higher performance LSI devices. The main challenges for future CMP technology can be summarized as follows:

1. Measures should be taken to prevent any reduction in mechanical strength that may occur due to the introduction of low-k materials.
2. Measures should be taken to improve CMP process control so that it can cope with the reduced film thickness that has resulted from micronization.
3. Measures should be taken to heighten the quality of CMP surface/interface in order to secure electric reliability in micronized interconnections.

Options in diverse specialist fields, plating and equipment technology and cleaning need to be considered when addressing these issues. In addition, consumable material technology (represented by slurry and pads), measuring equipment, and evaluation technology related to machines, precision, electronics, physics, chemistry, measurements and optics, must be thoroughly reviewed in order to better understand device process technology.

Figure 4.1 indicates the technologies and control factors of each unit in the CMP system structure that is used for the planarization of a wafer for an LSI device. As shown, there are a significant number of factors and elements to be considered for each process and unit.

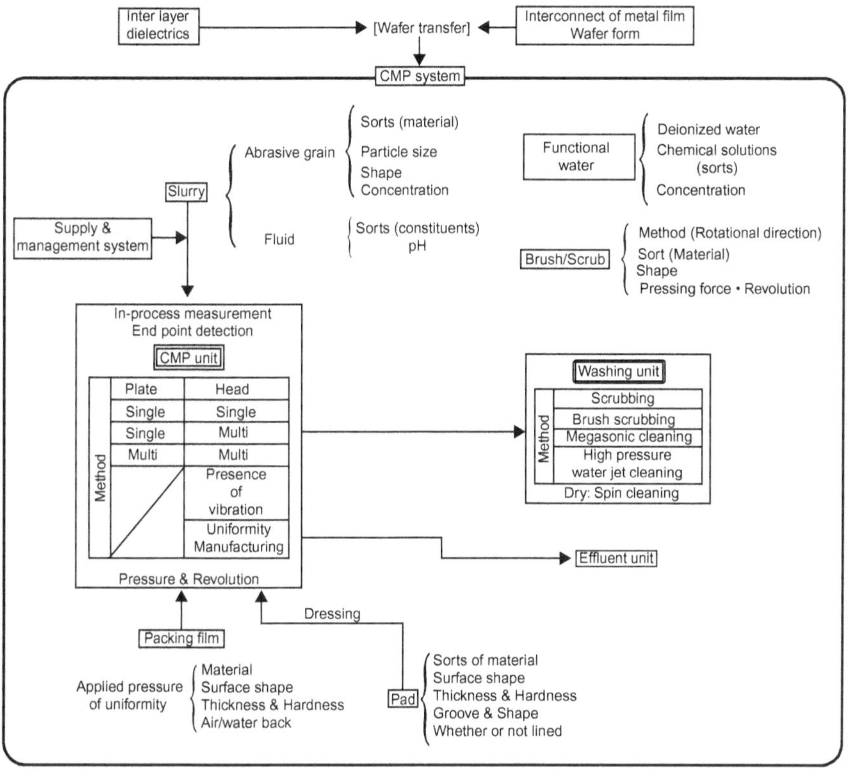

Figure 4.1 Element technologies and control factors of each unit in the CMP system structure

4.2 BASIC STRUCTURE OF THE CMP SYSTEM

Generally speaking, the most important properties required for a CMP system are:

1. The process system has to have ultra-precision and high rigidity, so that both planarity of wafer level and uniformity of residual film thickness can be reproduced in a stable manner for machining margins of less than 1 μm on the surface of a wafer with a diameter of several hundred millimeters.
2. The material removal rate from the wafer surface has to be high, in order to achieve high throughput.

Currently, a CMP system can achieve a productivity of 30 pcs/h/head by using a wafer with a diameter of 300 mm (12″) at a machining speed of over 0.5 μm/min. In view of running costs, the CMP system needs to be lighter and smaller. When considering using more than 10 CMP systems in a clean room, the overall footprint of the machinery needs to be minimized. Various options for CMP equipment and system structures that can achieve machining precision and productivity simultaneously have been discussed. Both one-by-one and batch systems have pros and cons, and it is difficult to say whether one is better than the other.

Polishing is the final finishing process for a bare silicon wafer, and efforts are continually made to improve it, improving wafer quality and precision by dealing with stricter requirements each year. For example, it has achieved a surface roughness below the nanometer level. Wet polishing has limited applicability in the wafer production process, because of concerns over contamination. New methods in device processing require realizing more precision, higher efficiency and more cleanliness in a process technology that is hard to control; only instinct and experience aid in handling new methods in device processing. Peripheral (circumferential) equipment and automation are indispensible. With this in mind, the basic requirements for design of a CMP system can be summarized by the following three points:

1. The peripheral equipment needs to ensure cleanliness in the polishing unit and be equipped with a FAB environment to deal with proper disposal of the slurry and waste fluid. It must also have a loading and unloading station for wafer cassettes and a handling system, such as a wafer moving mechanism.
2. In terms of automation, the equipment should have a monitoring and control system for the CMP process for distribution of machining pressure, finding the machining end, controlling the pH of the slurry, loading of pads, and controlling the temperature.
3. As a part of the semiconductor fabrication line, the CMP equipment should have a unified control and management system.

Table 4.3 summarizes the development of CMP equipment.

CMP is conducted on work piece materials for a diverse range of ultra-LSI devices, such as SiO_2 for interlayer insulating and embedding insulating films, connecting metals (copper, aluminum or tungsten) for interconnections and interconnection layers, barrier metals (Ta (Tantalum), TaN (Tantalum Nitride), TiN (Titanium), WN (Tungsten Nitride)) for preventing metal diffusion, polysilicon for forming capacitors in trenches, and Si_3N_4 for a stopper of machining and for insulating films. Above all it is essential to process work piece materials to perfect

Table 4.3 Points for Development of CMP System

Term \ Aim			Uniformize in wafer surface and between wafers	Planarization	Stabilization of rate High through put	Cleaning	Cost reduction	Note
Main construction and mechanism	Main processing mechanism part	Head part Platen part	Rigidity (anti-polishing resistance, anti polishing pressure), Method of wafer keeping and pressuring, Backing film (Wafer holding characteristics, Curve compulsion characteristics, Equable pressure performance, Vacuum chuck performance). Shape and size of retainer, Uniformize of trace density (Number of revolutions, Fluctuation, Vibration amplitude, etc.). Avoidance of resonance point, Long-term stability of various movement					End point detection monitoring technique (Low raising dust design, Air current control)
		Conditioning part	Same as above Uniform motion mechanism (e.g. Grind stone), Uniformize of trace density (Uniform processing of pad surface)					High accuracy high rigidity lightweight design Low thermal expansion materials design
	Wafer washing mechanism part		High efficiency slurry removal, Perfect removal of metal pollution and particle (Less or equal to 10^9 piece/cm^2), Single wafer processing washing and the stability, Washing process (Brush scrub, Megasonic, Wash solution, Temperature, Time, ...)					Anti-chemical resistance countermeasurement
	Wafer conveyance mechanism part		Loading/unloading station of wafer. Cassette to cassette, Inversion mechanism of wafer, High-speed movement. Raising dust prevention					System control technology
Consumption meterial	Slurry		Grain size distribution, Dispersibility, pH level. Additive, Stability, Variety and shape of grain, Temperature management					Waste fluid treatment technology Cost reduction, recycling Abrasiveless slurry
	Pad		Elastic deformation characteristics, Following characteristics, Surface texture and shape, Pressure distribution (Wafer entire surface : Uniformize, Local part: High pressure only in salient) High speed stability, Restraint of hydroplane phenomenon, Temperature countermeasurement					Cost reduction Long term stability Slurry-less pad

planarity (i.e. without steps or micro-roughness within a given machining margin) and to high quality without process defects. Of course, high machining efficiency (throughput) to meet production requirements is also necessary.

Figure 4.2 is a schematic diagram of the relationship between the slurry of solution-dispersed particles and the pad for CMP of a wafer for an LSI device, as well as the factors that affect the two. There are a large number of such factors; hence, a range of studies to optimize them, including CMP system and processing conditions, is required.

Various machining defects can potentially occur when machining a wafer with a pattern. Figure 4.3 shows machining defects that can occur during CMP of metal, including thinning, recessing, dissing, micro-scratches, and

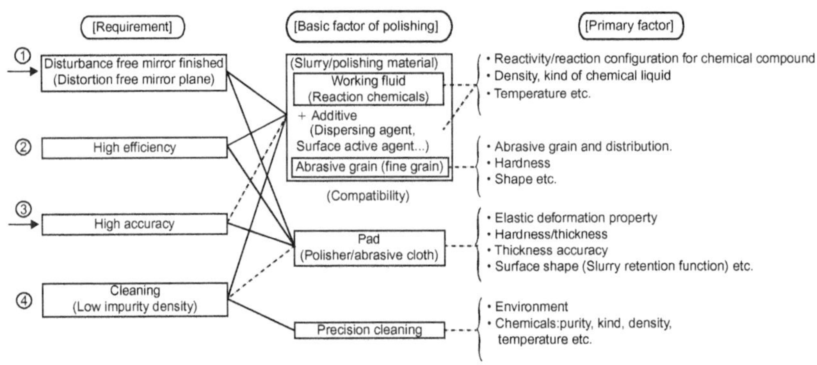

Figure 4.2 Relationship between requirements and basic factors in CMP

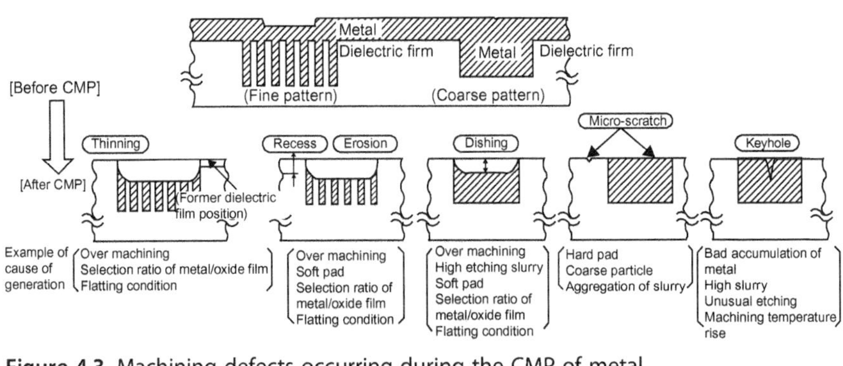

Figure 4.3 Machining defects occurring during the CMP of metal

keyholes. Their causes are mostly related to over-machining and the relationship between the slurry and the pad, as can be seen in the figure. Machining conditions must be correct, and the machined surface must be without defects to obtain planarity and uniformity.

The requirements for the planarization of a wafer by CMP can be summarized as follows:

1. Micro-roughness on a pattern must be planarized and flattened on a wafer.
2. Cleaning of the machined surface is necessary after CMP.
3. End of machining has to be clear.
4. High throughput, stable machining, and cleaning property are essential.
5. The system must be small and light.

Of these points, 1–4 are related to machining, 2 and 4 relate to cleaning, 5 relates to system design and 4 relates to the wafer carrier. Requirement 3 should be considered alongside measurement during machining, although stable machining is fundamental.

Design considerations for a CMP system cover details of machining, cleaning, carrying and controlling arrangements. The CMP system has basically the same structure as the polishing equipment used for bare silicon wafers, but, because extremely high precision is required, various plans and improvements are required. Because it is part of the semiconductor device fabrication process, various highly efficient functions have to be included, like full automation (cassette to cassette), a clean room, a close link between steps like machining and cleaning, measurement during machining (to find the end of machining), and communication with a host computer (telecommunication function).

Figure 4.4 shows a framework in which a CMP system is one part of the semiconductor production process, including important controlling factors. The CMP system consists of two main parts – the machining system and the cleaning system – together with peripherals, such as carrier, control, various equipment and tools. In terms of precision and productivity, the control and conditioning of the pad during the machining process are important; however, rough cleaning with pure water, polishing, finish cleaning and drying are important for removing particles and metal impurities during the cleaning process.

Figure 4.5 is a schematic diagram of the basic structure of the polishing mechanism, using the platen/rotary type CMP system as an example. As can be seen, the system is composed of a polishing head and its driving mechanism. The latter applies rotation and pressure to a wafer that is held

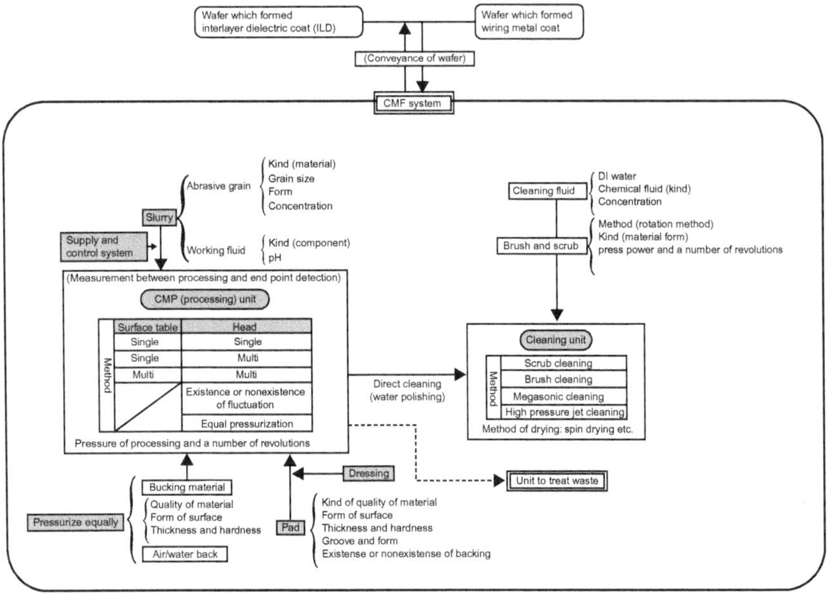

Figure 4.4 A framework of the CMP system

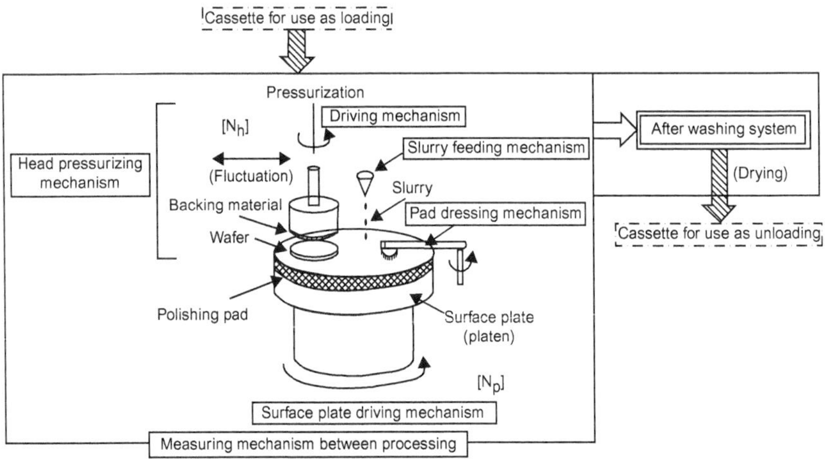

Figure 4.5 A schematic picture showing the basic structure of the polishing mechanism using the platen/rotary type CMP system

at the top. The equipment also contains a platen to which a polishing pad and its driving mechanism are attached, a conditioning (dressing) mechanism for the polishing pad, a cleaning mechanism for the wafer and chuck surface, and a slurry supply mechanism.

4.2.1 Polishing Head

In the planarization process, machining margins are as small as 1 μm and the required precision is below 10 nm, so the wafer must be held precisely in position, and pressure must be applied uniformly to the entire surface. Hence a polishing head that holds a wafer and applies pressure is considered as a central part of CMP processing. The polishing head consists of the following three parts:

1. A part that applies uniform pressure to the back of the wafer.
2. A retainer ring that prevents detachment of the wafer.
3. A chuck that holds the wafer while it is carried.

The part that applies pressure is the most important because it controls the uniformity. Additionally, the retainer ring prevents the wafer from detaching during machining, improves precision, and minimizes dull edges. The chuck holds the wafer within the retainer before machining, and releases it onto the pad. After machining, it holds the wafer until it reaches the next station, where air is applied to detach it from the chuck. These functions are stored in the chuck, and, while not directly related to machining precision, they are important actions for the firm holding and placement of a wafer. The polishing head can either rotate or oscillate.

4.2.2 Pad Table (Platen)

The platen, to which the pad is attached, rotates at several tens rpm, as shown in Figure 4.5, and the head rotates similarly. The surface precision of a platen is very important in polishing bare silicon wafers.

It should be noted that the temperature of the pad surface rises during machining. CMP takes less than 1/10 of the machining time for a bare silicon wafer, so the rise in temperature in this process is much smaller than that seen in machining, but it still cannot be ignored. The driving mechanism of rotation, the uniformity of relative movement between a wafer and a pad (platen), the density of motion track, the surface precision of a platen, and vibration control during machining are all important. This means that the use of belt and gear apparatus is avoided as much as possible for the rotation of a head and a platen, and the direct drive method is usually employed.

Very heavy loads are applied to the platen during machining. For example, a 300 mm (12″) diameter wafer to which pressure of 300 g/cm^2 is applied will experience a pressure of over 200 kg (Figure 4.6). Therefore the rotating platen needs to be extremely rigid in order to reduce any

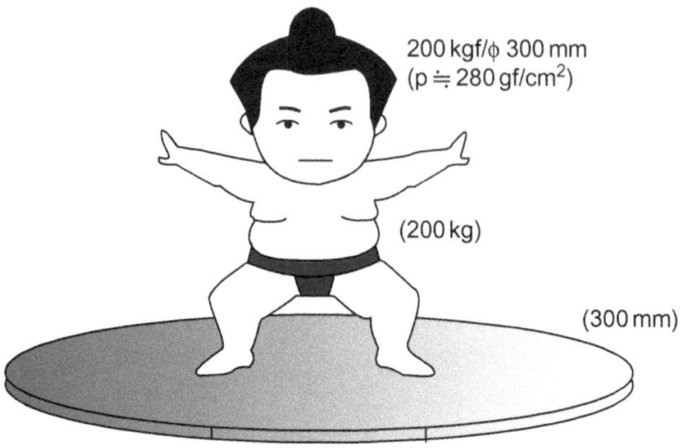

Figure 4.6 An example of polishing pressure

variation in applied pressure as much as possible. Large bearings must also be used for maximum machining precision (uniformity in particular).

4.2.3 Conditioning the Pad

Maintaining the machining ability of the pad is extremely important. This is particularly true when the end of machining is judged from the machining time. Conditioning (dressing) of the pad is carried out to restore its surface to prime condition, as shown in Figure 4.7. As polishing progresses, machining chips, product reactions, and solid bodies in the slurry produce small holes in the pad, which cause the machining quality to deteriorate. Conditioning also helps to maintain the planarity of the pad.

The main method of conditioning is to remove the pad surface layer by using a diamond wheel (D#100–#300). This is done after a given number of wafers have been processed. The conditioning function can be installed in the head in a retainer where the wafer is held. The dresser must uniformly cover the entire radius direction of a pad to ensure planarity, and must remove the material consistently from the entire pad surface.

Conditioning can be done by applying ultrasonic vibrations to the dresser at the same time as applying high pressure to the wafer to effectively recover from loading by removing the pad surface. A conditioning method to recover from loading, instead of removing material from the pad surface, is also proposed. The method proposed by Asahi Sunac Co. sprays ultrapure water at pressures between 10 and several 100 MPa (see Figure 4.8). This method is extremely useful for conditioning a pad made of non-woven fabric, such as Suba-type or synthetic leather (suede-like Politex)

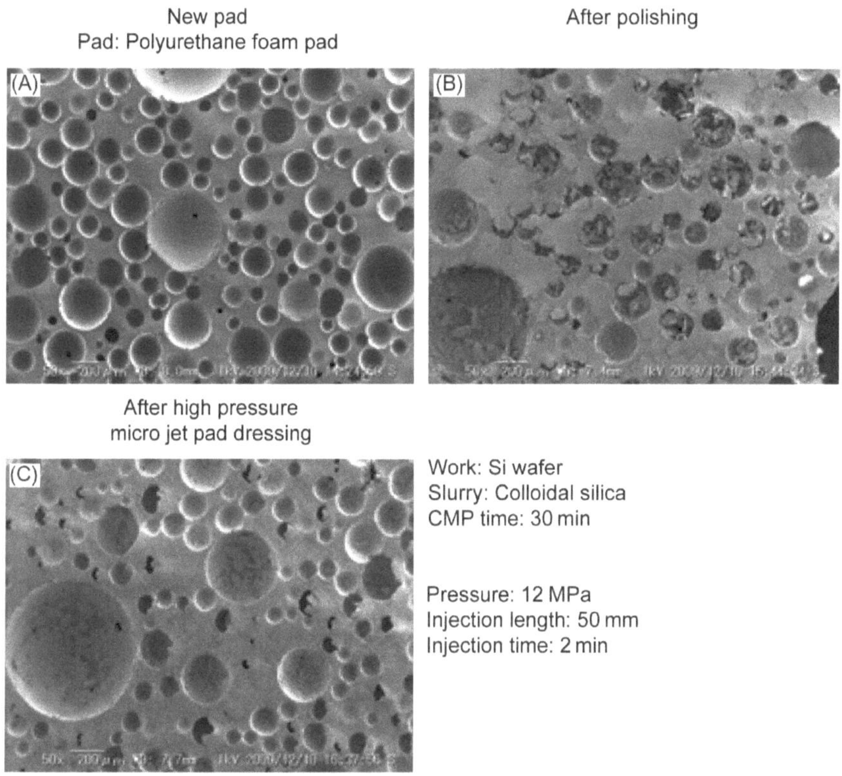

Figure 4.7 Conditioning (dressing) of a pad

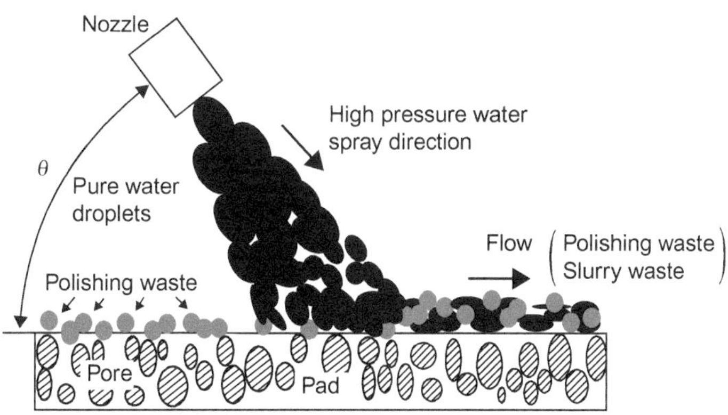

Figure 4.8 Spraying ultrapure water from an high pressure micro jet (HPMJ)

Table 4.4 Features of Each Conditioning Method of a Pad

Tool • Method	Removal of Pad Surface or Nothing	Efficiency
Fixed abrasive (diamond) tool	Existence (high)	High
Ceramics with groove	Existence (high)	Low
Diamond tool + ultrasonic	Existence (middle)	Middle
High pressure atomizing jet	Nothing	Middle

pads, but is not suitable for pads of IC1000. Table 4.4 shows features of the various pad conditioning methods.

4.2.4 Measuring Mechanism Used During Machining

In general, the CMP process has larger variations in machining properties than other LSI production equipment, and operating conditions are decided by repeated machining and measurement on many wafers prior to actual machining. This expends time and consumables, so CMP is carried out under much more stable machining conditions. Together with control over machining time, measurement during machining determines the end of machining.

Various measuring methods have been proposed and developed, but none are satisfactory in terms of safety. A lot of different measuring equipment is mounted in the CMP system:

1. Torque detection, which measures change in friction coefficient between the wafer and pad as a change in rotation torque of the holding head and plate.
2. The static electric capacity method, which detects the thickness of residual insulating films on a wafer.
3. The optical equipment, which monitors film thickness by using a laser beam in a rotating plate.
4. Vibration analysis equipment that analyses the frequency spectrum, as detected by a vibration sensor attached to a head or spindle.
5. The differential trance detector installed in a head.

4.2.5 Slurry Supply Mechanism

The slurry supply mechanism has a direct effect on the quality of the CMP process. For example, if the slurry is not treated and supplied properly, particles in it agglomerate and cause machining defects. During the CMP of oxidized films, the supply of the slurry is controlled so that

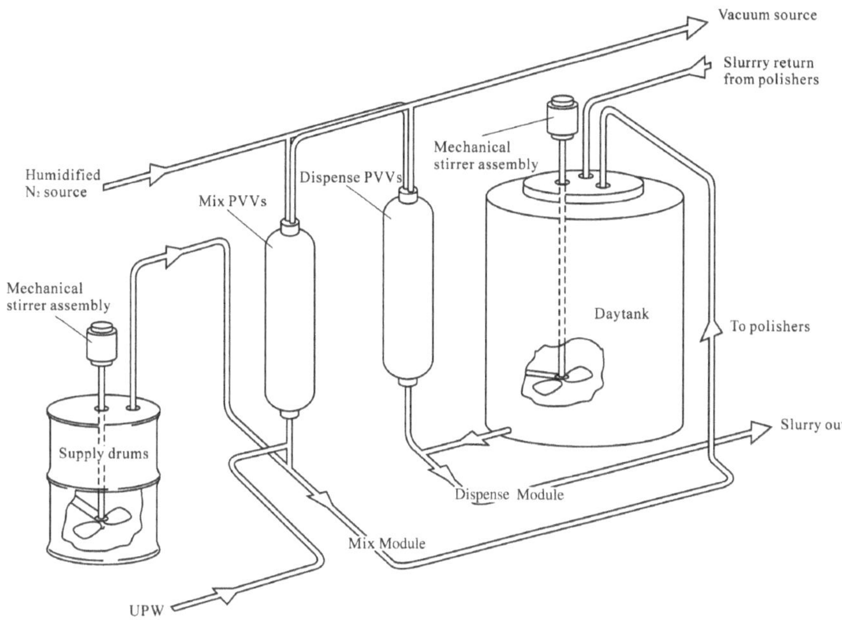

Figure 4.9 An example of the slurry supply system and its appearance

it does not get dry, as this causes machining problems. In the CMP of metals, the concentration of oxide should be maintained at a constant level, and drying and precipitation of the slurry should be avoided.

Slurry supply can use vacuum/pressure or a pump. Both systems supply slurry to the machining system by circulating it in a slurry pipe. It is assumed that the slurry may contact valves, orifices, pipes, vacuum or N_2 gas and stay there while turning over, so action should be taken to supply slurry in a stable manner at all times. Figure 4.9 shows one example of a slurry supply system.

4.2.6 Cleaning Mechanism

Particles and impurities left on a wafer during the machining process are removed by cleaning. The majority of these originate in the slurry, and a minority comes from the pad. The number of particles left on a wafer after CMP range from 1000 to 10,000/cm^2 (impurities 10^{14}/cm^2). These are generally removed by water polishing, before precision cleaning takes place in the cleaning station. The basic steps in the current cleaning process are:

Brush scrub cleaning → ultrasonic cleaning → spin dry

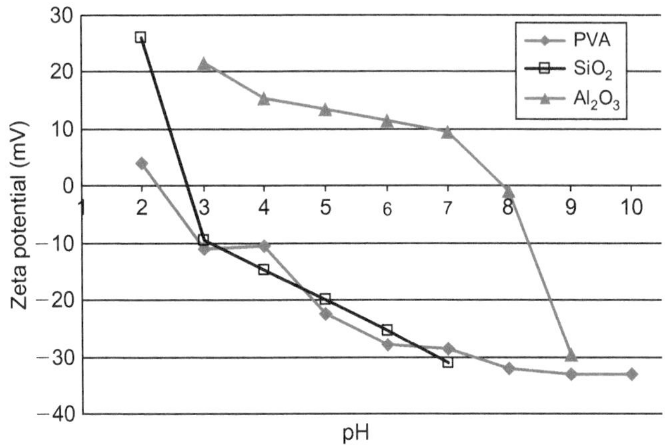

Figure 4.10 The correlation between pH and Zeta potential

Soft PVA (polyvinyl alcohol) in the form of a roll or a disc is used for brush scrub cleaning, because it is hydrophilic, high in tensile strength, and has modest softness. The Zeta potential of particles in the slurry means that pure water, an alkaline solution (like ammonia in water), and an acid solution (like diluted hydrofluoric acid) are selected as cleaning liquids. In short, the brush scrub cleaning agent is selected on the basis of the correlation between the pH and the Zeta potential, as shown in Figure 4.10. At pHs above 8, the Zeta potentials of Al_2O_3, SiO_2 and PVA are negative, so these particles are easy to remove from oxide films, and difficult to attach to a PVA sponge.

In supersonic wave cleaning, an ultrasonic wave of megahertz frequency is irradiated into the liquid. This is considered mega-sonic cleaning. By using megahertz frequency cavitations, the threshold rises, a high removal effect is obtained, and the occurrence of damage can be controlled.

CMP systems that include a fully automated cassette-to-cassette system, which integrates machining and cleaning into one system, have now been developed. In order for such a system to become established as standard semiconductor production equipment, dry-in/dry-out processing is required. Figure 4.11 shows a ground plan of the placement of each unit in a CMP system.

4.3 ELEMENT TECHNOLOGY

A CMP system which planarizes a wafer needs to meet certain targets in precision, quality, efficiency and cleanliness. To make a standard device

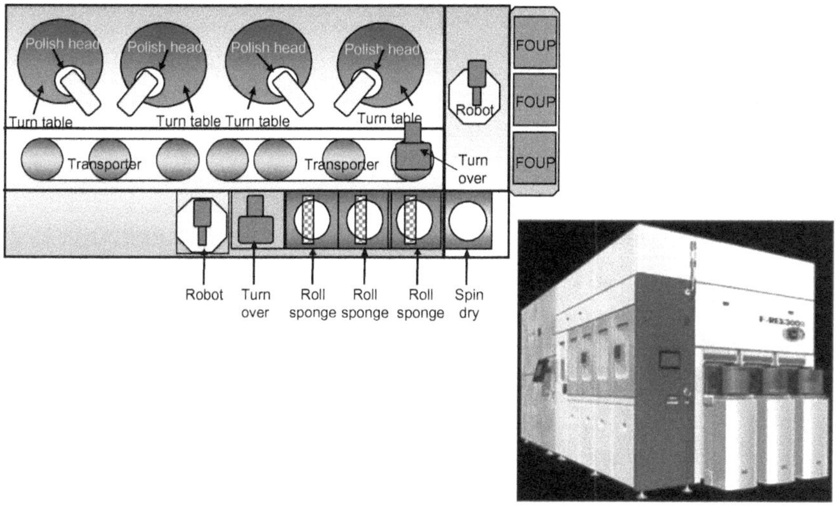

Figure 4.11 Photograph and plan of a CMP system

wafer surface (surface reference), uniform micro-removal of the whole wafer surface is carried out. To ensure uniformity on the wafer, Preston's equation is used:

$$M = \eta \cdot \rho \cdot \upsilon \cdot t \tag{4.1}$$

where M = machining amount, ρ = machining pressure, υ = relative speed, t = cycle time, and η = Preston coefficient decided by the machining condition.

The most important factor is the uniformity of $\rho \cdot \upsilon$ when machining the wafer surface. In order to achieve planarity, uniformity and machining efficiency, the slurry and pad are key factors and the most important consumable materials. Cleaning of the machined surface, which is also dependent on use of the proper consumable materials, is also important to ensure device quality. The mechanism of polishing movement, uniform application of pressure to the wafer, correct consumable materials and cleaning technology are all key factors of the CMP system, and will be discussed more thoroughly below.

In planarization CMP, it is also necessary to make the endpoint of machining very obvious, so trends in machining measuring methods will also be discussed. Considerable simulation research on planarization is taking place; if geometric precision, including defects like erosion and dishing, can be predicted by computer simulation, improvements in planarization precision can be made.

4.3.1 Movement Mechanism of Pad Tools

In CMP that involves chemical actions, M varies significantly by η in the Preston coefficient. The chemical composition of the slurry and the type of particles in it, the pH, temperature, surface condition of a pad, change of surface roughness of the device, and pattern density can all cause the value of η to vary significantly. Unless these factors are controlled, it is difficult to maintain consistent machining of a wafer. In addition, these problems become more critical as the wafer diameter gets larger. For example, the amount of slurry applied will vary between the wafer periphery and its center, so the movement mechanism is important.

When considering the movement mechanism in high precision CMP, attention should be paid to the relative motion between the wafer and pad, uniformity of track density, and vibration control during machining. The most typical CMP system has two rotating mechanisms; a head and a pad platen. If the rotation of the two (Nb, Np) are the same, uniform machining is in theory possible (Figure 4.12). However, in reality, subtle polishing marks are easily created that can cause the head or platen to oscillate. Because of this possibility, a direct driving method that does not have a rotation transmission mechanism, such as a belt or gear between each spindle, is sometimes introduced. Various types of movement mechanisms have been proposed to improve machining precision and efficiency, such as orbital (eccentric, small circular) movement, linear (straight line) movement, contact scan movement of some part of a hard pad in a

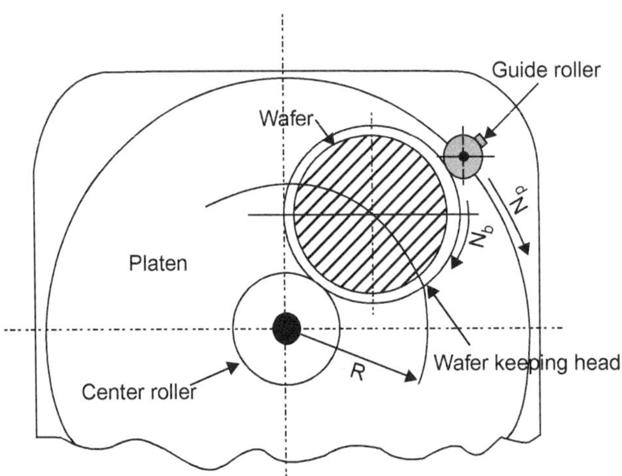

Figure 4.12 Mechanism of ring type CMP machine

donut shape, and dram (linear contact). Some of these are still at the research stage.

In 1990, Professor Doi's team proposed a system of eccentric, small circular movement with which uniform relative speeds could be obtained by taking the same small, circular track at any point and position on a wafer in terms of relative speed of a pad against a wafer. Later, IPEC (currently Speed Fam-IPEC) in the US introduced the orbital motion of a pad platen, as shown in Figure 4.13. Although the head mechanism of their system is different, the way of obtaining uniform, relative speed with small, circular movements is the same.

Ram Research Co. announced a linear (LPT) movement system (Figure 4.14) to improve removal rate and planarity by increasing the relative speed instead of lowering the pressure, as is done with conventional CMP equipment. This method pushes the wafer onto the polishing pad that moves in one direction and is similar to a conventional belt-type polishing method. When the spin speed of the wafer is reduced to 1/10 of the moving speed of the belt, the edge speed of the wafer becomes uniform over the entire wafer surface. A vertical model of this multi-type equipment that uses the two sides of a tool in the form of belt has been produced by Aplex in the US. These types of equipment attempt to gain machining distance by rotating a pad at high speed, and achieve planarity at low pressure.

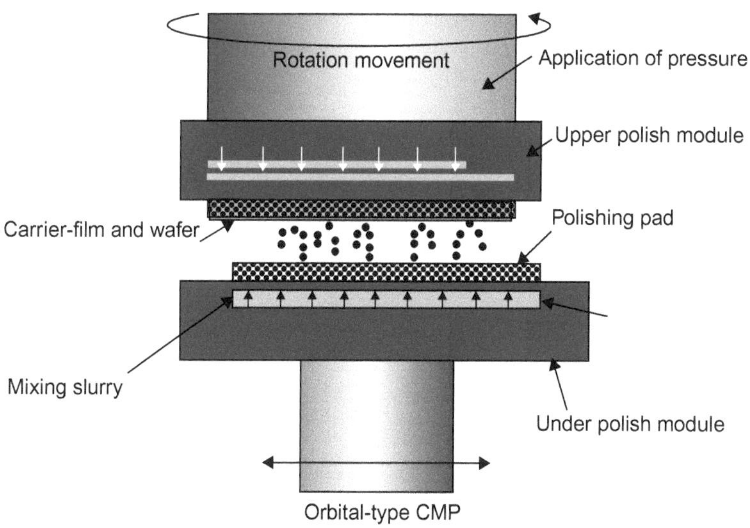

Figure 4.13 The orbital motion of a pad platen

Figure 4.14 A linear (LPT) movement system

Drum-like CMP equipment is also in operation, as proposed by Professor Doi's team. This system conducts polishing by using what is essentially linear contact; a rotating and oscillating drum, on which a long pad is rolled in a spiral form, is pushed onto a rotating wafer, while the slurry is supplied. This group has also recently proposed twin drum-type CMP equipment, and produced a prototype. This method is intended to give displacement between the spindles of two drums to obtain the same running distance at any position on a wafer. It is expected that the system will be highly efficient and precise during measurement machining, and will be important in handling the large diameter wafers expected in the future.

One other type of equipment that may be classified as orbital, and employs larger circular movements at the wafer head, has been announced. This equipment, launched by Obsidian Business Division in Applied Material (the old Obsidian Co.), introduces oval movement of the wafer head that can move independently on the X and Y axes on a special slurry-free, bonded particle pad (3M Co). Figure 4.15 shows the movement mechanism of this system and its appearance. Table 4.5 summarizes the available pad movement types in CMP systems, both on the market and in R&D.

4.3.2 Element Technology Related to Uniform Machining of a Wafer

When machining work piece materials, it is necessary to define a surface reference for machining. In the planarization of a wafer for an LSI device, the front surface is the surface reference, as machining is conducted from the front. This is because the machining margin is extremely small, and the precision of the substrate itself (particularly thickness precision) is not

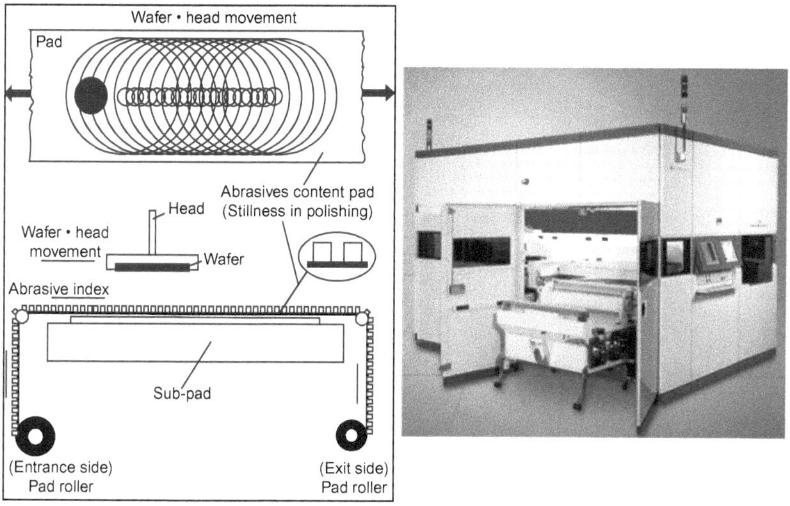

Figure 4.15 The movement mechanism of the Obsidian CMP system (AMAT) and its appearance

trustworthy. On the other hand, the machined amount of a bare silicon wafer during single polishing is 10–0 μm; thus, having the back side as the surface reference is convenient to get good TTV (Total Thickness Variation).

Figure 4.16 shows two ways of defining a surface reference during wafer polishing. In the past, substrate wafers were mounted by either wax mounting or carrier mounting. The carrier method places the wafer in a pocket formed on the carrier head surface. If the wafer is wax mounted, its back side can be finished relatively easily by controlling the planarization precision of the head itself (for example, by using a high precision surface ceramic plate) and the thickness of the adherence layer (wax layers that become thin pieces). This approach is widely used for finish polishing bare silicon wafers. However, this method cannot be used for planarizing a wafer during device production. With this method, only the convex part on the wafer surface on which subtle roughness is formed, like the device surface, has to be uniformly polished. In order to do this, polishing has to cope with warp on the surface and thickness variation to remove a very small amount of machining margin. Therefore, the surface reference should be the front surface, and the back side of a wafer is pushed uniformly onto a pad to follow the entire pad surface while correcting warp. The polishing method in which a pad is adhered to the wafer surface and polishes according to wafer surface geometry can be called surface reference polishing in theory.

Table 4.5 A Summary of Pad Movement in the CMP System

Movement Form of Tool Pad		**Direction of Tool**	**Character**	**Manufacturing Company (Institution)**
Disk tool (platen)	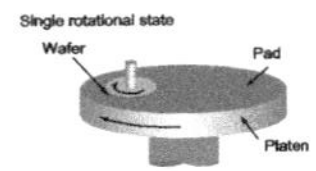	Up or down (or horizontal)	Popular mode, high speed mode	Ebara Corporation, Toshiba Machine Fujikoshi Machinery Corp., Lapmaster SFT Corp., (Okamoto Machine Tool Works Ltd), Peter Wolters Tokyo Seimitsu Co., Ltd, Speedfam, Mitsubishi Materials, Sumitomo Metal Industries, Ltd, Applied Materials
	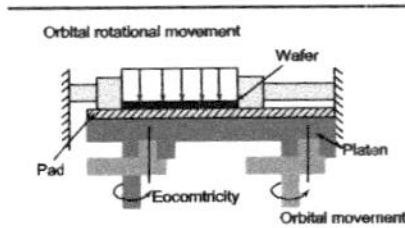	As above	Miniaturization, saving slurry, large diameter measure, uniform pressure jig	Fujikoshi Machinery Corp., Saitama University Inst. of Physical and Chemical Research
Belt shape tool	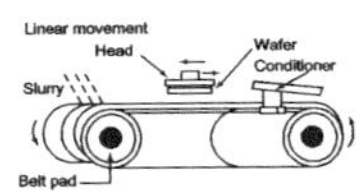	Horizontal or vertical	Low pressure high speed uniform pressure jig	Lam Research Corporation (horizontal type)
				Applied Materials (pad taking up type)
Hollowing-out type tool	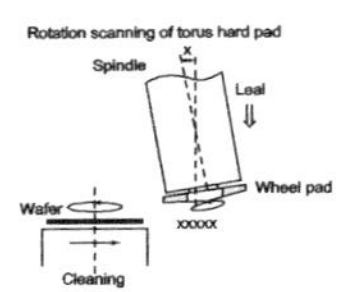	Down (or option)	Partial contact scanning processing of torus semifixation hard pad, control of pressure turnover number, effect of slimy and fixation abrasive	Sony
Cylinder tool	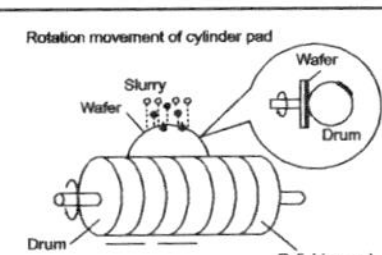	As above	Line contact type processing simple measurement during processing large diameter method miniaturization	Fujikoshi Machinery Corp., Saitama University (Okamoto Machine Tool Works Ltd.)

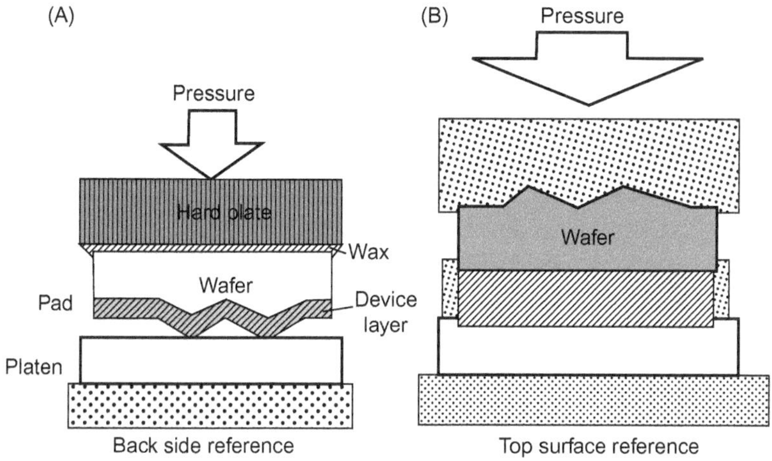

Figure 4.16 Method of surface reference and wafer holder

In contrast, the free mounting method fixes (water adherence) a wafer by using the surface tension of water by pumping soft foamed polyurethane films (backing material) in a carrier. Because the geometry of the back side of a wafer is absorbed by deformation of the soft elastic films, the surface can be removed uniformly. Using backing materials like the above to apply a uniform pressure has become the norm.

4.3.2.1 Evolution of Carrier Head Structure and Effects of Uniform Pressure

It is not an exaggeration to say that developments in wafer chucks and pressure mechanisms make improvement possible in the CMP of oxide films. It is essential to push the entire wafer surface onto the pad surface, and to apply pressure uniformly in machining to the surface reference. The wafer chuck and pressure mechanism are key components for CMP equipment; hence, each CMP equipment maker has their own design. The basic mechanism is to apply pressure vertically from a machining head, and additional pressure from the back side of a wafer. A retainer ring is attached to the outer periphery of the chuck. For uniform application of machining pressure, ρ, the warp and thickness variation on a wafer have to be considered. ω_{max} occurs when uniform load is applied as described below:

$$\omega_{max} = (5 + \upsilon)\, \rho a^4 / 64\, (1 + \upsilon)\, D \qquad (4.2)$$

where $D = Eh^3/12(1 - \upsilon^2)$, υ = Poisson's ratio ($\upsilon_{Si} = 0.262$), ρ = load per unit surface area (kg/cm^2), D = bending rigidity, E = Young modulus ($E_{Si} = 1.73 \times 10^6$ kg/cm^2), h = thickness (cm), a = radius (cm).

For a silicon device wafer with a diameter of 200 mm (8″), the warp can be as small as 100 μm (device face is convex) and the value of ρ to correct it is below 10 g/cm^2. This can be ignored when compared with the machining pressure in ordinary polishing, so the influence of pressure due to warp correction is not considered.

Figure 4.17 shows an example of the basic structure of the surface reference most widely used. In this case, the head consists of four elements; the wafer chuck that holds the wafer, the retainer ring that prevents the wafer from detaching, the backing film that keeps the surface reference, and the machine housing that applies pressure while these are being held. Application of pressure and rotation are basic functions of the machining head. Generally, a rotation spindle is placed at the center of a head, and pressure is applied along it. In contrast to the dead weight method, in this case the pressure on the wafer tends to be highest at the lower part of the spindle center, and lower towards the edges, so some measures need to be applied to obtain a more uniform pressure distribution in the head housing. The air/water back pressure application method designed by Professor Doi's team in 1991 is one such measure.

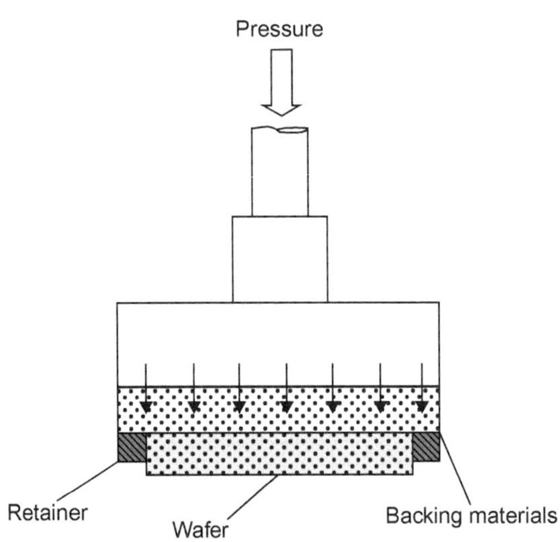

Figure 4.17 The basic structure of the surface reference

Figure 4.18 is a schematic drawing of a uniform pressure application method, in which a bag made of thin elastic rubber, and filled with air or water, is pushed onto the back side of the wafer. This method achieves much greater uniformity than those using a backing film (Figure 4.17). Alternatively, a membrane pressure application method, using rubber films, can be applied as shown in Figure 4.19. This membrane method is used by Applied Materials Co., Syvec Co. (Mitsubishi Material Co.), and Speed Fam IPEC Co.

These methods push the back side of a wafer through backing materials (carrier film) or thin rubber films. Backing materials are not consistent in form and are consumable, so variation in uniformity on the machined

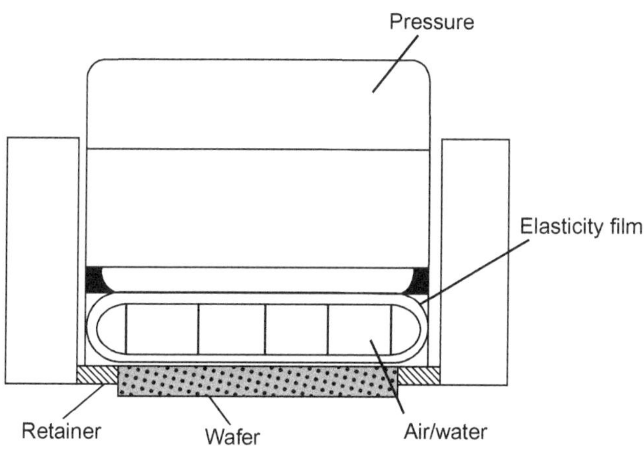

Figure 4.18 Uniform pressure application method

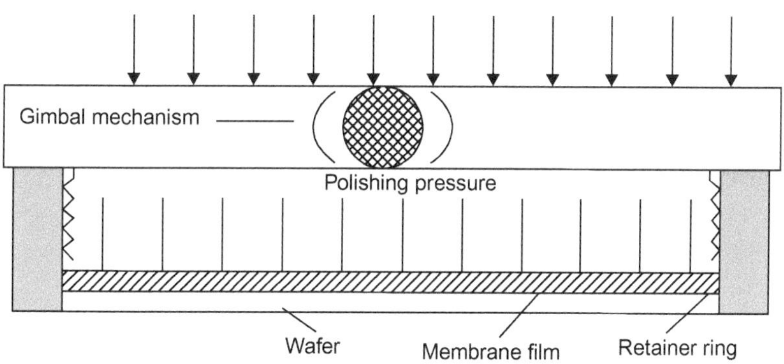

Figure 4.19 A membrane pressure application method using rubber films

surface occurs. The membrane pressure application method also has its limits, even though it can improve precision. A uniform fluid pressure application method has also been proposed which may improve further on existing techniques.

Figure 4.20 shows the principle of the direct air back pressure method, which applies uniform pressure directly by applying air as fluid. By avoiding the use of solid bodies, such as backing materials and membranes, and applying pressure directly with an air layer (10–100 μm), stable machining and high uniformity can be achieved. This direct air/water back method (or air/float method) has been tested in the form of an air float carrier head (Figure 4.21) in CMP systems by Tokyo Seimitsu Co., and favorable results have been achieved. However, it is easier for the slurry to penetrate to the reverse of the wafer, and for air to leak around the retainer. Water can also be used instead of air.

Figure 4.22 shows the variation in the residual oxide film thickness after CMP to remove 1 μm from an oxide film wafer by the direct air method shown in Figure 4.21. Very good uniformity (520 ± 10 nm) in the oxide film thickness is obtained, except at the wafer periphery where the variation is 3 mm.

An alternative to the above options is a method in which the pad follows the wafer surface as much as possible when applying pressure. One proposed example is the linear CMP system, developed by Ram Research Co., which incorporates a mechanism which applies uniform pressure to the wafer carrier head and the lower part of the pad

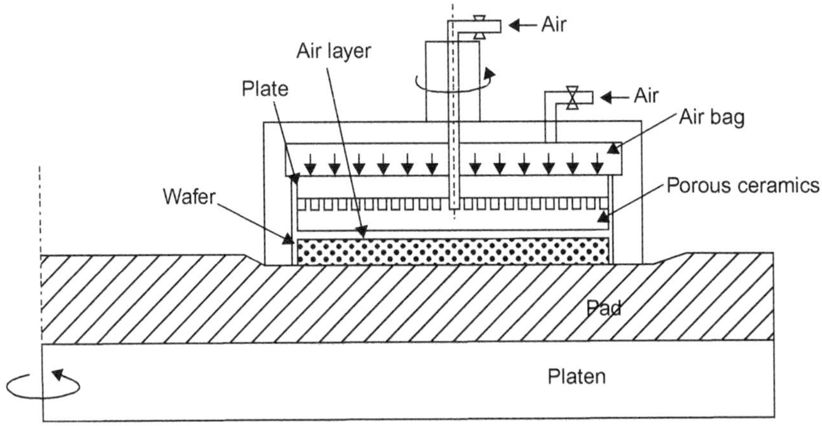

Figure 4.20 A schematic drawing showing the principle of the direct air back pressure method

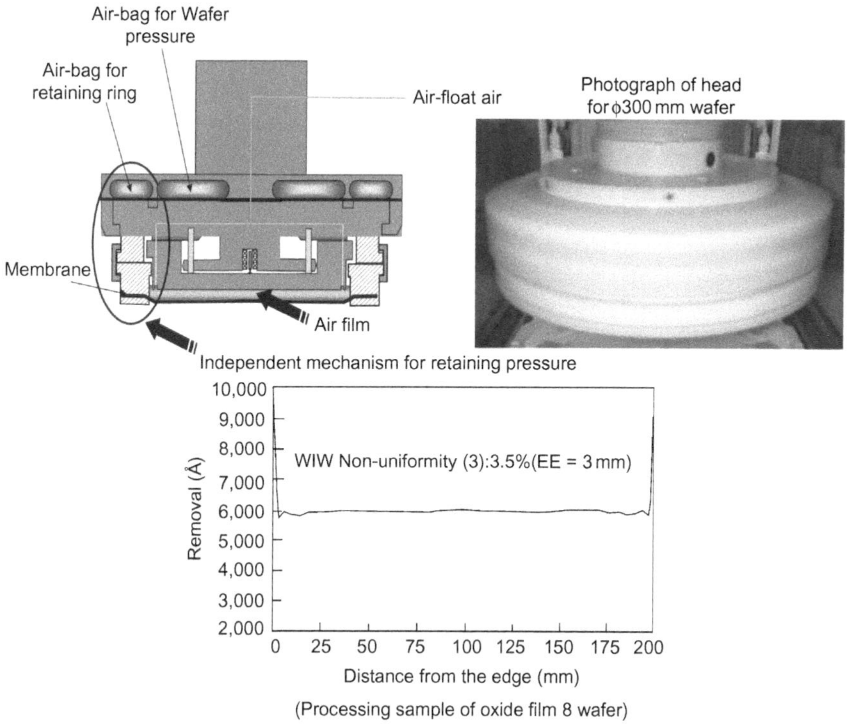

Figure 4.21 Air-floating system head of Tokyo Seimitsu (drawing: courtesy of Tokyo Seimitsu Co., Ltd)

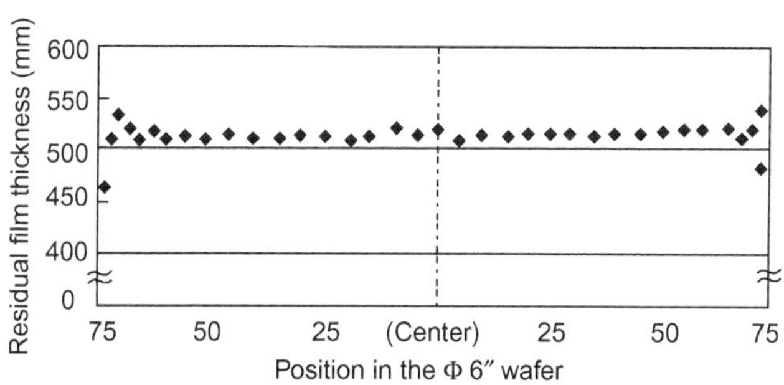

Figure 4.22 Variation of the residual oxide film thickness

(Figure 4.14) by spraying air from the fluid bearing platen on the lower part of the pad so that the pad follows the geometry of the wafer surface. The improvement in uniformity is obtained by introducing a uniform pressure application method with backing films on a wafer carrier head.

4.3.2.2 Effect of Retainers

If the wafer is simply pushed onto the pad surface to apply polishing pressure, a dull edge is created at its periphery. The elasticity of a pad like IC1000 that is generally used in planarization CMP is more than 10 times that of a polishing pad for a bare silicon wafer. The nominal elasticity of IC1000/SUBA400 in a two-layer structure is 1/2−1/3 of a single layer pad; so, although varying with machining conditions, the dull edge of a wafer can be as large as 10 mm.

When a rigid body is pushed against an elastic body, the pressure increases exponentially from center to periphery, according to Boussinesq's formula in elasticity theory. In addition, the pad will deform due to the applied pressure, as shown in Figure 4.23. At the wafer periphery, a pad changes from a deformed state to a deformation-free state, and pressure is applied both perpendicular to a wafer and to the wafer edge. Therefore, the machining amount during polishing is proportional to the applied pressure, according to Preston's equation. Thus, in the early stage of polishing the wafer periphery is removed preferentially, resulting in the creation of a dull edge.

In order to get as many semiconductor tips as possible from one wafer, uniformity over as large a surface area as possible is desired. When assessing uniformity of the wafer surface after CMP, the width of

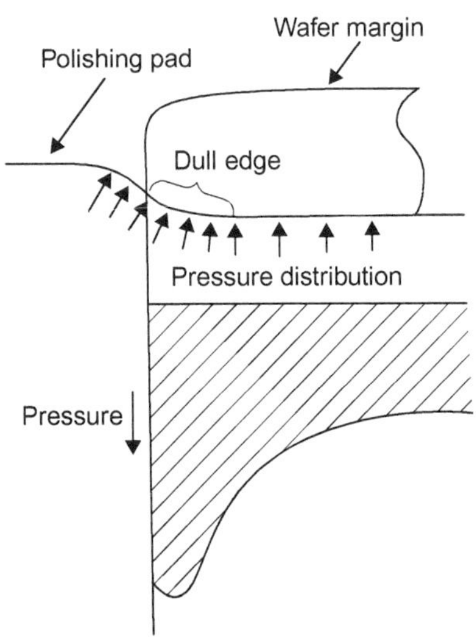

Figure 4.23 Pad deformation due to pressure applied to a wafer during polishing

the wafer periphery that is out of the evaluation region is called edge exclusion (EE). The EE is generally about 3 mm, but recently this has been required to drop to 2 mm.

Placing a dummy around a wafer held in a carrier head can be used to minimize edge exclusion. Usually a retainer (ring) is used to hold the wafer firmly during machining. Figure 4.24 shows a retainer ring at the same position as a wafer held as a dummy. When a retainer ring that has not been adjusted is used, the machining amount varies at the wafer periphery; when an adjusted ring is used, this problem is reduced. An unadjusted retainer ring causes a reverse force on the pad, raising it several millimeters away from the edge of the retainer ring.

A mechanism that applies pressure to the retainer ring independently can have the same effect as adjusting the width of the retainer ring. Figure 4.25 shows a retainer ring in a carrier head (titanium head) made by Applied Material Co. By this mechanism, the pressures P1 and P2 are applied independently to the retainer ring and the wafer (membrane), so the edge profile of the wafer can be controlled. Applying pressure to the retainer ring independently is one method that manufacturers are adopting to achieve the goal of an EE of 2 mm.

Hard-wearing materials with low friction coefficients (for example, polyimide-based fluorine resin) are used to make retainer rings.

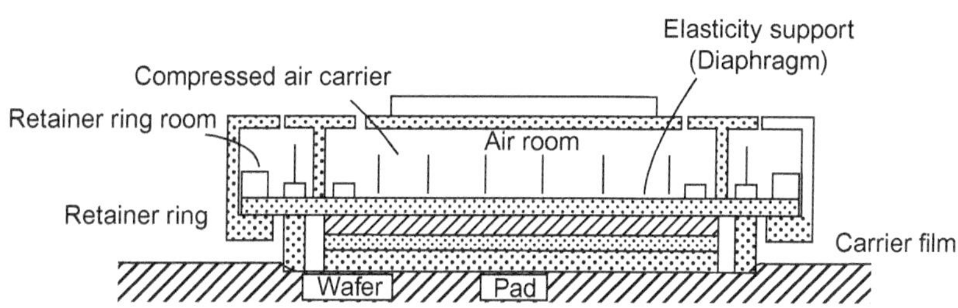

Figure 4.24 A schematic drawing of wafer head with retainer ring

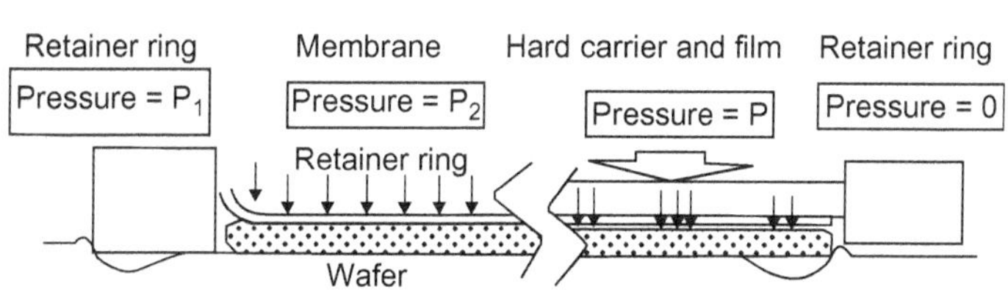

Figure 4.25 Independent pressure of a retainer ring in a carrier head

4.4 ROLE OF SLURRY IN CMP

Machining properties depend on the consumables used, including the pad and the slurry. CMP processing costs are divided between CMP/ cleaning equipment, consumable materials, and labor. Of these, consumables account for about 70% of the total, and include the pad, slurry, backing material, pure water and reagents. The pad and the slurry are the most expensive of these, so it is quite understandable why they are considered so important.

4.4.1 Development of Slurry

Slurries used as abrasive agents used to be made of natural particles dispersed in water. In this section, the evolution of the slurry and polishing history will be introduced.

Polishing technology has progressed rapidly. A technique similar to the one used in the 17th century before Newton could achieve a flat mirror surface with surface roughness below 100Å and an order of several 10Å relatively easily. In order to understand this history more fully, conventional polishing theory will be discussed.

First, the polishing mechanism will be explained by taking glass as an example. This can be broadly classified into the following:

1. Micro-removal approach, by Rayleigh in 1901.
2. Flow view by Beilby in 1903.
3. Chemical action view by Grebenschikov in 1931.

Micro-removal is an approach in which machining a flat mirror progresses by removal of the convex part of the surface by the micro-cutting action of particles. Flow is the view that flow of the surface created during machining plays a big role in flattening, and chemical action is the view that the chemical action of a solution (i.e. machining fluid) creates gel layers of silicic acid on the glass surface, which are then removed. Those who have experience in mechanical polishing are aware that all of these are reasonable approaches. In chemical action, chemical reactions between work piece materials (glass), machining liquid, particles and the pad play an important role in making the glass surface flat. According to a paper by Kaller, material removal by chemical reaction together with filling of concave parts by silicic acid gel both play a role in flattening.

CMP is the ultra-precision polishing technique most often used for producing semiconductors. It employs both mechanical and chemical removal. J. Regh and his team, in the US in the early 1960s, wrote the paper outlining the basis for current CMP technology. It uses a liquid mixture of ammonium fluoride (NH_4F) and copper nitrate ($Cu(NO_3)_2$) and utilizes an exchange reaction of a Cu ion, as follows:

$$2Cu^{2+} + Si \rightarrow Cu + Si^{4+}, Si^{4+} + 6F^{-}, SiF^{2-} \qquad (4.3)$$

This method, however, was not used for silicon wafer processing because copper ions remained on the surface, and the surface roughness was poor – although the machining efficiency was more than ten times higher than ordinary polishing. E. Mendel of IBM in the US proposed the CMP of single crystal silicon via a chemical reaction in an alkaline solution, such as NaOH, and created the basis for today's bare silicon polishing process:

$$Si + 4NaOH \rightarrow Na_4SiO_4 + 2H_2 \qquad (4.4)$$

4.4.2 Basic Science of the Slurry

The basic principles of the CMP slurry will now be discussed, paying special attention to its behavior and the machining mechanism.

4.4.2.1 CMP of Oxidized Film (ILD, STI) and Study of its Mechanism

The machining of interlayer insulating films (ILD) and shallow trench isolation (STI) is conducted by SiO_2-CMP. The polishing mechanism has developed from that of glass and crystals. When the newly activated surface exposed by polishing water comes into contact with, hydration occurs instantly. If alkali is present in the water, soft layers of water glass (i.e. layers of silicic acid gel) are created on the surface. In fact, if pure water without particles is rubbed with a pitch, glasses can be polished, as confirmed by Kaller. Therefore, it is reasonable to assume that soon after polishing begins, the soft layers covered by a hydrate film on the glass surface are removed by particles.

In 1960, Kaller and his team showed from tests of glass polishing that the hardness of particles had little effect on polishing; it was their chemical activity that was significant (Figure 4.26). They found that multivalent metal oxides (e.g. CeO_2, Fe_2O_3, SnO_2 and MnO) were effective, and pointed out that the simultaneous action of particle, pad and water was also important.

The glass surface that has been polished is covered with a thin, soft layer from which alkaline metal comes out and bonds with OH base, making the mechanical removal action of particles favorable. He also explained that ion dissolution and ion exchange from the glass surface

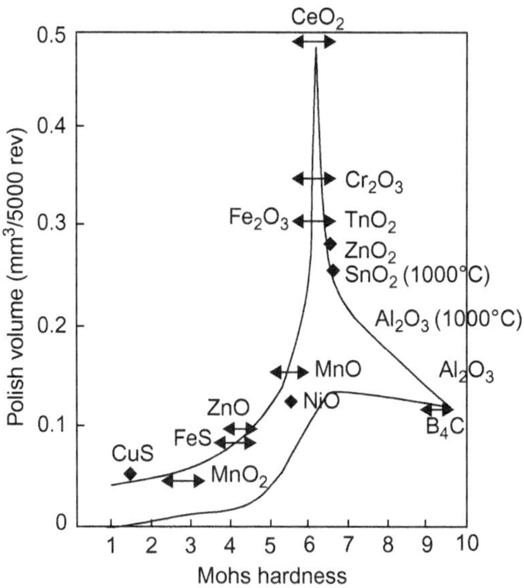

Figure 4.26 Polishing volume by various abrasives (Kallar)

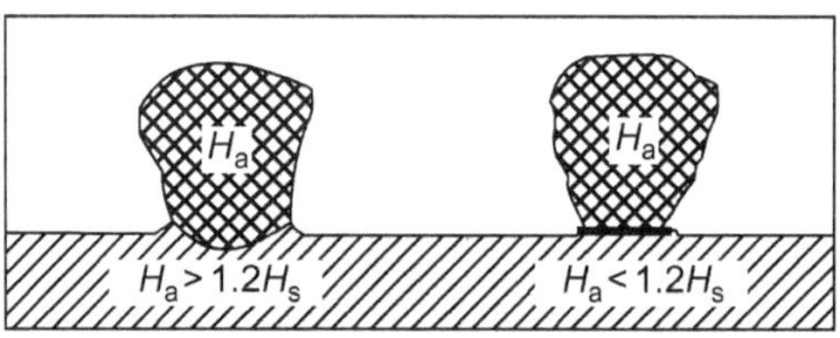

(a) Transformation of flatten surface (b) Transformation of abrasive
(H_a: hardness of abrasive, H_s: hardness of surface)

Figure 4.27 Plastic deformation between abrasive and surface

could be done better in the conditions where hydrate fluid chips were adhered to the glass surface than the glass in direct contact with water.

On the other hand, in the CMP of oxidized films with a slurry that contains SiO_2, the machining is performed on material of the same chemical composition, and the oxidized film wears by plastic deformation. Under static compression, SiO_2 particles deform plastically more easily than oxidized films (Figures 4.27 and 4.28), and they have a much larger surface area that can be cooled down effectively by the surrounding water. Therefore, as shown in Figure 4.28, oxidized films increase in temperature due to friction, and so decrease in hardness, making them softer.

Figure 4.29 shows the mechanism of CMP of oxidized films by oxidized hydrate particles. Plastic deformation is accelerated by frictional

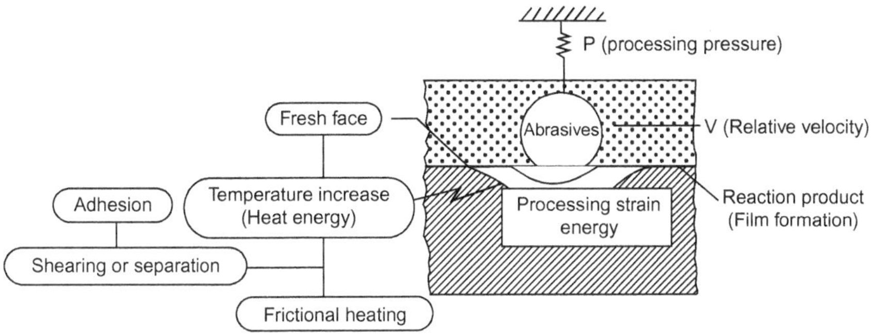

Figure 4.28 Proceeding factors of CMP mechanism

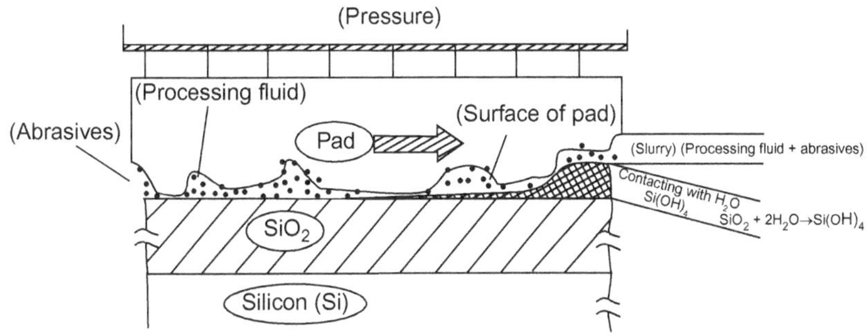

Figure 4.29 Polishing mechanism of oxidized films CMP

heat, and the removal action is faster on soft surface layers containing water. A chemical process is also taking place, but increasing its rate does not necessarily speed up the machining rate. For example, a high machining rate can be achieved by using a slurry with a high pH; however, changing the pH has other effects. A slurry with a low pH can achieve a high machining rate of oxidized films, thanks to agglomeration of SiO_2 particles due to reduced surface electrical charge.

4.4.2.2 CMP of Metal (Tungsten, Aluminum, Copper etc.) and its Mechanism

Metals can be polished by chemical etching, but in general the machining efficiency is not high and the quality of the machined surface is compromised. Currently, an oxidizing agent and particles are added to an acidic machining solution. The oxidizing agent is added because hard, brittle materials are rather easy to machine, and the oxidizing agent makes

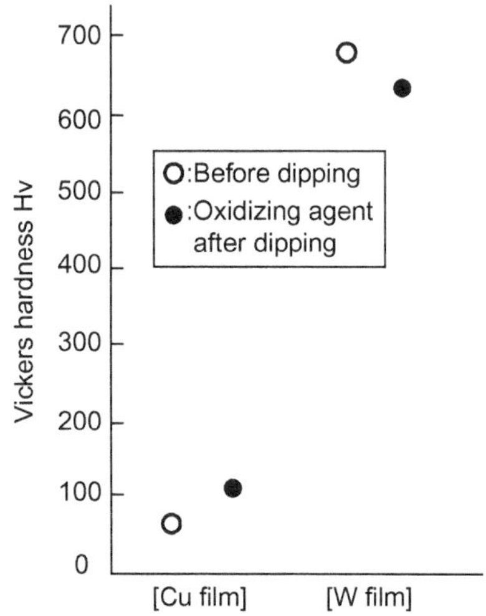

Figure 4.30 Hardness of a metal surface before and after being immersed in an oxidizing agent

the surface layer brittle by oxidizing the metal. Figure 4.30 shows a comparison of the hardness of a metal surface before and after being immersed in an oxidizing agent (H_2O_2). In contrast to tungsten, copper becomes 1.7 times harder after this treatment. This is because tungsten is subject to a higher etching effect with hydrogen peroxide than other oxidizing agents (i.e. iron nitrate).

If a metal (Cu) surface is oxidized with an oxidizing agent in the slurry, the oxidation films so formed are passivated and the oxidation reaction does not progress. The passivated oxidation film that is formed is brittle, and can be removed by a pad and the particles in the slurry, thus exposing a new metal surface. This is again exposed to the oxidizing agent, hence a new passivated oxide film is formed on the surface. By repeating this process, metal CMP progresses. This machining mechanism also occurs in aluminum (Al) and tungsten (W). Figure 4.31 is a schematic drawing of the general machining mechanism for the CMP of metals.

Cu is more easily etched with various acids, so a corrosion-resistant agent (BTA, $C_6H_5N_3$) is added. However, because copper oxide has a weak protective effect, CMP is more difficult for Cu than for W. The chemical reaction between Cu and the slurry therefore needs to be considered, for which the Pourbaix diagram is often used. An example diagram is shown in Figure 4.32, showing the equilibrium potential between the metal surface and the solution in the etching region, the passivation

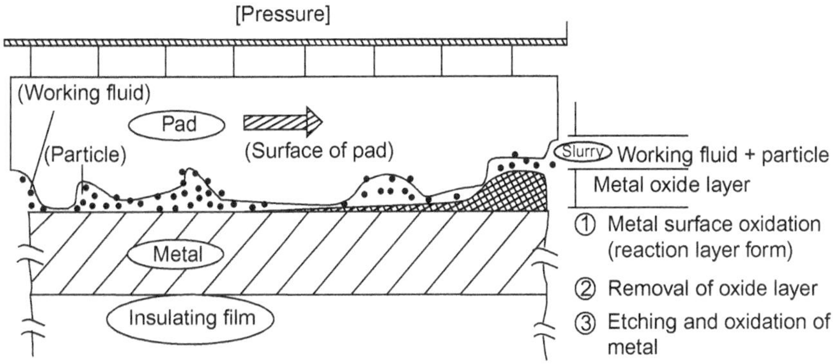

Figure 4.31 A schematic drawing of the machining mechanism of metal CMP

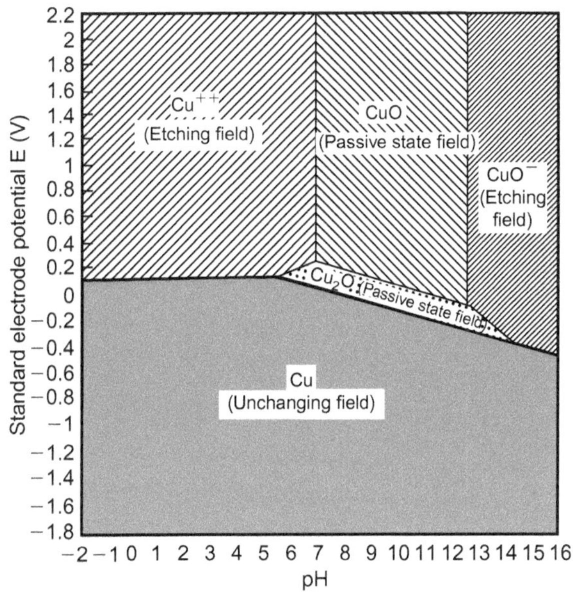

Figure 4.32 Pourbaix diagram for Cu

region, and the non-metamorphosis region. In the Damascene process, the pH of the slurry should be at the border between the etching region and the passivation region. Because Al and Cu have etching regions at both acid and alkaline pHs, many additives can be used. Both metals are soft and scratches are easily produced, so special care needs to be taken in controlling the amount of oxidizing agent, in order to produce conditions in which the passivation film is fully formed.

4.4.2.3 Dispersing Ability and Stability of the Slurry

When mixing and dispersing fine particles in a solution, dispersed particles tend to agglomerate, so measures to prevent agglomeration are necessary. Defects, like scratches on the machined surface, must not be produced in the CMP of interconnection metal and insulation films. Agglomeration of particles in the slurry causes defects on the machined surface, so it is essential to control both the dispersion ability and stability of the slurry. When preventive measures are taken and the initial dispersion state lasts long enough, "a good dispersion slurry" is present.

Silica particles have silanol (SiOH) groups on the surface that are bonded to each other by relatively weak hydrogen bonds; they have a thixotropic tendency, in which particles can be detached by a small external force (e.g. vibration or agitation) (Figure 4.33). In microscopic studies, it is found that in a slurry of silica particles dispersed in an alkaline solution, OH^- groups form on the particle surface, and it becomes negatively charged. The amount of this negative charge depends on the pH. Positive ions, like K^+ and Na^+, are attracted by this negative charge, hence a positively charged solution surrounds the particles, and electrical double layers are formed (Figure 4.34). The potential of the outermost face (sliding face) of the electrical double layers, (i.e. potential of solution (dispersant) and silica particles in the slurry) is called the Zeta potential.

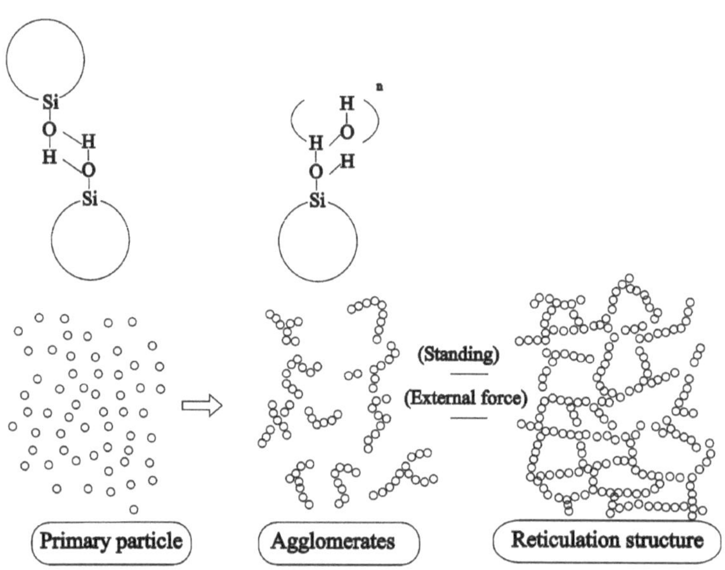

Figure 4.33 Silica particles and the interaction between them

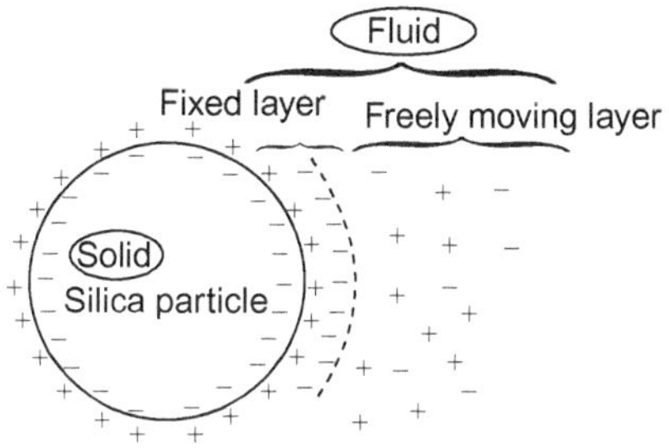

Figure 4.34 Model of electrode of particles, and electrical double layer

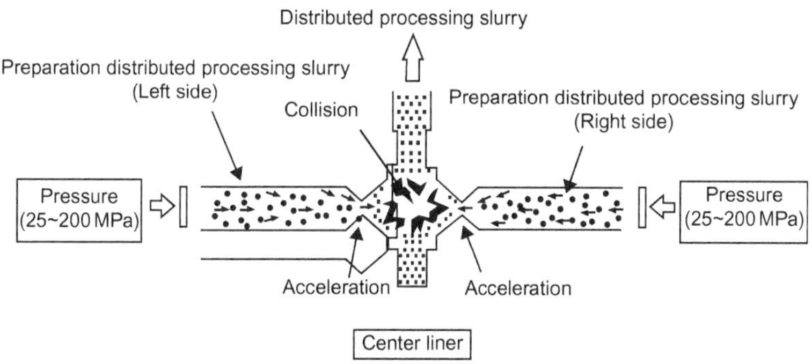

Figure 4.35 Opposite collision by high pressure slurries from two nozzles

When this is small, the attractive force between the particles increases and agglomeration is likely to take place. Therefore, the Zeta potential should be increased to above the critical value by adjusting the pH of the slurry.

Addition of surfactant is also useful for obtaining good quality CMP by controlling the occurrence of scratches. This is one way of preventing agglomeration, as stated above. It is an example of using repulsive forces by making the particles absorb macromolecules. It should be noted, though, that if the particles have a wide size distribution, crystal growth occurs during storage, producing coarse particles. Dispersants can control crystal growth by absorption.

Methods of dispersing particles in the slurry for an extended length of time without using a dispersant have also been proposed. This approach makes the particles collide at twice the speed of sound by propelling them at high pressure from two nozzles, as shown in Figure 4.35. The dispersion mechanism is not yet fully developed, but it is an interesting approach, as the effect is displayed in the slurry containing dispersant.

4.4.3 Classification and Production Methods of CMP Slurries

Figure 4.36 classifies slurry by particle type. In this section, representative slurries containing silica (SiO_2), ceria (CeO_2) and alumina (Al_2O_3) will be introduced.

4.4.3.1 Silica-Based Slurry (SiO_2)

Silica Particles With Silicic Acid Sodium ($NaSiO_2$) As Raw Material

Precipitated silica and colloidal silica are produced in the process shown in Figure 4.37. Since the original raw material is silicic acid sodium, relatively inexpensive SiO_2 particles are obtained. Precipitated silica ((a) in the figure) is not a single particle but rather a body of several precipitated particles. Colloidal silica, produced at point (b) by ion exchange, grows as

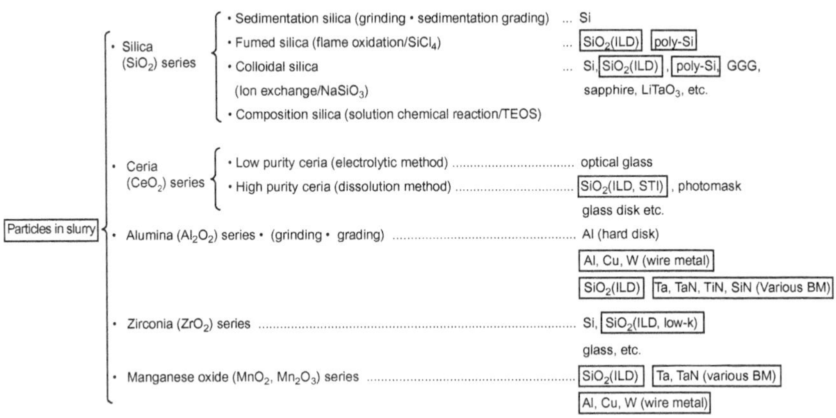

Figure 4.36 Particles (abrasives) in slurry

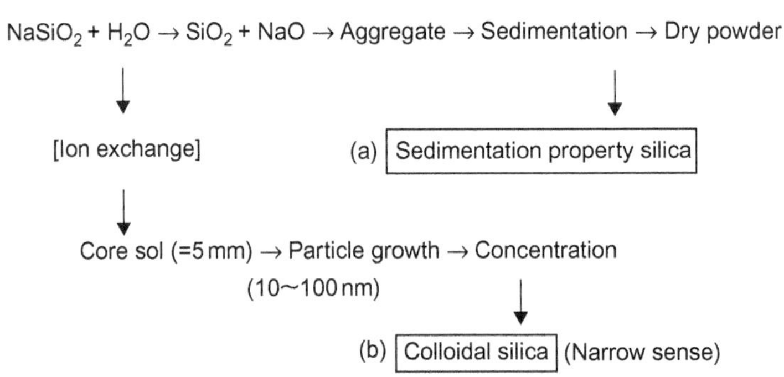

Figure 4.37 Process of producing precipitated silica and colloidal silica

a single crystal with fine particle of 5 nm as core in liquid. The product available on the market is in the colloidal state.

A considerable amount of sodium remains in a precipitated silica slurry, because the raw material is silicic acid sodium. In the original polishing processes, dry precipitated silica was mixed with water to obtain the slurry. Currently, however, because alkaline metal contamination causes problems, this method is rarely used.

Colloidal silica that is adjusted and produced by ion exchange is low in metal contamination compared with precipitated silica, and this is now used as a polishing slurry for the primary and secondary polishing of bare silicon wafers for LSI. In general, colloidal silica has a siloxane structure and a colloidal form, in which negatively charged silica particles are dispersed in water. Ordinary silica is alkaline, with a pH of 8–11, and is stable. In many cases it matches the dispersant in terms of negative ion and non-ion. TEM photos of colloidal silica are shown in Figure 4.38; the particles are round and their size is consistent. It is possible to get various particle geometries, and size ranges of 20–200 nm, by controlling the reaction temperature and reactant concentrations.

Silica Particles with Organic Silicate as the Raw Material

This ultra pure colloidal silica is obtained from organic silicate by wet hydrolysis. Colloidal silica ranging from 100 to several 1000 Å can be obtained by hydrolysis of organic silicate in an organic solvent, such as alcohol. The finished product is obtained by distillation and precision filtration. Colloidal silica from Fujimi Incorporated Inc., Planerlite series, is a typical example and is about 99.999% pure. This colloidal silica can be used for CMP of

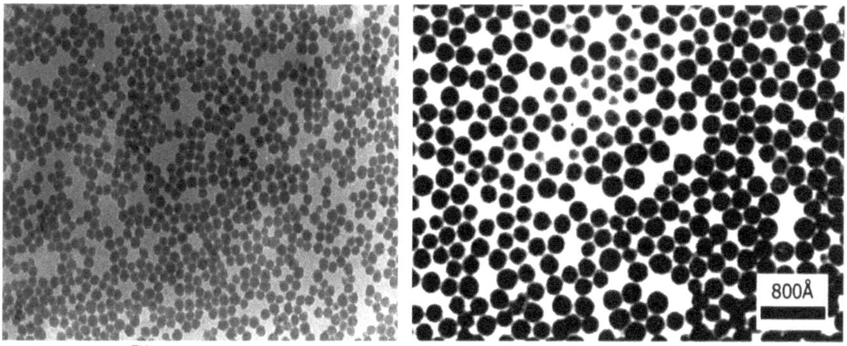

(a) Klebosol™-Clariant (50-70 nm) (b) Nalco 2350

Figure 4.38 TEM images of colloidal silica particles

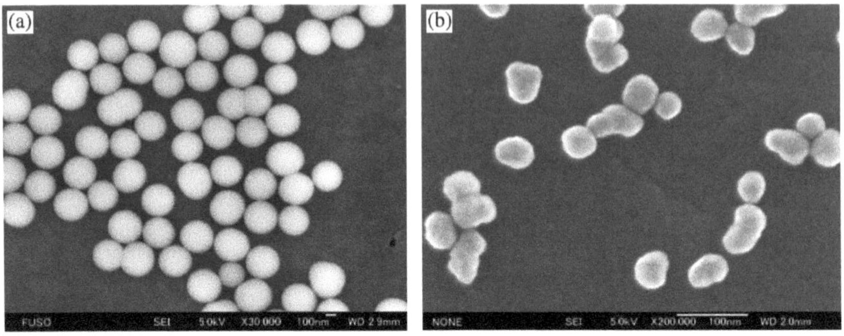

Figure 4.39 SEM photos of high-purity colloidal silica (Fuso). (a) Spherical shape colloidal silica (f 200 nm) (b) Cocoon shape colloidal silica (f 30 nm)

poly-silicon. Figure 4.39 is an SEM photo of highly pure colloidal silica made from organic silicate. The cost of this material should decrease in the future, but currently the emphasis is placed on its purity.

Silica Particles with Silicate Tetrachloride as the Raw Material

This material is called fumed silica, and is produced by oxidation in a flame, followed by desalting refining. It is considered to be highly pure and relatively inexpensive; hence it has been used for polishing silicon wafers for some LSI devices. Silica slurry for CMP of interlayer insulating films is produced from this type of silica. Oxidation of SiO_2 particles is conducted in a flame, literally fumed, so the particles have agglomerated to some extent when produced. Hence, it is of the utmost importance to keep the size of this secondary agglomeration consistent and stable, to obtain stable, reproducible machining performance.

4.4.3.2 Ceria (Oxidized Cerium) Based Slurry (CeO_2)

This slurry stands almost in the middle in terms of pH, and features a high machining rate in the neutral region, while a silica-based slurry in general achieves a high machining rate in the alkaline region.

Ceria achieved high-quality results in glass polishing in Europe in the early 1930s. The raw material is a rare earth chloride, obtained when extracting thorium from monazite, and CeO_2 accounts for 45% of all rare earth elements used as raw materials. Conventional ceria production adds chlorine and NaOH to the rare earth chloride, to turn elemental Ce into the oxide in order to separate it from other rare earth elements. This is then sintered and milled to get fine particles; but, depending on the sintering condition, different polishing properties are obtained. Sintering ceria at low temperatures achieves a low machining rate but results in a nicely machined surface with fewer scars. Ceria sintered at high temperatures can give a high machining rate, but scratches are easily produced on the machined surface.

There are various views of the machining mechanism of a ceria-based slurry. Some conclude that the particles are worn and crushed to fine particles, resulting in a larger active specific surface area, and thus achieving a higher machining rate than would be expected from the initial particle size. This is called micro-mechanical action. However, because both cerium and silicon are tetravalent, some chemical interaction, such as transfer of electrons between SiO_2 and CeO_2, may be included in the mechanism.

Ceria is generally used as a glass polishing slurry as:

CeO_2: about 85%

M_nO_m: about 15% (oxide of rare metal)

Average particle size: 1.0–2.0 μm

This type of ceria has problems in purity, particle size, and precipitation occurs due to the high specific gravity of CeO_2; it also has poor dispersion ability. This means that it is difficult to use it as it is for ultra-precision polishing of semiconductor grade material.

Recently, sub-micron-sized fine ceria has been developed, by micronizing CeO_2 particles with an ultrasonic dispersion machine and ball mill, in an effort to improve both purity and dispersion ability. This is now used as the slurry for interlayer insulating films and STI. The GPL series from Showa Electric Industry, the GPX-HS series from Hitachi Chemical Co., and Ceria from Nissan Chemical Ind., are representative products of this type, are almost colloidal in form, and are good in their dispersion abilities.

4.4.3.3 Alumina-Based Slurry (Al_2O_3)

Alumina-based slurry plus an oxidizing agent is currently used for the CMP of metal for interconnections, like tungsten (W), copper (Cu) and aluminum (Al). This type of slurry is still at the development stage, except for the slurry for tungsten. Major problems to be overcome include developing a slurry without oxidizing agent, and preventing scratches on soft metal generated by hard and embedded particles. Developments in alumina-based slurry have occurred; for example, a colloidal alumina-based slurry is available.

The slurry used for interlayer insulating films and metal films is different. Figure 4.40 summarizes the slurries currently available and those being developed.

CMP object material		Slurry		Manufacturer • sales agency
		Abrasives	Working fluid/addition agent	
Oxide film (ILD)		SiO_2	KOH, NH_4OH, amine, etc.	Nittahaas, Cabot, Fujimi, JSR, EKC, etc.
		CeO_2	yes	Nittahaas, Showa Denko, Hitachikasei, Nissan Chemical, EKC, Mitsui Kinzoku, etc.
		ZrO_2	yes	Nissan Chemical
		$Al_2O_3(\gamma)$	yes (organic dispersion)	Sumitomo Chemical, etc.
		Mn_2O_3		Mitsui Kinzoku
Metal	W (plug)	Al_2O_3	H_2O_2, $Fe(NO_3)_2$, KIO_3	Nittahaas, Cabot, Fujimi, Sumitomo Chemical, Baikowski, EKC, etc.
		MnO_2		Mitsui Kinzoku
	Al (wire/plug)	SiO_2	H_2O_2, $Fe(NO_3)_2$, KIO_3	Cabot, Nittahaas, Fujimi, etc.
	Cu (wire/plug)	Al_2O_3	The same as above, quinaldinic acid, glycine, BTA, etc.	Nittahaas, Cabot, Fujimi, Shibaura Mechatronics, EKC, Sumitomo Chemical, etc.
		NO	Oxidizing agent, resolvent, anticorrosive	Hitachikasei

Figure 4.40 Slurries for oxide film and metal CMP

4.5 ROLE OF PADS IN CMP

4.5.1 Basic Principles of Planarization and Pad Behavior

The ideal planarization of wafers preferentially removes the convex part of any micro-roughness that exists on the wafer surface.

The dynamics of CMP and of general polishing have been discussed above, and can be described by Preston's equation, as follows:

$$\delta H/\delta t = Kp \cdot L/A \cdot \delta s/\delta t \qquad (4.5)$$

where H is the height of the convex pattern, t is machining time, L is load, A is surface area, s is moving distance, and Kp is a constant given by the machining conditions. Removal speed mainly depends on the partial pressure and is decided by load, contact surface area, and the speed of the pad that passes across the wafer surface.

In order to understand the basic factors affecting pads, it is important to understand how pads relate to surface precision in silicon wafer CMP. Figure 4.41 is one example, using a silicon wafer as a starting material. It shows the dependence of the planarity on the machined amount, including the dull edge, of a wafer with a lapped surface. Non-woven fabric is used in a pad. When the machining starts, the dull edge phenomenon appears first, and, as machining progresses, the wafer becomes increasingly planar. However, as machining progresses this increase in planarity levels off. The final planarity (Sf) increases as machining pressure increases. This can be explained by the pressure distribution that is seen when an elastic body (pad) is pushed onto a rigid body (wafer). Figure 4.42 is a schematic picture of the machined geometry of a wafer, and the change in pressure

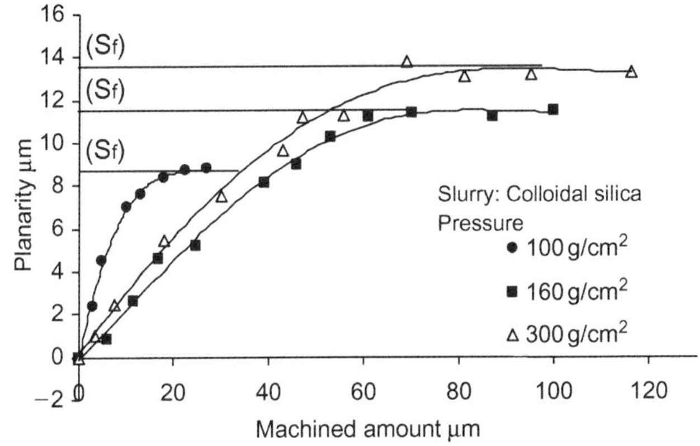

Figure 4.41 An example of dependence of the machined amount on the planarity

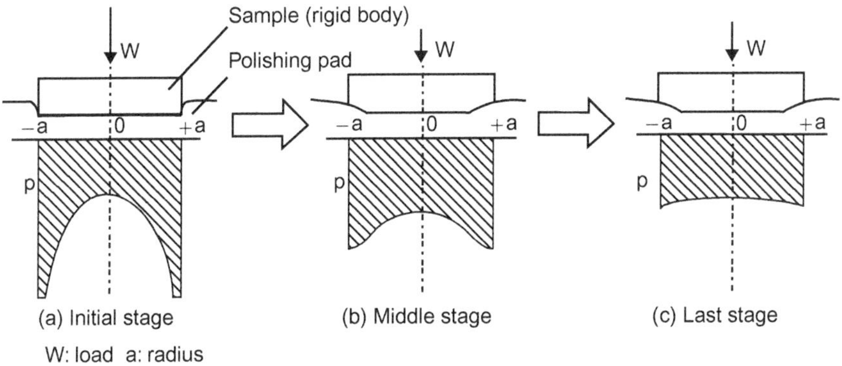

Figure 4.42 Schematic picture of the machined geometry of a wafer and change in the process of pressure distribution on the wafer surface

distribution across the wafer surface. When the final surface planarity Sf adapts to the deformed geometry of a pad, the deformation of the pad (δ) can be given by the difference between the deformed amount at the bottom of the wafer center (δ) r = 0 and the deformed amount at the bottom of the periphery (δ) r = a:

$$\delta = \frac{(\delta)_{r=0} - (\delta)_{r=a}}{E} = \frac{2(1 - v^2)\rho a(1 - 2)}{\pi E} = K \cdot 1 \cdot \rho \qquad (4.6)$$

where E = Young's Modulus, v = Poisson's ratio, ρ = pressure, and K = constant.

From the above equation it can be seen that the final surface planarity, Sf, depends on the reciprocal of the Young's modulus of the pad,

1/E, nd the machining pressure, ρ. Therefore, in order to achieve a planar surface with a reduced dull edge (i.e. small EE):

1. Use a small, thin pad with less deformation.
2. Apply low machining pressures.
3. Make sure there is uniform distribution of pressure on the wafer.

Both (1) and (2) closely relate to the machining quality and efficiency. In terms of (3), close attention should be paid to pressure application, and a method to create grooves of some kind on the pad surface is proposed.

Like the above, deformation due to the pad's thickness, hardness and applied pressure has a big influence on surface precision. This is demonstrated by the example of polishing a bare silicon wafer with a pad made of non-woven fabric.

CMP planarization of a wafer with a rough pattern on the surface will now be considered. Figure 4.43 shows the basic planarization mechanism using a relatively soft pad. It can be observed that the physical properties of a pad, such as its flexibility, hardness, roughness, and interaction with particles in the slurry, influence each other and are important factors in planarization CMP.

On the area adjacent to the convex part of wafer patterns, the pad will deform elastically to some extent. However, even in such regions, the pad does not make direct contact with the concave part of a device pattern, and the possibility of a concave part being machined with a convex part is low. If the deformed region is larger than the space between device patterns, initially only the step height of the convex part is

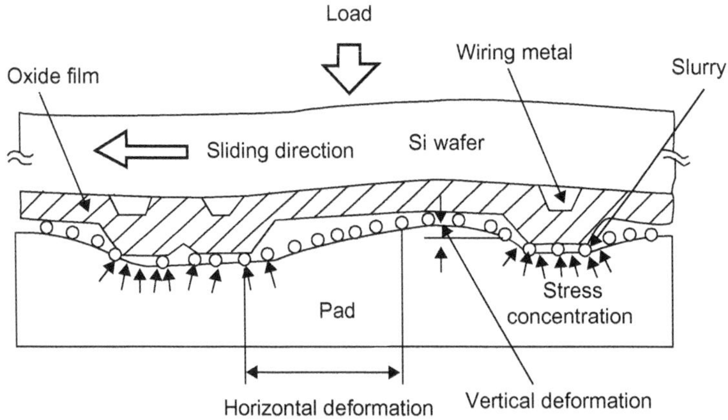

Figure 4.43 A schematic of a basic planarization mechanism using a relatively soft pad

machined and removed by the action of the pad and slurry. Then, machining gradually begins on the concave part. A change of size that is considered as a whole part of planarization starts to become smaller, when films in the limited area to each convex part and films in the concave part are started to be machined. If uniformity in machining speed becomes less uniform in a relatively large area, like in the level of the whole chip or wafer, the same mechanism is considered to be taking place.

It may be preferable to use a pad with high rigidity and less deformation horizontally, which can achieve very good planarity on a tip, while the uniformity across the entire wafer is inferior. On the other hand, a soft pad has a narrower deformation region horizontally; hence, the planarity of a single tip is inferior, but the uniformity across the entire wafer is improved. This is a general property and relates to the nanotopography or nanotopology of silicon wafers, which is increasingly becoming a topic of discussion.

This nanotopography is waviness on the micro-scale, which causes problems, particularly on bare silicon wafers in STI CMP. Figure 4.44 shows that when there is nanotopography (change in height) of 20 nm over 0.5 mm (distance), or ~100 nm (change in distance)/5.0 mm (distance) on a bare silicon wafer, oxidized films are accumulated along this feature. Different effects will result from these features, depending on whether a hard or soft pad is used. When using a hard pad, the oxidized film surface is planarized, but there is a variation in film thickness. When using a soft pad, the film thickness is consistent, but the surface is not planarized. In either case, high precision is essential on a bare silicon wafer, and is a challenging subject for the future.

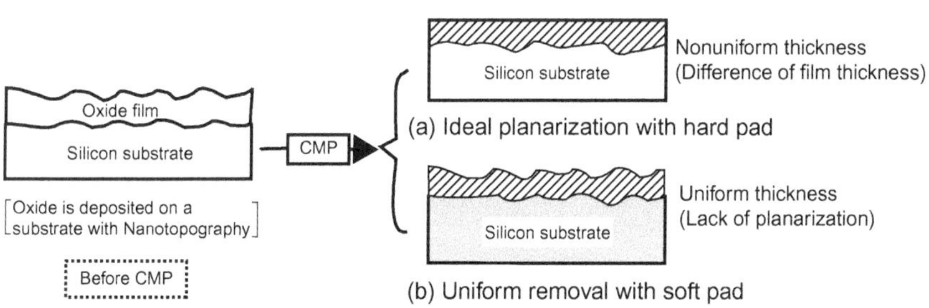

Figure 4.44 Two cases: CMP process on oxide films deposited on silicon substrates with nanotopography

4.5.2 Pad Requirements and Classification

The previous section discussed the importance of the elastic deformation of pads during machining. The variation in pad thickness should be minimized, but suitable pad materials must also be selected on the basis of their hardness and thickness. The basic factors covering pad selection for CMP can be summarized as follows:

1. It should have high efficiency (to retain slurry).
2. It should achieve planarity and high quality (with appropriate rigidity (hardness) on the pad surface).
3. It should ensure uniformity (pad surface can follow wave and warp of a wafer).
4. It should allow reproducible machining (should have waste disposal parts for machined chips and reaction products).
5. It should ensure cleanliness (fewer impurities in pad materials).

In the above list, (1) is essential, since enough slurry must be supplied to the work piece surface for machining. Generally, this is achieved by fine holes and grooves in some pattern on the pad, which also allow disposal of machined chips ((4) above). When grooves are formed on the pad surface, the pressure on a wafer is uniformly distributed, and a good, flat machined surface can be achieved over the entire wafer. This is why a conventional pitch (pad), has lattice marks on its surface for pitch polishing of lenses and prisms.

The previous sections indicate that there is a trade-off between (2) and (3). Various difficulties arise when considering what type of material should be selected, how thick the pad should be, and what profile structure it should have in order to obtain planarity, quality and uniformity. Also, as mentioned in (5), in order to introduce CMP to the LSI device process, close attention must be paid to the purity and quality assurance of the slurry. Table 4.6 shows analytical results showing the impurity levels of metals in various pads available on the market. It is noticeable that consideration is given to high purity of the planarization CMP pad in particular.

In general, fine holds/grooves are created on a pad, and machined chips and reaction products fill them. If all these holes are completely filled with chips (this is called "loading"), the machining property deteriorates. It is extremely difficult to recover the initial state of the pad after loading simply by digging out the chips, so a diamond wheel is used to remove the loaded surface, which is called dressing. This operation grinds off the

Table 4.6 Analytical Results of Impurity Metals in Various Pads Available in the Market

Metallic element \ Pad	Nonwoven fabric	Artificial leather	Independent formed structure
Al	<2	0.6	0.7
Fe	1	5.3	1.7
Mg	10	4.3	0.3
Cu	<1	<0.1	0.1
Ni	<1	<0.1	3.1
Ca	41	20	1.8
Zn	12	0.2	0.7
Na	220	1.0	6.4
Cr	<1	0.2	1.8
K	62	1.6	0.7

loaded pad surface, and the initial condition of the pad surface is restored so that machining can continue.

Table 4.7 shows pads available in the market classified by structure. In general, pads are either foamed or non-foamed. Foamed pads are again divided into independent and continuously foamed structures.

4.5.2.1 Pads with an Independent Foamed Structure

In independent foamed pads, the slurry exists only in the gap between the wafer and the pad, or on the contact interface, so reduced slurry flows at the same machining pressure are preferable. Figure 4.45 shows SEM photos of the structure of IC1000 from Dow Chemical Corp. and Nitta Haas Inc. (formerly Rodel Inc. and Rodel-Nitta). As can be seen in these photos, foamed pores form in an almost perfect circle, filled with gas generated by the foaming process. Deformation of the IC1000 pad in the vertical direction displays excellent physical properties with less delay in stress/deformation in the general viscoelastic behavior.

However, because the slurry does not penetrate into the inside of this type of pad, it is difficult to remove machined chips and reaction products, and the pad surface becomes loaded after a relatively short machining time. To solve this problem, the slurry flow is made active and grooves are created to support the disposal of machined chips on the pad surface. After a given time, a suitable dressing must be applied to the pad surface to retain stable machining speeds, obtain a consistent wafer surface, and extend the pad life.

Table 4.7 Pads Available in the Market Classified by Structure

The Morphology of the Foam (1)			The Morphology (2)	Filler or Not	Hardness etc.	Rodel-nitta Company
(1)	Closed-cell foam type (dry foam type)	Foam by additive	Bubble-free type	Without filler particle	No shell type	IC1000
		Form by foaming agent	Containing bubbles type	Without filler particle	Soft	MH-S 35 A
					Hard	MH-S 15 A
				With filler particle	Soft	MH-C 14 A (celium oxide)
					Hard	MH-C 15 A
(2)	Open-cell foam type (wet foam type)	Nonwoven fabric type	Primary resin		Soft	Suba 400
					Semisoft	Suba 400 H
			Second resin		Soft	Suba 600 D
					Semisoft	Suba 600
					Hard	Suba 800
					Very hard	Suba 650
		Suede type	Bonding type		Semisoft	Rodel 205
			Direct form type	Carbon reinforcement	Soft	Politex RN-H
					Semisoft	DPC 5350
				Carbon free	Soft	WHITEX-B
					Hard	SPM 3100
(3)	Two-layered composite type (lamination type)	Composite pad by bonding with the above (1 and 2)			Deformable type	IC1000 (032 = 800 μm)
					Undeformable type	IC1000 (050 = 1250 μm)
		Composite pad by bonding with sponge layer				IC1400 (050)

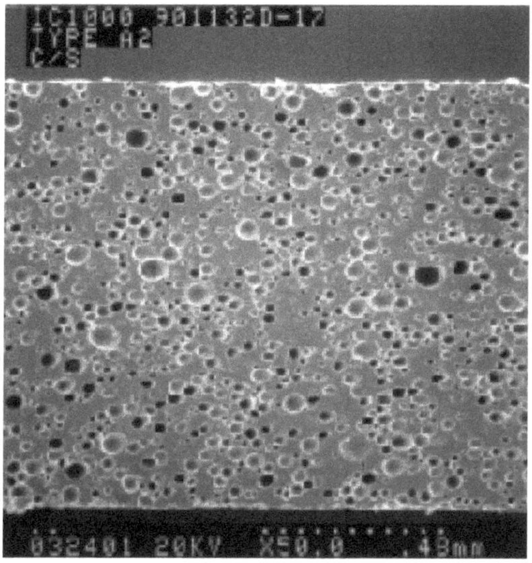

Figure 4.45 Cross-sectional view of an IC1000 pad

4.5.2.2 Continuously Foamed Pads

A continuously foamed pad generally has a non-woven fabric as its raw material, and various resins embedded in the fiber lattice act as binders. Figure 4.46 is an SEM photo of a pad of this type, made of a typical polyester, non-woven fabric. Because this type of pad can achieve a high machining rate and a smaller dull edge, it is widely used for the primary polishing of bare silicon wafers.

Figure 4.47 is an SEM photo of a suede-type pad, one example of a pad made of synthetic leather used for finish polishing bare silicon wafers. This pad is produced by creating openings by removing the surface layer after forming a foamed layer in polyurethane that is coated on the back side with non-woven fabric. This is classified as a continuously foamed body pad, and the slurry is retained in the nap layer (concave at several 100 μm) formed on the surface. This pad is soft in structure, so a high-quality machined surface without scratches can be obtained, and it is often used as a cleaning agent (water polishing) after CMP.

A pad with this continuously foamed body structure is easily penetrated by the slurry. The slurry penetration rate and the disposal of machined chips change according to factors that affect the total pad structure, like the existing ratio of fiber and resin per unit volume length, fiber diameter, and foam diameter of the resin layer and its density. Both the

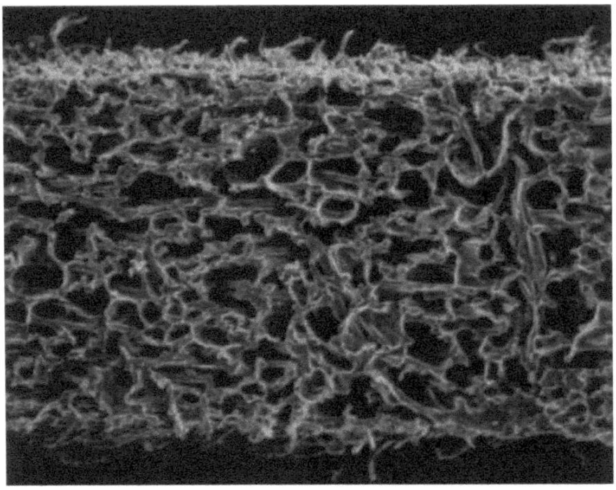

Figure 4.46 Cross-sectional view of a typical polyester, non-woven fabric pad

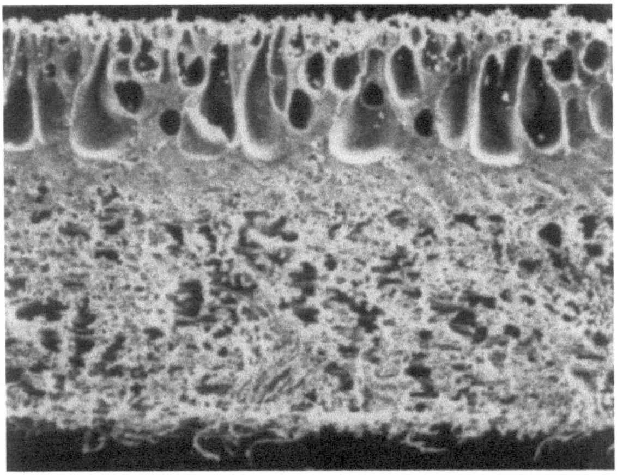

Figure 4.47 Cross-sectional view of a suede-type pad

dynamics of a pad and its profile structure are important factors in pad function.

4.5.2.3 Two-Layer (Complex) Pads

Before detailing the two-layer structure of a CMP pad, such as IC1000/Suba400 or IC1400, the ideas and concepts of the two-layer pad structure designed by Professor Doi will be explained. Figure 4.48 shows a two-layer pad designed and proposed in the late 1970s to achieve high precision and a high-quality machined surface on bare silicon wafers. The

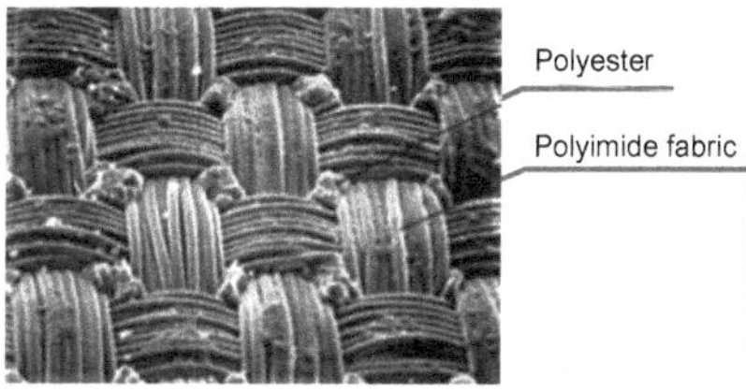

	Materials	Hardness	Thickness (mm)
Upper layer	Polyimide fabric	98	0.08
Lower layer	Polyurethane	65	0.5

(a) Top view

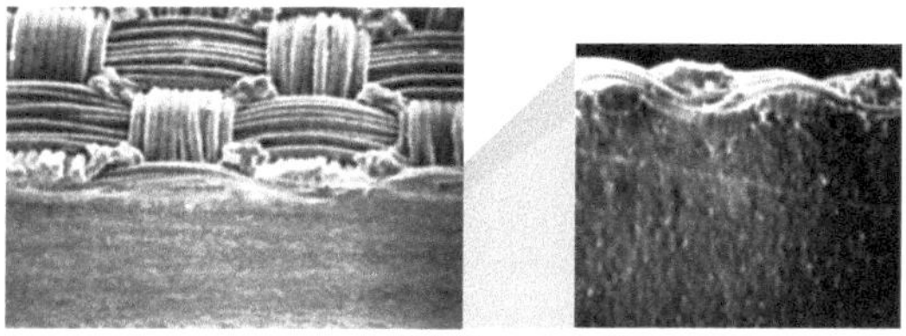

(b) Cross-section

Figure 4.48 Double-layer pad (Prof. Doi)

basic idea was to produce a pad of complex structure, in which a hard, thin polyimide fabric with less deformation was used as the top layer to secure high precision. A softer, elastic body was used as the bottom layer to control machining defects, like scratches, and to achieve high quality. It is important to completely bond the top and bottom layers together in order to achieve both objectives.

Polishing with this two-layer pad achieves a highly precise flat surface without a dull edge, as shown in Figure 4.49. In terms of surface precision, it is clear that this pad is far superior to the one made of synthetic leather in more general use. However, when this two-layer pad is used for wafers which have micro-roughness, on which the machining margin is extremely low, it is difficult to achieve flatness, partly due to roughness of the fabric. Nevertheless, the idea of this two-layer pad structure was introduced in the IC1000/Suba400. In order to get satisfactory performance in both planarity and uniformity, which are in a trade-off relationship, various combinations of materials are used. Figure 4.50 is a cross-sectional SEM photograph of the IC1000/Suba400 pad.

The IC1000, which has significantly improved hardness, consistency, purity and particles, is a hard-foamed polyurethane body with independent air foam. It is a highly functional pad developed specially for CMP, which has a direct effect on a wafer surface that contains steps. However, use of the hard IC1000 alone is not enough to simultaneously achieve local/global planarity, and, in order for a pad to profile the wafer surface with warp and wave, it is made into a viscoelastic body by placing a shock-absorbing layer

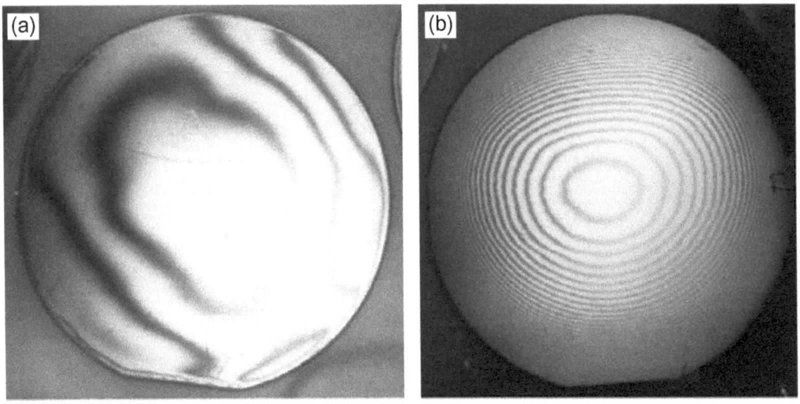

Figure 4.49 Effect of the double layer pad proposed by Professor Doi. (a) Fringe pattern by double layer pad (b) Fringe pattern by suede-type pad (polishing pressure 100 g/cm^2)

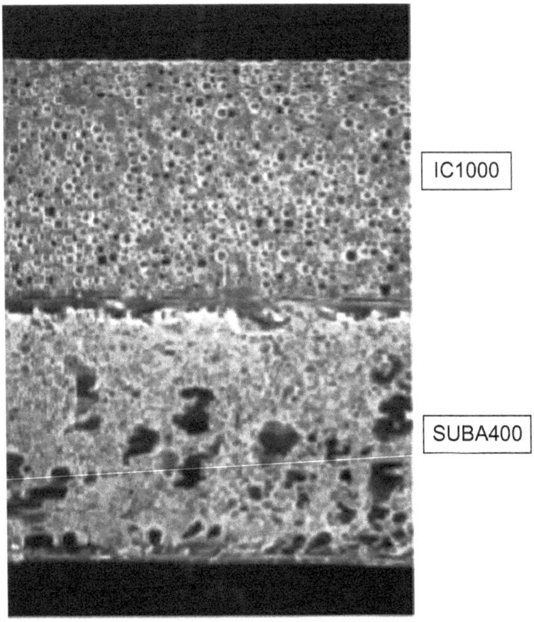

Figure 4.50 Cross-sectional view of an IC1000/Suba400 pad

on the substrate, as stated above. Among the pads in the Suba series based on non-woven fabric, the Suba600 and Suba800 are used most often.

Table 4.8 shows the physical properties of polishing pads IC1000 and IC1400, each used in a single and two-layer body. The IC1400 pad has a two-layer structure, and Type A based IC1000 has a highly crosslinked product on top, and a uniform urethane foam pad on the bottom (Figure 4.51). Because the pad is highly pure and water does not invade the substrate (i.e. adhering tape and plate), the uniformity of the wafer surface, and, therefore, the pad stability, has improved.

It is general practice to machine special grooves or holes on the pad surface for IC applications. There are various types of pads, including one in which small holes of 1–2 mm are created on the surface to allow good slurry supply and prevent the wafer from adhering to it, and another in which lattice, concentric circle or spiral-like grooves are created on the surface to increase slurry flow, in order to improve the flatness and uniformity of the wafer surface. Figure 4.52 shows four types of pads with which various machining methods are carried out.

Table 4.8 The Physical Properties of Polishing Pads IC1000 and IC1400

[Value in ( ) Is Physical Property m Monolayer.]

Pad	Compressibility (%)	Elastic Recovery (%)	Hardness (Asker-C)
IC 1000	(1.1)1.6	(75) 88	95
IC 1400	(1.1) 1.5	(75) 90	95

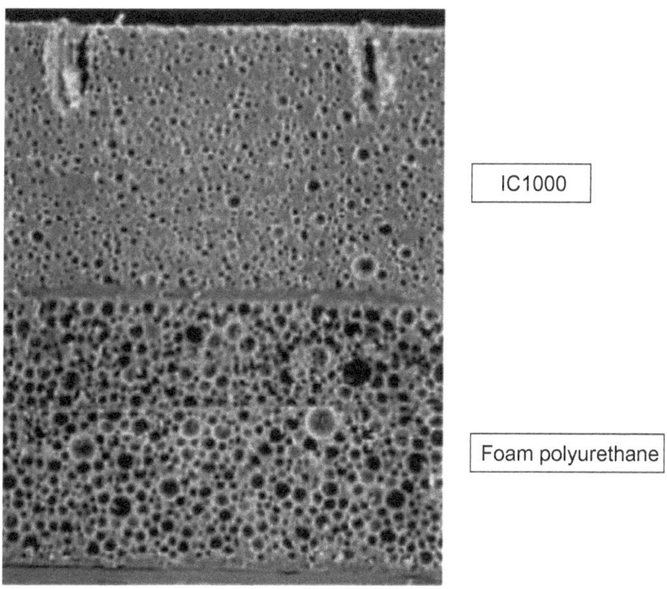

Figure 4.51 Cross-sectional view of a IC1400 pad (IC1000/ polyurethane)

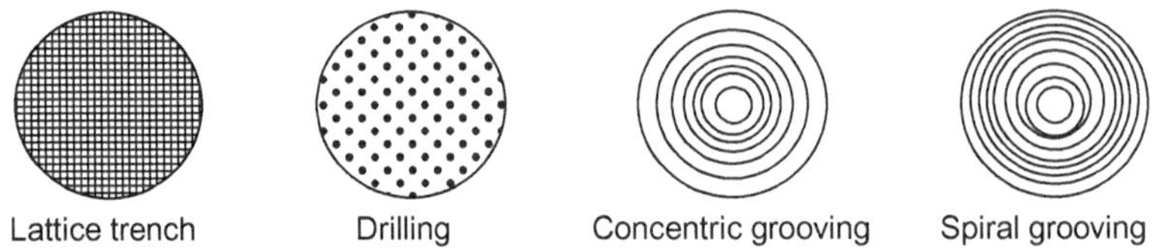

Figure 4.52 Four types of pads

4.5.2.4 Non-Foamed Pads

This type of pad is made from hard, synthetic resin, like a plastic plate. This material is not only hard, making it difficult to generate elastic deformation, but it also has a flat, smooth surface with no roughness, so it can planarize a wafer with micro-patterns. However, it does not retain slurry on its surface, and, thus, grooves are created on it. Without the grooves, the pad cannot achieve a high machining rate, and there are problems with machining precision.

Because this type of pad does not have foamed micron-holes, the loading phenomenon is not observed. On the other hand, it cannot deal with micro-waviness/nanotopography, which is a weak point. In order to mitigate these weaknesses, a thinner, non-foamed material with grooves can be added at the top, and an elastic layer at the bottom. This type of pad resembles the aforementioned two-layer pad. Figure 4.53 is a cross-section of IC2000 urethane from Dow Chemical Corp. and Nitta Haas Inc., as an example of a two-layer structure of a pad. It is not available on the market at present.

Next, the dynamic properties of pads will be touched upon. Figure 4.54 shows the typical relationship between load and deformation rate for independent and continuously foamed pads across their thickness, which is equivalent to a stress/deformation curve. Compared with the quick response of deformation to machining pressure for an independent foamed body, a continuously foamed body shows a considerably delayed response, and in some cases 100% elastic recovery does not take place, which causes permanent deformation at some level. In the case of a continuously foamed pad, both loading on the pad surface and a change in economic dynamics property must be considered.

During machining, machining efficiency (machining speed) is a fluctuating parameter that can be easily measured, and stable machining can probably be maintained if the loading of the pad surface is removed by dressing. However, a change in the dynamic properties of a pad not only changes the global planarity of the wafer but, in severe cases, also affects the

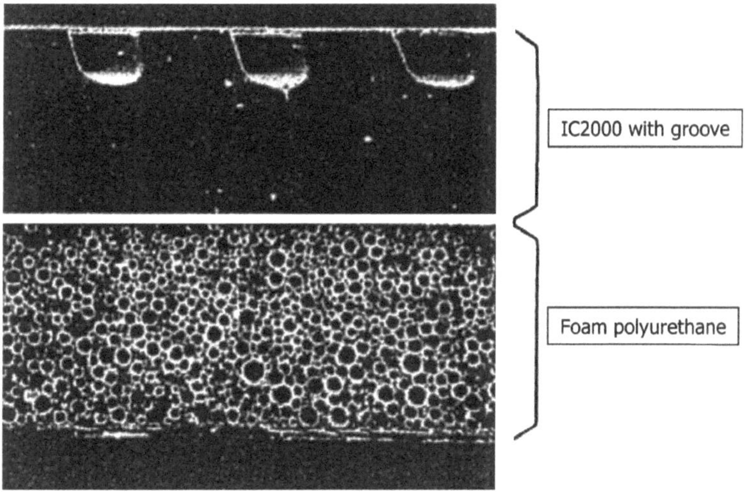

Figure 4.53 A cross-section picture of IC-2000

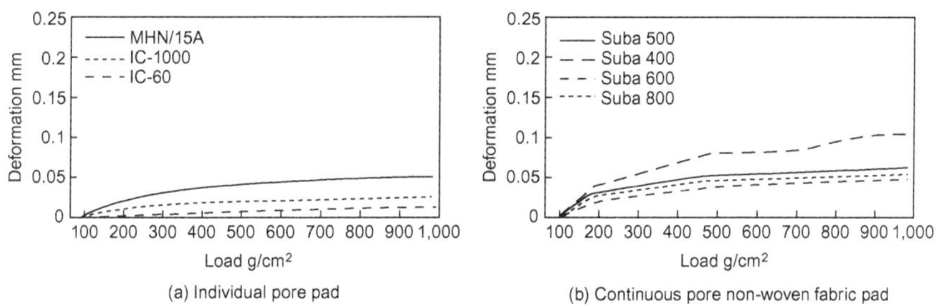

Figure 4.54 The typical relationship between load and the deformation rate. (a) Individual pore pad (b) Continuous pore non-woven fabric pad

roughness of the machined surface. It is therefore necessary to monitor fluctuations in speed during machining and to pay attention to loading.

Table 4.9 shows CMP pads and their features using Dow Chemical and Nitta Haas Inc. products as examples.

4.5.2.5 Bonded Particle Pad for CMP

3M Co. in the US have developed and announced a bonded particle, slurry-free pad, shown in Figure 4.55. It has a thin film sub-pad as the bottom layer. The top layer contains a resin in the form of square micro-pyramids or trapezoids mixed with particles of ceria, silica or alumina. The most suitable for the job is selected, along with the resin, for oxidized films and materials. Claims of a surface finish with dishing of

Table 4.9 CMP Pads and Their Features

The Stock Number	Objects	Structrures and Features
IC1000	SiO_2, W, Cu, poly-Si, Al	Normal CMP pad Surface: IC1000 Foundation: the bilayer consisted of Suba400
IC1400	SiO_2, W, Cu, poly-Si, Al	Foundation doesn't have permeability. Elastic recovery is fast. Surface: IC1000 Foundation: the bilayer consisted of UF

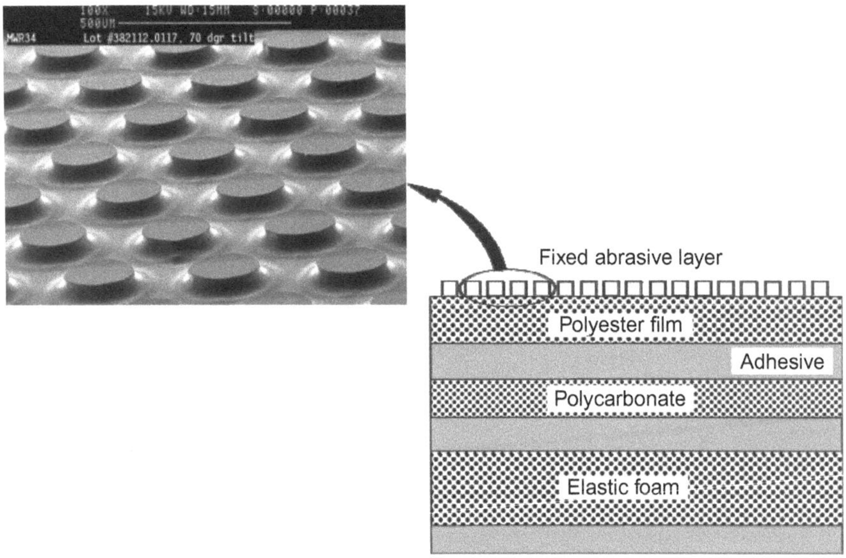

Figure 4.55 The structure of the slurry-free pad from 3 M

40 nm, oxidized film erosion 20 nm, and surface finish R6 86Å have been made for copper CMP. However, the machining ability of this type of pad is low for Ta/TaN (Tantalum/Tantalum Nitride), so this is used for the first step of CMP. Research papers have been published on the CMP of a device wafer with these pads, and its future development should be interesting.

The major future concerns in this area are machining efficiency, price, machined surface quality, and stability. For machining efficiency and machined surface quality, dynamic machining conditions and additives need to be reviewed. Scratch prevention is an important subject, as are measures to prevent fine particles from being pulled out.

4.5.3 Methods for Measuring the Physical Properties of Pads

The physical properties of a pad which are measured include thickness, compression rate, compression recovery rate, hardness and density. Some of the measuring methods involved are summarized below:

1. Measuring the physical properties of the current pad.
   Thickness, hardness and density are measured by the equipment (e.g. durometer) commonly used for rubber and plastic (JIS L1096). Measurement of compression and compression recovery rate is, in general, carried out using automated compression rate measuring equipment. Example results are shown in Figure 4.56.
2. Points for measuring physical properties to be reviewed in the future.
   The measurements mentioned above concern only the static, physical properties of a basic machining pad. However, it is possible that static, physical properties do not correlate with machining properties. Therefore, it is necessary to measure physical properties in a state that more closely resembles actual CMP, such as dynamic viscoelasticity, the change in compression of a pad over time by repeating the compression and release cycle, and the compression rate in the wet state. Dynamic viscoelasticity measurements are often made. A model has also been proposed in which the surface condition of a pad is measured – surface roughness, in particular – to actually monitor CMP. At any rate, standardization has not been achieved.

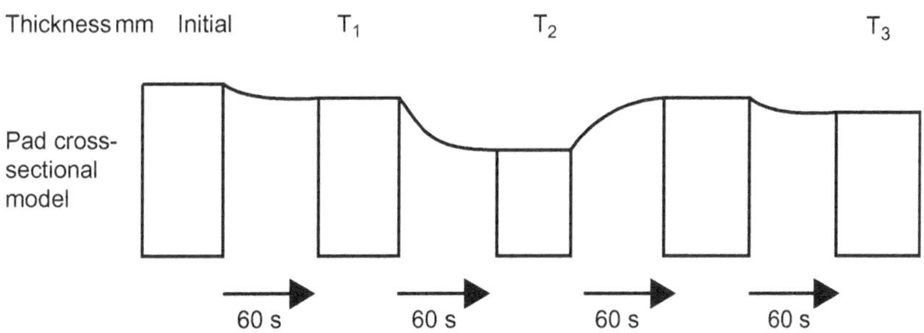

Compressive ratio (%) = $(T_1-T_2)/T_1 \times 100$

Compressive elasticity ratio (%) = $(T_3-T_2)/(T_1-T_2) \times 100$

Figure 4.56 Automated compression rate measuring result

4.6 ADVANCED EVALUATION OF PAD SURFACE TEXTURE

4.6.1 In the Beginning

Progress in science and technology depends, in part, on innovation in the semiconductor industry. Silicon wafers for use in device fabrication need extremely flat surfaces. As a result of the continuous shrinking of device dimensions, planarization requirements are reaching the order of nanometers. Hence, increasing demands are being made on CMP techniques.

In actual polishing conditions, the polishing rate and accuracy are affected by the real contact area between the polishing pad and the wafer. Specifically, it has been pointed out that the real contact area between the two is small, even though the apparent contact area is equal to the surface area of a large wafer.

Moreover, the surface of the pad deteriorates as polishing progresses, which leads to a decreased polishing rate and accuracy. Consequently, a diamond dresser (conditioner) is normally used to recover the pad surface. However, no effective quantitative evaluation methods or parameters for pad surface characteristics have been established.

Professor Uneda's group has developed a technique to evaluate the surface condition of a pad. The proposed method is based on contact image analysis using an image rotation prism. This section will describe the basic evaluation method, and experimental results for the contact ratio, number of contact points, spacing of contact points and peak width of the spatial FFT spectrum.

4.6.2 Experimental Apparatus and Method

Figure 4.57 shows the experimental apparatus and an illustration of the dressing method for a polishing pad. The pad is placed on a steel plate 300 mm in diameter, which is rotated clockwise at 30 min^{-1}. The diamond dresser is then rotated clockwise at 2 min^{-1}. The distance between the center of the polishing pad and the center of the diamond dresser, referred to as the offset, is 125 mm. Experiments were conducted with four different mesh sizes (#80, #100, #500, #1000) for the diamond dresser. The main experimental conditions are shown in Table 4.10.

4.6.3 Measurement Method Based on Contact Image Analysis Using an Image Rotation Prism

Figure 4.58 show the proposed analysis method using contact images. The images are obtained from 50 different locations for each measurement and

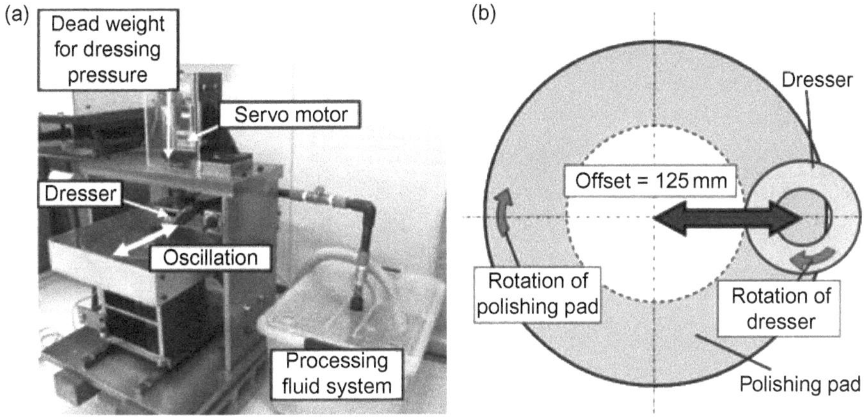

Figure 4.57 Experimental apparatus and dressing method. (a) Photograph (b) Schematic drawing

Table 4.10 Experimental Conditions

Mesh size of dresser	#80, #100, #500, #1000
Outer/Inner diameter of dresser	100 mm/ 82 mm
Dressing time	60 min
Dressing pressure	150 gf/cm^2
Dresser rotation speed	2 rpm (CW)
Pad rotation speed	30 rpm (CW)
Offset of dresser	125 mm
Pad	MH-S15A

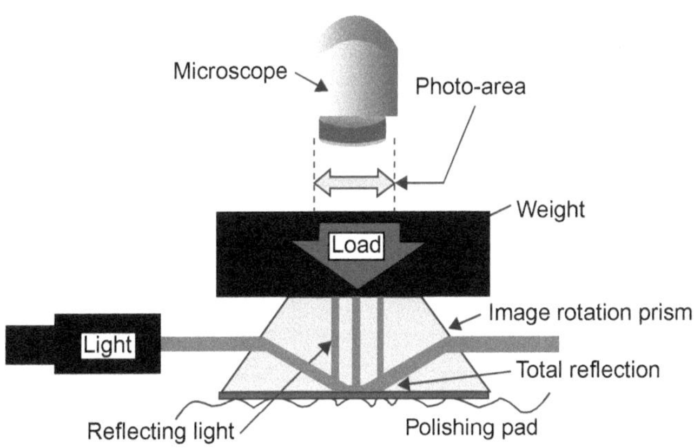

Figure 4.58 Schematic drawing of photography method of contact image between image rotation prism and polishing pad

the evaluation parameters are then extracted from these data. The average and standard deviation of the resulting distribution are then calculated.

In an image rotation prism, incident light from the side of the prism becomes totally reflected at the prism bottom when nothing is in contact with it. On the other hand, when there are contact points at the bottom surface of the prism, the light is diffusely reflected by the contact points.

The proposed method applies pressure from above to the image rotation prism, and the contact image between the polishing pad and the bottom surface of the prism is recorded by the microscope/camera system. The microscope can photograph a 7.3 mm × 5.5 mm area at a resolution of 1200 × 1600 pixels with a color depth of 16 bits.

First, we examined the threshold amount of the contact images. Figure 4.59 shows the relationship between threshold amount and the contact ratio. The contact ratio will be defined later. We photographed an image with nothing in contact with the bottom surface of the prism – i.e. conditions with an ideal contact ratio of 0. However, the image is slightly noisy. Consequently, we investigated the threshold amount when the contact ratio is less than 0.001%. The results in Figure 4.59(a), show the contact ratio decreases from the increment of the threshold amount. The results in Figure 4.59(b), show that a threshold amount of 6000 concentrates at a contact ratio of less than 0.001%. Figure 4.60 shows representative examples of observed contact images for different dressing conditions. The white areas on the images indicate the contact regions. From these images, it is clear that before dressing, all contact regions appear to be large, but the dressing process causes the size of the contact regions to decrease, and the number of contact regions to increase.

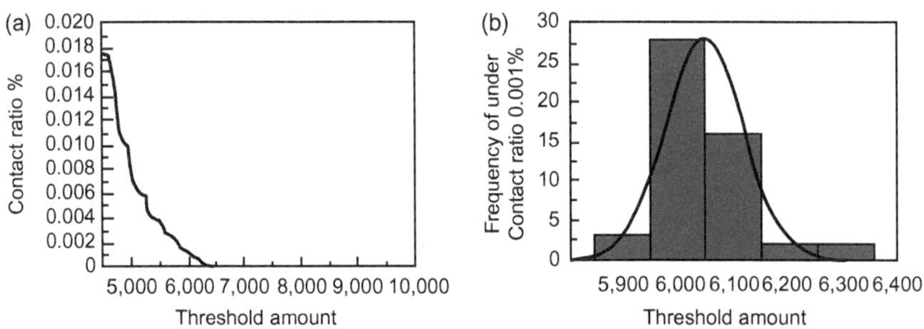

Figure 4.59 Relationship between threshold amount and contact ratio (a) Contact ratio of threshold amount (b) Frequency of threshold amount

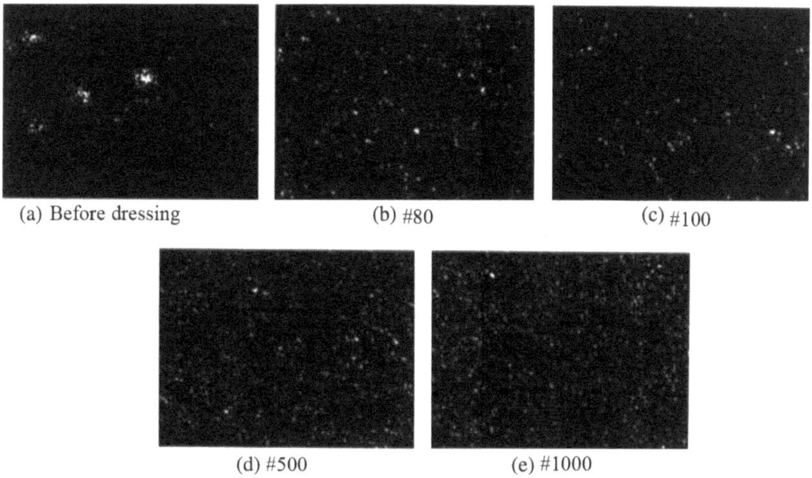

Figure 4.60 Representative examples of observed contact images under each dressing conditions

4.6.4 Results and Discussion

Figure 4.61 shows the relationship between mesh size of the dresser and the number of contact points. Note that the latter refers to the number of peaks in the spectrum of the contact image; such peaks indicate a contact point between the bottom surface of the image rotation prism and the polishing pad. The number of contact points increases with the mesh size of the dresser, with the most dramatic increase observed between mesh sizes of #500 and #1000, where the number of contact points increase by 2 and 3.5 times, respectively.

The reason for this increase in the number of contact points could be planarization and micro-roughness created by the polishing pad as the grain size of the dresser reduces. Thus, it seems that the effect becomes larger with increasing mesh size.

Figure 4.62 shows the relationship between the mesh size and the contact ratio. The contact ratio is the ratio of the apparent contact area to the real contact area. In this case, the apparent contact area is the area photographed by the microscope, and the real contact area refers to the total area of pixels for which the spectrum becomes larger than the threshold amount in the contact image. The contact ratio increases with increasing mesh size, in a similar manner to Figure 4.61. Specifically, the contact ratio becomes remarkably high for mesh sizes of #500 and #1000.

Figure 4.63 shows the relationship between the mesh size and the maximum value of the minimum spacing between contact points. The maximum value of the minimum spacing between contact points is the maximum in a set of the minimum spacing between contact points. It is seen to decrease as the mesh size becomes larger, and consequently as the

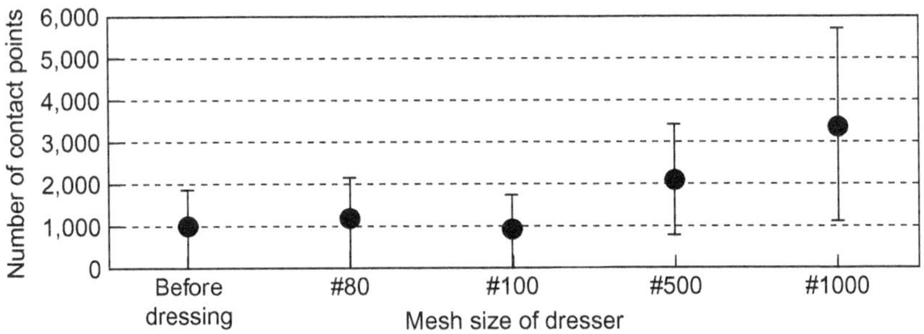

Figure 4.61 Relationship between mesh size of dresser and number of contact points

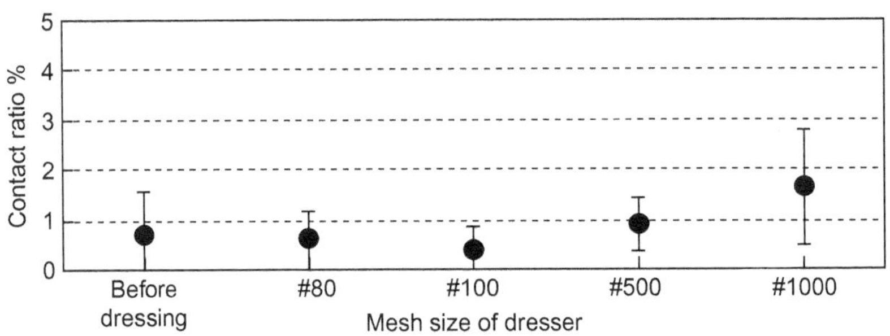

Figure 4.62 Relationship between mesh size of dresser and contact ratio

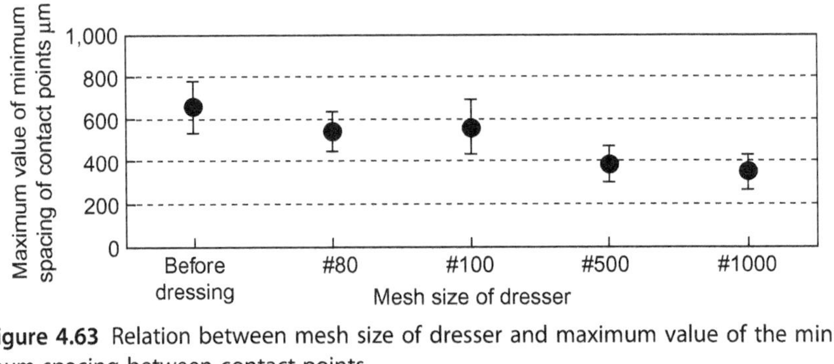

Figure 4.63 Relation between mesh size of dresser and maximum value of the minimum spacing between contact points

number of contact points increases. In addition, an opposite relationship is found for the results shown in Figures 4.61 and 4.62.

Figure 4.64 shows contour diagrams of the spatial FFT results for the contact image. From this image, the contour spectrum before dressing is seen to be concentrated in the center compared with that after dressing. Figure 4.65 shows the relationship between the mesh size of the dresser and the half width of the peak for the spatial FFT result. This parameter seems to be a quantitative measure of the cohesion of the contact points,

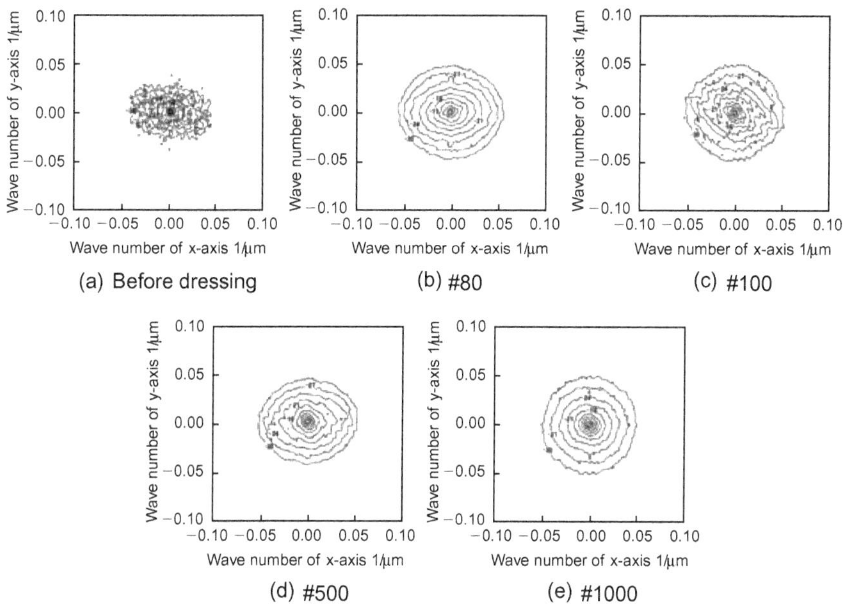

Figure 4.64 Contour diagrams of spatial FFT results of contact image

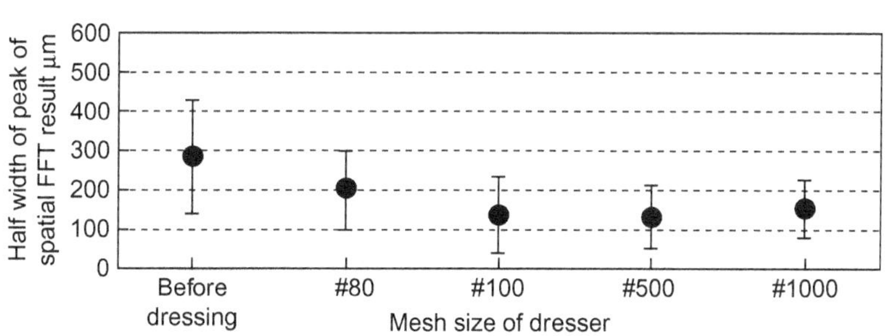

Figure 4.65 Relationship between mesh size of dresser and half width of peak of spatial FFT result

and decreases most dramatically from the before dressing case for a mesh size of #80.

Therefore, it is found that that the number of contact points, the contact ratio, the maximum value of the minimum spacing between contact points, and the half width of the peak of the spatial FFT result can be used as parameters to evaluate the characteristics of the polishing pad surface.

4.6.5 Summary

A novel method based on contact image analysis was used to evaluate the characteristics of a polishing pad surface, and the following results were obtained:

1. A pad could be evaluated quantitatively by using the number of contact points, the contact ratio, the maximum value of the minimum spacing between contact points and the half width of the peak of the spatial FFT result as parameters.
2. As the dresser mesh size increased, the number of contact points and the contact ratio were found to increase, whereas the maximum value of the minimum spacing between contact points and the half width of the peak of the spatial FFT result were found to decrease.

Figure 4.66 shows the the prototype equipment (PSP-100) from Fujikoshi Machinery Corporation.

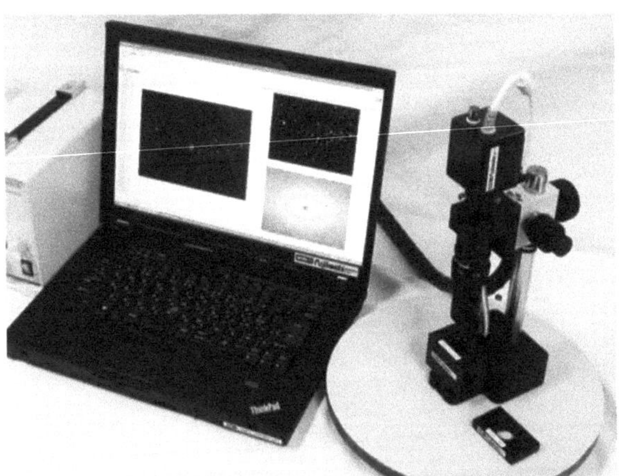

Figure 4.66 The external appearance of the commercially available equipment (PSP-100) from Fujikoshi Machinery Corporation

4.7 CLEANING AFTER CMP

Because metals such as copper, and also inorganic and organic-based low-k materials, are not chemically resistant, and not particularly hard, the cleaning liquids and methods used for silicon cannot be applied. Dedicated cleaning technologies are required.

4.7.1 Principles of Cleaning after CMP

4.7.1.1 Principles of Wafer Cleaning

Methods to remove contaminant particles adhered to the wafer are classified as physical removal, lift-off and dissolution, as shown in Figure 4.67. Physical removal includes brush scrubbing (Figure 4.67(a)) and sonication (Figure 4.67(b)). The brush scrubbing method removes particle contamination by scrubbing the wafer surface with a sponge brush while applying a cleaning liquid. Ultrasonic cleaning applies ultrasound (~MHz) to a cleaning liquid, such as a pure water/alkaline mixture, to detach particles from the wafer surface by cavitation.

The lift-off method removes contaminant particles and metals attached on the surface by slightly etching the surface layer, as shown in (Figure 4.67 (c)). Dissolution cleaning (Figure 4.67(d) selectively removes contaminants only; it does not affect the surface layer of a work piece. In this method the

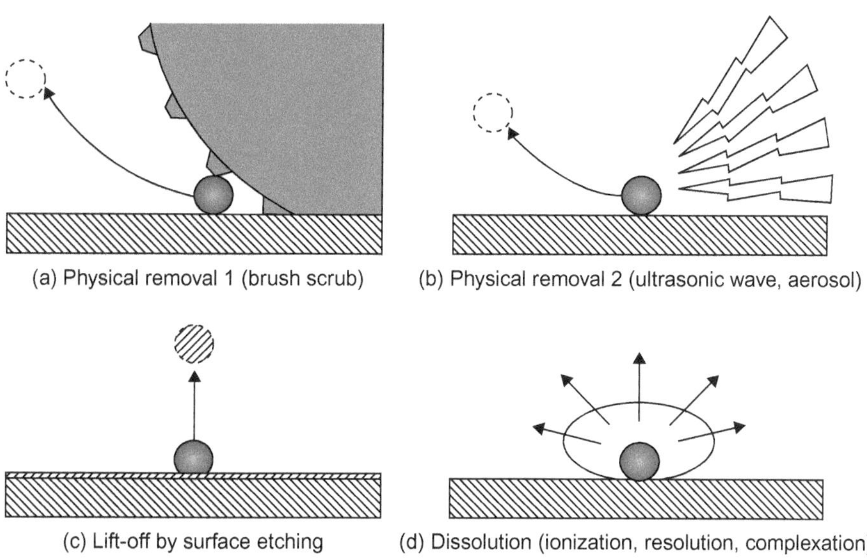

Figure 4.67 Removal methods of contaminant particles

contaminant metals become ions by reaction with acid and alkaline reagents. Organic films also react, to form soluble metal complexes.

4.7.1.2 Cleaning Processes after CMP

The removal of contaminant particles using a brush is illustrated in Figure 4.68. In the first step, particles adhered to the wafer are detached by physical force with a brush, or lift off by the slight etching of the surface layer. During the second step, contaminants are removed from the wafer by not re-attaching onto the brush or the wafer, and the end of the process begins. Electrical repulsion prevents re-attachment; hence the electrical potential on the wafer, the brush, and the contaminant particle surface must be such to make them repel each other. The electrical potential on the surface generally has the same value as the Zeta potential, and is affected by pH. In general, under alkaline conditions, many particles, like silica and alumina, have a negative Zeta potential. A negative Zeta potential means that the SiO_2 surface and the PVA brush surface each generate electrical repulsion. Therefore, the particles detached from the wafer are effectively removed from it, and do not re-attach to the brush or the wafer.

The first requirement for cleaning after CMP is the removal of polishing particles and polished chips. More than 10,000 contaminant particles sized 0.2 μm or larger remain attached to the surface of a 300 mm (12″) wafer after polishing. Removal of these contaminant particles using an appropriate agent (chemical liquid) and brush scrubbing is required.

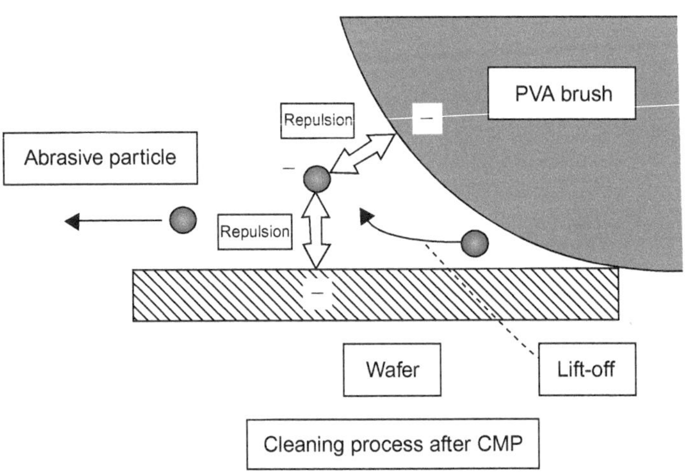

Figure 4.68 The removal mechanism of contaminant particles

Metal contaminants must also be removed. There is a risk that abrasive particles may be contaminated with metals during their production. High purity abrasives are available, but even in these, metal impurities at the ppm (parts per million) level can be left on the wafer. This is a very high contamination level, compared to the pure water and chemicals used for semiconductors, which are controlled to ultra-high purity, ppt (parts per trillion), levels. In addition to copper and tungsten, various metal contaminants like iron (Fe) or potassium (K) can enter the slurry. In many cases, the surface density of these metal contaminants (the number of impurity atoms per cm^2) can reach over 1×10^{12} atoms/cm^2; a threshold of 1×10^{10} atoms/cm^2 is acceptable after cleaning.

4.7.2 Cleaning Methods and Equipment

The main cleaning method is scrubbing with a brush. Brush types include:

- The roll brush, where the wafer is held horizontally and the brush rolls from top to bottom (Figure 4.69(a));
- The disc brush, where the wafer is placed between circular brushes (Figure 4.69(b)); and
- The pen brush, which scans the wafer surface with a pen-like brush (Figure 4.69(c)).

These brushes are made of a sponge of PVA with nylon and mohair fibers.

There is also a roll brush where the wafer is inserted between brushes, as shown in Figure 4.69(a). This method can create more space in the cleaning room. Pen brushes are often used for higher precision cleaning after cleaning with a roll and a disc brush. In ultrasonic cleaning, ultrasonic waves are applied through pure water or a cleaning liquid by a nozzle equipped with a ~MHz ultrasonic wave vibrator (Figure 4.69(d)).

After polishing, the wafer should not dry before cleaning, and each wafer should be cleaned individually. When CMP was first used, separate standalone polishing and cleaning equipment was needed, but now this type of equipment is only used experimentally. In the CMP of copper and other materials, the Cu interconnection starts to erode if cleaning is not carried out soon after polishing; therefore, it is necessary to have both cleaning and polishing functions in one piece of equipment or to build a dry-in/dry-out system.

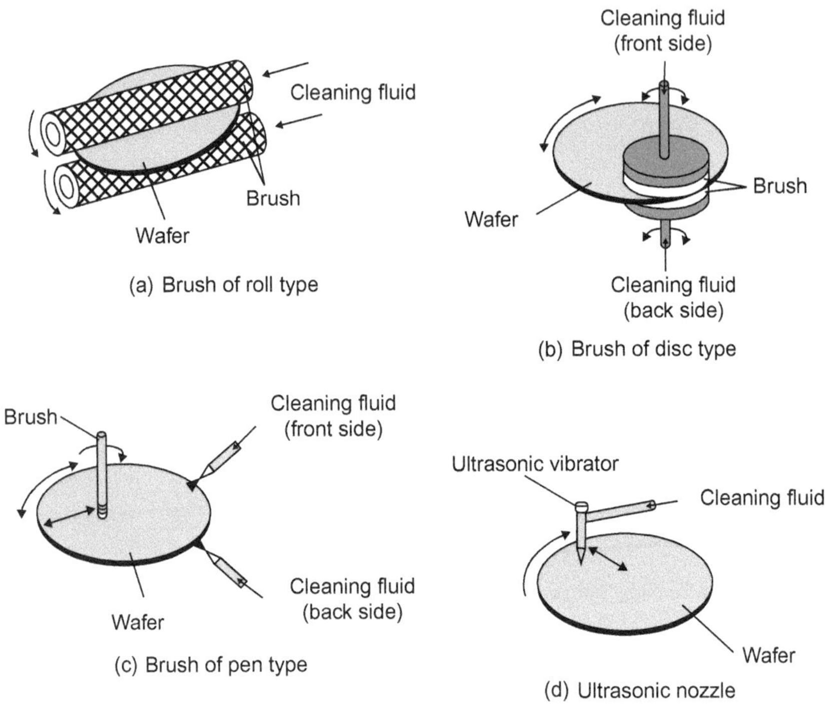

Figure 4.69 The main cleaning methods used in cleaning equipment after CMP

4.7.3 Cleaning after CMP

The fluid used varies with the film to be cleaned. Table 4.11 summarizes the abrasive type, dispersant and chemicals used to remove particle and metal contamination during cleaning.

4.7.3.1 SiO_2 Films

SiO_2 occurs in interlayer insulating films and STI. When SiO_2 films are polished, a silica-based slurry is mainly used, and KOH (potassium hydroxide) and NH_4OH (ammonium hydroxide) are used as dispersants. If potassium is present in the CMP slurry, potassium contamination remains on the SiO_2 films and silicon substrate. It is possible to remove silica particles relatively easily with pure water and dilute ammonia solution, and metal contamination like K is generally removed by lift-off with DHF (Dilute Hydro Fluoric acid). Also, as shown in Figure 4.70, in the case of STI-CMP (Shallow Trench Isolation) with Si_3N_4 (Silicon Nitride) as a substrate film, an oxidized cerium (CeO_2) based slurry that has high selectivity is sometimes used. After

Table 4.11 Representative Example of Cleaning Methods After CMP for Various Coats

Polished Coat	Abrasive Grain	Dispersant	Particle Removal (brush)	Metal Decontamination
SiO_2	Silica, cerium oxide	KOH, NH_4OH	DI water, dilute ammonia water, DHF (→APM)	DHF (→HPM)
W	Silica, aluminum	H_2O_2, $Fe(NO_3)_2$	dilute ammonia water	DHF, organic acid
Al	Silica, aluminum	H_2O_2, $Fe(NO_3)_2$	DI water, reduced water	organic acid
Cu (barrier)	Silica, aluminum	Oxidizer	DI water, reduced water, DHF	DHF, organic acid
low-k	Silica	KOH, NH_4OH	dilute ammonia water, reduced water (use water vibrated by ultrasonic waves at the same time)	DHF, organic acid

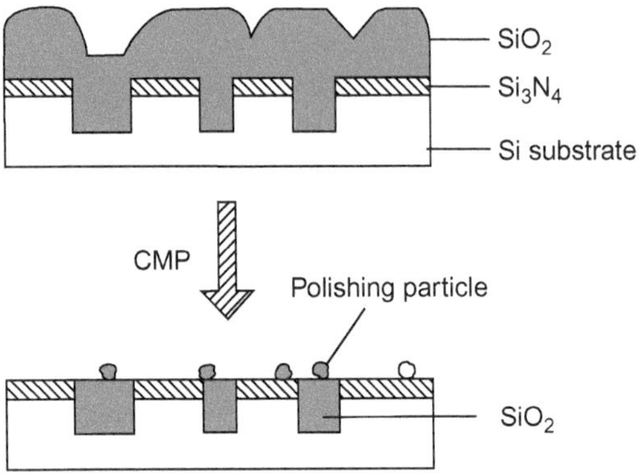

Figure 4.70 Cross-section of STI-CMP

STI-CMP, much higher cleanliness is required. After this step, gate oxidation takes place, so great care and control are required in order to avoid metal contamination inside the oxidation furnace. After STI-CMP, particle removal is often conducted by cleaning with APM (ammonia and hydrogen peroxide solution) and metal contamination is removed with HPM (hydrogen peroxide and hydrochloric acid solution), which produces a cleaned surface with few particles and metals.

4.7.3.2 Tungsten Films

In tungsten (W) CMP, a slurry is used that contains alumina particles to which an oxidizing agent, such as iron nitrate and hydrogen peroxide, has been added. Removal of alumina particles is effectively done with diluted ammonia water (~2%) and a brush scrub. The same method is used for the removal of silica particles. However, in the case of W CMP, as shown in Figure 4.71, keyholes are sometimes created near the center of the W plug, and removal using only a brush is not sufficient, because abrasive particles clog the holes. In this case, it is useful to remove particles from the holes by using ultrasonic waves.

When slurry containing iron nitrate is used, a large amount of Fe contaminants are left on the surface, so cleaning with DHF and organic acid is said to be useful. W CMP is mainly used for forming a W plug, and the girth of the W plug is covered with barrier films of titanium (Ti) or titanium nitride (TiN); therefore, Ti and TiN films come to the surface during cleaning after CMP. If DHF is used as the cleaning liquid, the Ti part is etched and the treatment must be very rapid. If an organic acid,

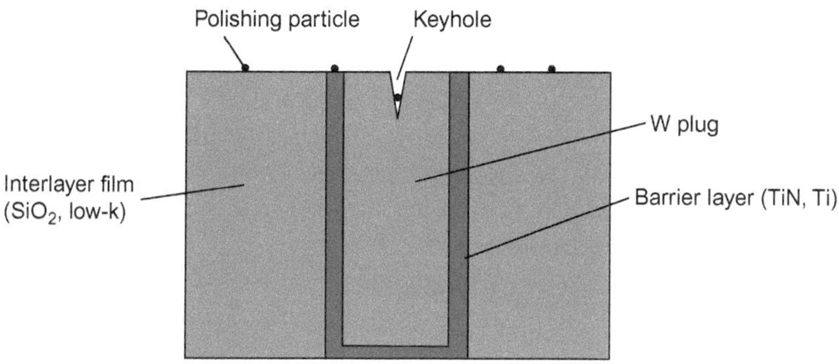

Figure 4.71 Schematic of W CMP

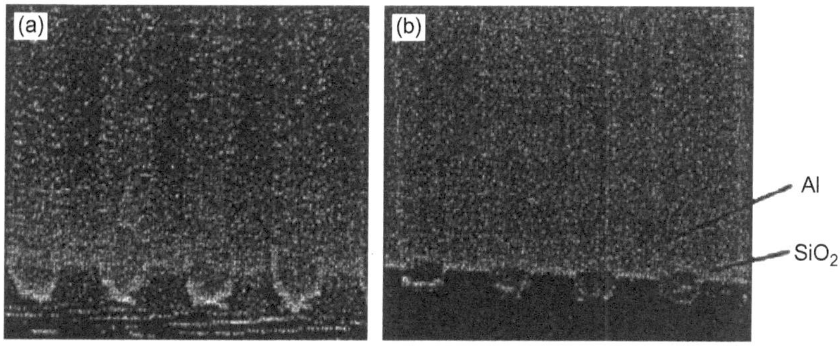

Figure 4.72 An SEM photo of the wafer surface cleaned with (a) pure water and (b) reducing water (pure electrolytic cathode water) after Al CMP

such as citric or oxalic acid, is used, it is possible to remove only Fe contamination without affecting the W, Ti or TiN on the surface.

It should be noted that W films are not dissolved by acid or alkaline solution, but are dissolved by acid and alkaline liquids containing H_2O_2; thus, APM (Ammonia Hydrogen Peroxide mixture liquid) and HPM (Hydrogen Peroxide chlorate Mixture liquid) cannot be used for STI-CMP.

4.7.3.3 Aluminum (Al) Films

Al CMP was much discussed until Cu interconnections became commonplace. Because Al reacts with acid and alkaline solutions, chemical cleaning is strictly avoided, and brush cleaning is conducted with pure water. However, this makes particle removal difficult. Residual particles in the form of interlayer films can be removed relatively easily, but particles on the Al interconnection are difficult to remove. Figure 4.72 is an SEM photo of the wafer surface cleaned with (a) pure water and (b) reducing water (pure electrolytic cathode water) after Al CMP. Fine particles (particle diameter below 0.2 μm) that cannot be removed by pure water on the Al interconnection are effectively removed by the second technique. It is thought that in the CMP process particles are removed from the surface by etching a thin AlO_x layer on the Al surface with pure electrolytic cathode water, which has a reducing property. The production of reducing water and its cleaning effect will be detailed later.

4.7.3.4 Cu Films

Because Cu films easily form ammonium and amine complexes, they can be etched easily, but this makes it difficult to use ammonia solution as a cleaning liquid. DHF does not etch Cu films, but care must be taken with treatment time and concentration because the surface gets rough.

One proposed cleaning method uses reducing water (electrolytic cathode water) that is close to neutral pH to removes particles, and an organic acid (oxalic acid) to eliminate metal contamination. Figure 4.73 is an SEM photo showing a wafer after cleaning with electrolytic cathode water followed by Cu/HSQ CMP. The fine particles on the Cu surface are effectively removed, as they are for Al interconnections.

Oxalic and citric acids are the organic acids in most common use. Citric acid is not as good at removing metal contamination as oxalic acid, so a chelating agent is added to the reagent on the market. An alternative cleaning method involves cleaning the surface with a brush set up to be an electrolytic anode after CMP with abrasives containing MMC and quinaldate that can remove both particle and metal contamination is reported. In addition to these methods, trials are taking place to improve

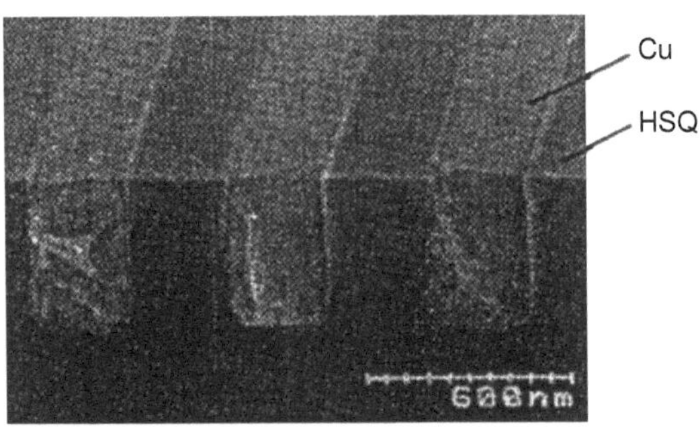

Figure 4.73 An SEM photo showing the state after cleaning with electrolytic cathode water follow by Cu/HSQ CMP

cleanliness using a cleaning liquid to which surfactant is added, but the surfactant must be carefully selected, because any residual organic matter in it may cause poor adherence at the interface between the interconnection and the top layer.

Particles are difficult to remove from Damascene interconnections because of the recess-like step height between the interconnection parts and the interlayer films, so the use of ultrasonic waves to detach particles is useful. However, if the adherence between the interconnections and interlayer films is not strong enough, there is a risk of the interface between Cu and low-k films peeling off due to the impact of ultrasonic waves, so the strength of ultrasonic waves and cleaning time should be carefully considered.

4.7.3.5 Low-k Films

Low-k films may be damaged if they come into contact with chemical liquids after CMP. When low-k films are added as planarization layers, CMP of the interlayers (low-k films) is necessary, because a global step height is created due to insufficient planarization of the overall pattern, as shown in Figure 4.74(a). In this case, low-k films are polished directly, and the cleaning liquid comes into contact with the film. If copper layers are overpolished is carried out in CMP, the high polishing speed may remove the covering SiO_2 films at the periphery of the wafer, hence exposing low-k films to the cleaning liquid (Figure 4.74(b)). Therefore, it is necessary to consider the resistance of low-k films to chemical liquid.

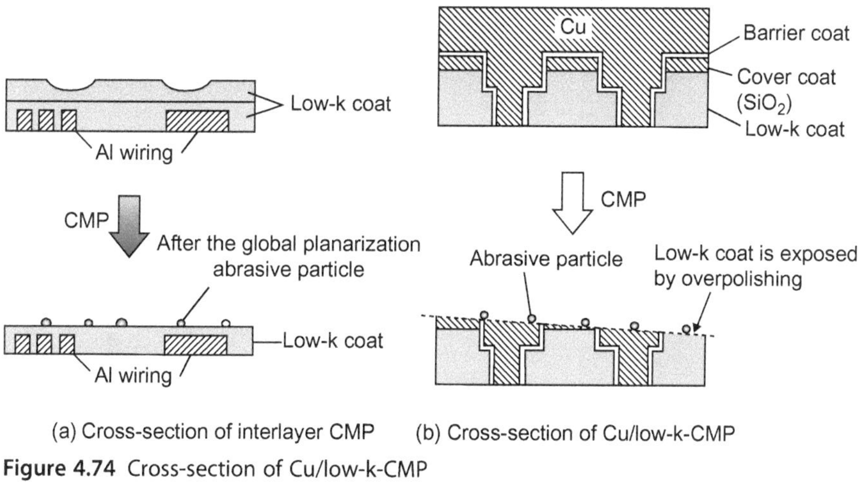

Figure 4.74 Cross-section of Cu/low-k-CMP

In addition, because most of the low-k film surface is hydrophobic, the cleaning property deteriorates compared to a hydrophilic surface, like the SiO_2 surface. This is because the Zeta potential of the surface surrounding contaminant particles and the Low-k film surface being hydrophobic is in two ends and both are easy to be attached. Therefore, after CMP, sufficient removal of contaminant particles from the low-k film surface is not possible with conventional brush cleaning. The removal effects with ultrasonic wave cleaning have been considered, but have to be fully tested. Low-k films are soft compared with SiO_2 films, and concern arises over the degrading of film quality from the impact of ultrasonic wave.

The influence of various cleaning liquids and the ultrasonic wave cleaning used on low-k films is discussed below.

Organic Low-k Films (SiLK™)

Organic films like SiLK are chemically resistant, and are not degraded as they are not etched by DHF or ammonia water. However, layers with improved adherence between SiLK film and substrate layers are easily etched with HF solution. Therefore, when sections of SiLK film at the edge of the wafer are left exposed, precautions must be taken for the possible delamination at the contact area of DHF.

Figure 4.75 compares the etching rates of films treated by ultrasonic waves in pure water. SiLK films are etched at the highest rate, 12 nm/min. Ultrasonic waves create OH radicals in water, which can oxidize and dissolve organic matter, and since SiLK is an organic compound, it is etched. Films like SiLK are etched by ultrasonic waves in water, but changes in the film quality are not observed, since the etching state differs from the film forming condition.

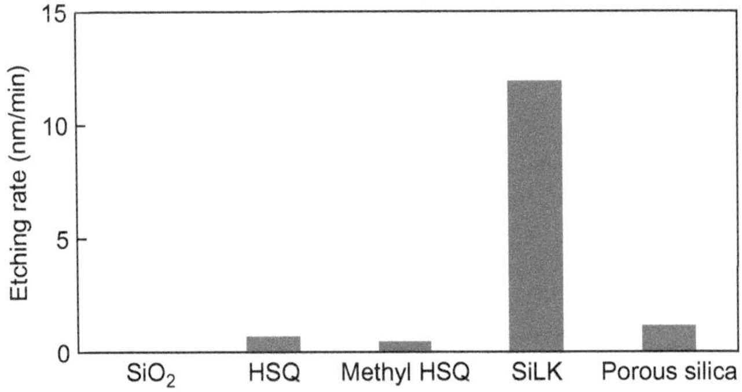

Figure 4.75 The etching rate of low-k films treated by ultrasonic wave pure water

Porous Films (Porous Silica Films)

Porous films are attractive in terms of their low dielectric constant, but there are various problems when cleaning them. The etching rate of plasma SiO_2 films by DHF is about 20 nm/min, while porous silica over 1 μm thick is etched 50 times faster. This is because porous silica films are Si-O-based, and are porous.

The surfaces of porous silica films are hydrophobic because of the presence of terminal CH_3 groups. After treatment with ammonia solution, the Si-CH_3 groups are reduced by half, and the number of Si-OH groups increases. Therefore the films change in character, and absorb water more easily, which tends to increase their dielectric constant. Furthermore, after ultrasonic treatment, their insulation resistance tends to be reduced, because porous silica films are not strong and are easily affected by shock waves.

4.8 SURFACE DEFECTS AND INSPECTION TOOLS IN CMP

The CMP process is essential for device miniaturization, higher integration, increased yield and reliability. The processes applied vary from STI, ILD, Via Plug, interconnection implantation and many others. The materials used in each process also vary, and include Si, SiO_2, SiN, W, TiN, Cu, and low-k. As the integration in LSIs becomes higher, the number of defects in the CMP process that are regarded as killer defects, which reduce yield, similarly increases. As the number of parameters affecting the process increases, and the chase work not only for the process but also for the sources of defect generation becomes more complicated, extremely accurate defect inspection becomes essential.

4.8.1 Types of CMP Defects

The defects caused by CMP processes include particles, slurry residues, micro-scratches, hazes, cracks, recess, dishing, erosion, void, film thickness variation and so on. It is necessary to detect these defects and take appropriate action in order to maintain a high yield.

4.8.1.1 Defects Occurring After Oxide Film CMP

The major defects that follow oxide film CMP are shown in Figure 4.76. Currently, scratches remaining on the processed surface after polishing are the most problematic defect. The severity of these scratches ranges

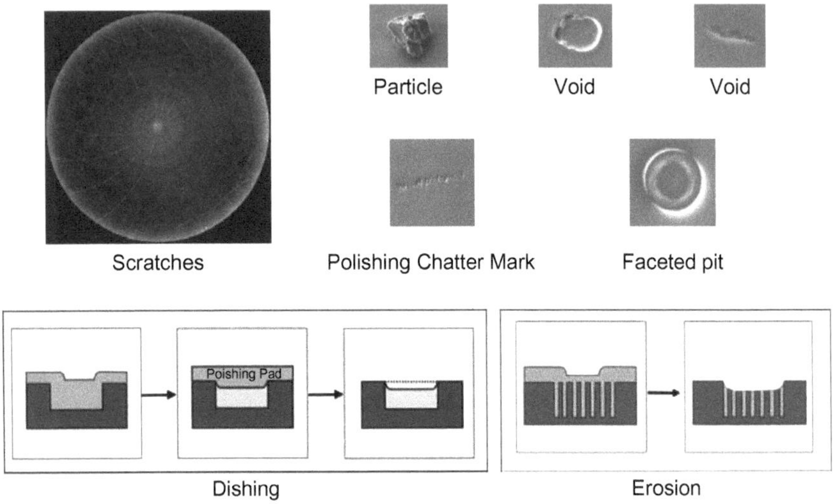

Figure 4.76 Major defects after oxide film CMP

from those visible to the naked eye to micro-scratches observable by optical microscope, a wafer surface quality inspection system, or AFM or SEM. Various shapes such as flocculating flaws, continuous track marks, and dents are also commonly reported. Even micro-scratches can contain W and Cu residues in their concave parts, which causes short-circuiting in the interconnections. These micro-scratches are caused when dents are created by such means as abrasion by slurry aggregation, dropping out of the abrasive of a diamond dresser, foreign matter in the polishing pad, etc. Since this defect grows in size during cleaning and wet etching, its detection becomes easier. Information on the location of the defect, and its length, width, and depth, etc. is reported.

It is considered that defects such as slurry residues and particles are mainly generated by insufficient cleaning. A particle is generally a tiny foreign body that affects the yield of semiconductor devices. Though the clean rooms used for fabrication are designed to minimize airborne particles, dust generated from the human body, the moving parts of the manufacturing equipment, or the cassette and other fittings, cannot be avoided. Once this dust adheres to and aggregates on the wafer, it can cause short-circuits and disconnects in the circuit patterns in the lithography and dry etching processes. Particle counters and/or surface analyzers are commonly used for the measurement of particles.

Haze is delocalized light scattering on the mirror surface of a polished wafer, resulting from micro-roughness or defect clusters on or near the surface. These are often detected visually, even now, by use of a strong light in a darkroom.

Other defects such as dishing, rip out and cracks in the low-k material may also become problems. Dishing is a phenomenon observed in the Damascene process or in STI fabrication in which two kinds of thin films are used. In the Damascene process, the center of the metallic interconnection becomes thin; however, in STI fabrication, the center of the insulation film in the trench area becomes thin. Since the stopper nitride film at both ends of the isolated domain is less easily polished than the oxide film in STI, it causes a deflection in the polishing pad and as a result the oxide film is overpolished. The amount of dishing is determined by the mechanical properties of the polishing pad, the width of the interconnection and the isolation trench, slurry properties, polishing conditions etc.

4.8.1.2 Defects Occurring After STI Insulation Film CMP

Defects similar to the ones that occur during oxide film CMP also occur in the CMP process for STI fabrication. Additionally, silicon nitride films may not be uniform in thickness after CMP, due to minute asperities on the wafer surface. These variations cause variations in the shape of the device isolation trench, which makes the isolation imperfect. To avoid this problem, it is necessary to improve the manufacturing process of wafers and to inspect the nanotopography of the wafer surface. Nanotopography is the pattern of minute asperities ranging between flatness and micro-roughness, and it is evaluated as short wavelength asperity after the waviness of the entire wafer is removed.

4.8.1.3 Defects Occurring After Metal CMP

Defects similar to oxide film CMP occur in the case of tungsten film CMP, and void defects, slurry residue, and micro-scratches are found in copper CMP. Measurement of the scratches on the surface of the metal film is extremely difficult, so it is necessary to distinguish scratches caused by corrosion (mentioned later) from those caused by particles.

A void defect is a gap that occurs when the Damascene process performance deteriorates during the formation of the metal interconnection. It is not directly caused by CMP, but needs to be detected in order to improve the device yield.

Micro-roughness causes rough surfaces and corrosion on Cu films, and appears at the post-cleaning stage. Corrosion occurs in the interconnecting metals, and could be caused by the metal dissolving when residual impurities in metal films or on the surfaces are ionized. It tends to occur in CMP when the balance between mechanical and chemical reaction is lost, and the chemical processes dominate.

In W films, the keyhole phenomenon occurs when a seam (a joint created between films in the deposition process) becomes too wide, especially when H_2O_2 is used as an oxidant.

In Cu CMP, corrosion is observed only when chemical etching advances after the surface is polished to beneath the insulation and mechanical action will not work when the polishing pad cannot reach the Cu surface.

4.8.2 Defect Inspection Tools

4.8.2.1 Inspection of Unpatterned Wafers

Inspection tools are used on unpatterned wafers to measure shape (wafer geometry) by using a dummy wafer, and to monitor the level of cleaning in mass production lines. Wafer geometry includes flatness, nanotopography, edge roll off, etc., and optical interferometry is mainly used for such measurements. Light scattering is generally used for cleaning inspections, which check size, location, quantity, haze, etc. of particles and scratches.

Wafer Geometry

Since high spatial resolution, stable reproducibility and less uncertainty difference among tools are required for the current wafer measuring tools in accordance with miniaturization of chip fabrication, the principle of optical interferometry is more commonly adopted.

Optical interferometry uses a CCD camera to measure phase shifts in interference fringe patterns (PSI: phase shift interferometry). These are generated by the interference between the light reflected from the optical reference surface and that from the wafer surface.

WaferSight 2® by KLA-Tencor (Figure 4.77) uses interferometry for double-sided polished wafers. As high performance is required, Fizeau interferometers are arranged at both the front and rear of the wafer. Moreover, the interferometers are equipped with large optical reference surfaces to simultaneously measure the whole surface of a 300 mm wafer, and a common semiconductor laser is used as the light source. Additionally, they are installed in a seismic isolation structure to enable

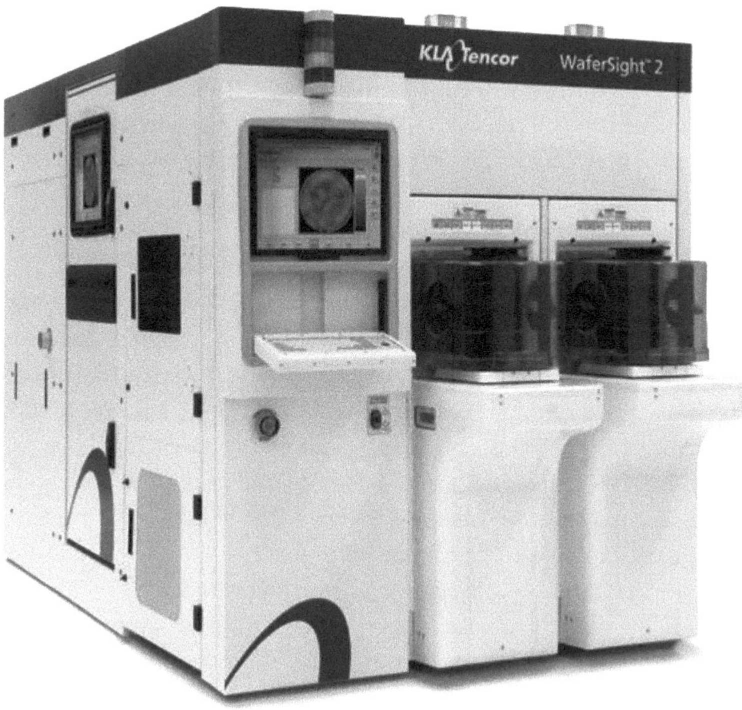

Figure 4.77 WaferSight 2® by KLA-Tencor

high precision measurement. Observing and checking both front and rear enables the measurement of wafer flatness and nanotopography.

(a) Flatness: Wafer flatness influences the depth of focus (DOF) during exposure. In the case of immersion steppers, which are widely used in advanced device fabrication, the depth of focus has improved remarkably, but the demands of miniaturization exceed this.

In general, defocusing occurs at the wafer edges. Furthermore, it is difficult to maintain the uniformity of chucking among photolithography scanners. The amount of defocusing could easily be different even if the same wafer is mounted and exposed using another chuck of the same type. To reduce this problem, the flatness of the wafer surface is measured in the stick-mounted state immediately before exposure is executed by using the focus adjustment function of photolithography scanners.

(b) Edge Roll Off: Every device manufacturer is trying to narrow the edge exclusion width to increase productivity. The yield could increase by 2–4% if edge exclusion were narrowed from 3 mm to 1 mm per wafer, depending on chip size.

Currently, flatness is not a problem at the wafer center, but it will become a serious problem at the edge if the edge exclusion length reaches 1 mm. SEMI (Semiconductor Equipment and Material International) has already standardized four types of measuring method for measuring geometry at wafer edges. Since each method has its own features, they suit different applications.

(c) Nanotopography: Nanotopography measures asperity components on wafer surfaces which have a wavelength of 0.2 mm to 20 mm. Although nanotopography problems have decreased thanks to recent improvements in fabrication processes, its measurement is used to detect local short-wavelength defects on wafer backs. In the case of wafers that require heat treatment for film formation, such as epi-wafers, abnormal growth can occur during this step, causing asperity. The laser marks imprinted on the edges of the wafer reverse and could cause defocusing. It is necessary to measure precisely the size and height of any defects on the wafer back in order to maximize chip yield.

As wafer fabrication processing advances, measuring technology is required to be more accurate, more stable and have higher resolution. In addition, instrumental error in measuring tools has become a big problem, since it affects mass production. Preparation of standard wafers to provide reference values for wafer geometry parameters such as flatness cannot keep up with the rapid advance of processing. Ensuring calibration and traceability of measurements still remains an important issue.

Particles, Scratches and Haze

In contrast to features of large-scale geometry, such as flatness and nanotopography, light scattering methods are mainly used for inspecting tiny defects such as particles, scratches and haze.

A bright field optical microscope is generally used for optically inspecting small objects. It obtains the images from wavelength and intensity changes in the light that is transmitted through a sample. Based on the light wavelength limit and numerical aperture (NA) of the lens, its resolution is limited to about 0.2 μm. This means it is not possible to detect tiny defects, such as micro-scratches and nano-particles.

For extremely tiny defects, the dark field method, using light scattering, is used. Its greatest advantage is that it is possible to observe extremely tiny objects without any pretreatment to a sample using scattered light instead of direct light. This makes it possible to observe objects 1/20 to 1/30 the size of the wavelength of the light source (a few tens of nanometers) – much higher resolution than an optical microscope. The dark field method works by preventing direct light from illuminating an object lens.

Although it is possible to evaluate length and shape information by the dark field method, it is extremely difficult to evaluate the width. This is because the image observed by the dark field method is merely formed by scattered light, and it is not an actual image. Because the scattering intensity is dominated by various factors such as diameter, shape, refractive index, angle of incident light, and wavelength, it takes considerable effort to directly evaluate the size of the object by this means. The S/N (signal to noise) ratio must be very high to deal with tiny objects, thus it is necessary to increase the intensity of the scattered light. Lasers are therefore used, because they emit very bright light and induce scattering efficiently. In addition, a photo-multiplier is needed to detect low levels of scattered light. Using the dark field method of oblique illumination for wafer inspection enables sensitive measurement of surface defects only, as it is difficult for the light to penetrate further, even into transparent layers like oxide films.

The Surfscan SP2XP® by KLA-Tencor (Figure 4.78) is used as a typical wafer defect inspection tool in the 45 nm generation because it has high sensitivity and rapid throughput.

The bright field channel method is an alternative to the conventional dark field channel of oblique illumination and vertical illumination, for detecting defects that are otherwise difficult to measure. This operates simultaneously to the dark field channel and the differential interference contrast function by using laser light phases installed to recognize the size and depth of a defect. Furthermore, the tool can scan the objects with both oblique and vertical illumination of the dark field channel in one step. The apparatus is also equipped with channels for gathering narrow and wide light ranges for each angle of incidence, thus enabling data construction based on five different optical components together with the bright field.

Use of ultraviolet laser light can minimize errors caused by embedded defects underneath the wafer surface. Such an inspection tool makes it possible to detect shallow CMP scratches, meaning defects such as surface roughness, water marks, slurry residues and small asperities that scatter light only weakly can also be measured. Void defects can be distinguished from relatively harmless particles and foreign materials by reprocessing the measured data.

Haze is a low frequency, low-amplitude component of the light scattered from the wafer surface, and is expressed as the ratio of the average

Figure 4.78 Surfscan SP2XP® by KLA-Tencor

surface scattering intensity to the intensity of the laser incident light. In the case of bare ware, haze is closely related to surface roughness. A high S/N ratio is important even for wafers with rough surfaces when inspecting for haze. Surfscan SP2 is equipped with smaller spot sizes than the conventional ones, so improving the sensitivity for wafers with rough surfaces, due to smaller amount of background included in the image. This machine can achieve remarkably low detection levels for fake defects, as the scanning threshold can be set at a higher level than haze. It can achieve the required sensitivity at high throughput even for high haze wafers thanks to the stepped-up sensitivity and background noise reduction functions.

4.8.2.2 Inspection Tools for Patterned Wafers

Inspection tools using short wavelength light, such as ultraviolet, are needed for inspecting wafers with patterns, to analyze the small-scale pattern shapes and related defects. The tools are further refined by combining them with SEM review tools, automatic defect-sorting systems, detailed shape measurement by AFM, etc. Any problems need to be dealt with within a short period of time, to minimize the time before market release.

It is also necessary to measure shapes defects, such as plug, Damascene, STI recess, erosion of oxide film, and scratch depth, which cannot be evaluated by AFM due to its narrow field of view. Although the 3D function of SEM enables depth to be determined by using stereo viewing, a wafer surface geometry measuring tool capable of measuring height is required for wider scope inspection in the air. A surface geometry measuring tool capable of high precision measurement based on the tactile probing method has been used in the past, but non-contact measurement systems, such as the confocal laser microscope and the white light interferometry type surface asperity measuring tools, are currently used. The following section explains the principles of the other inspection tools that are combined with the previously mentioned light scattering detection methods.

SEM (Scanning Electron Microscope)

This is the most widely used electron microscope for surface observation and analysis. When a sample is irradiated by electrons, the incident electron energy is mostly converted to heat, but the repeating collisions excite secondary electrons and the emission of various electromagnetic waves (X-rays, ultraviolet, visible and infrared radiation; see Figure 4.79). Some electrons are reflected from the surface of a sample, and scattered, either elastically or non-elastically, outside the sample (also called backscattered electrons, a few keV to a few tens of keV). Their intensity changes are influenced by the asperity of the sample surface.

Among the electrons emitted from a sample, those with low energy of 50 eV or less are called secondary electrons. Figure 4.80 shows the principle of surface measurement by SEM. Secondary electrons have various features which are not observed in backscattered electrons and they are the most important signals in SEM. The contrasts created in the secondary electron image are caused by different amounts of emergence (efficiency) of secondary electrons. Only the secondary electrons generated within 10 nm of the surface are emitted from a sample and form the image. Secondary electrons are the emitted electrons which are excited and generated by incident electrons.

The features of SEM can be summarized as follows:

a. Low acceleration voltage (electron gun normally has acceleration voltage from 1 to 30 kV) and low current make generation and collection

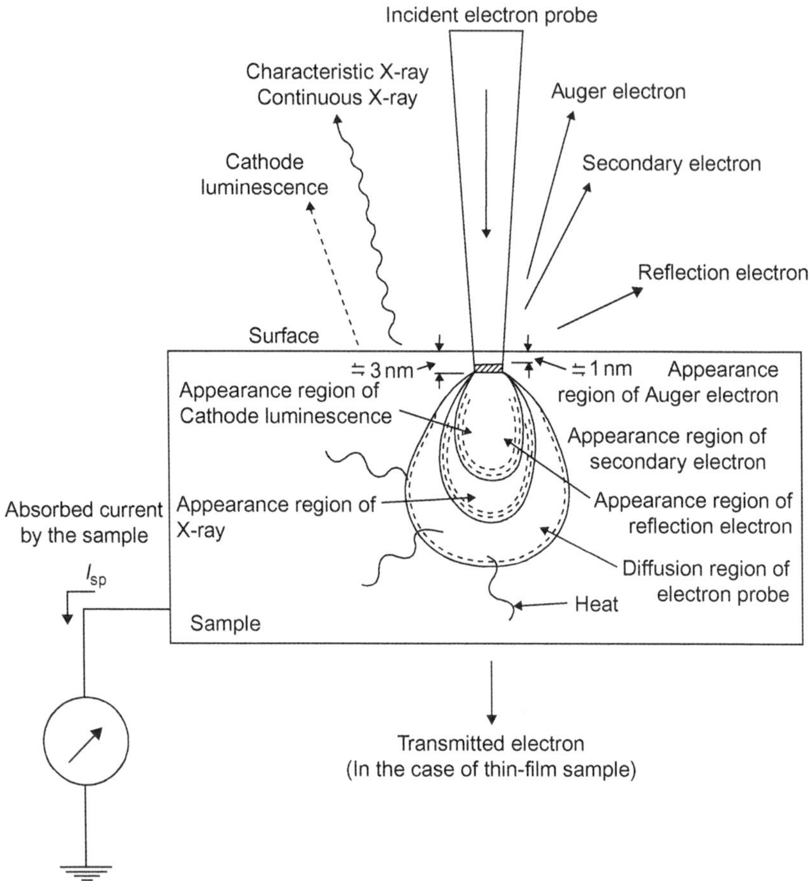

Figure 4.79 Interaction and secondary information by electron probe

efficiency so high that it is suited for surface inspection of samples, and detection of organic residues which are weak against electron irradiation. Unlike TEM, the sample does not need to be a thin film, and it is possible to measure bulk materials, powders, etc. However, the samples must have conductivity. Conductive treatments such as deposition and sputtering are commonly applied to non-conductive materials, but it has recently become possible to obtain a clear image without conductive treatments on a sample by preventing charging phenomenon using a low vacuum environment.

b. Because a large DOF can be used, very rough micro-asperities and dents can be observed in 3D. SEM has 100 times larger DOF than that of an optical microscope; for example, an optical microscope can

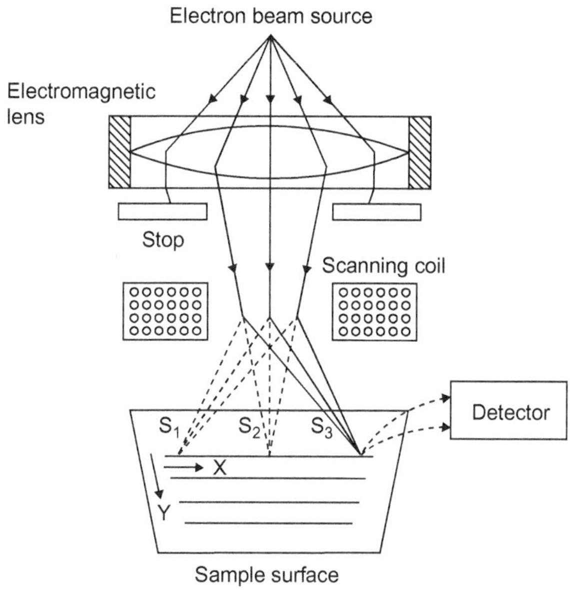

Figure 4.80 Principle of surface measurement by SEM

observe 10 μm in depth at 100 times magnification, while SEM is capable of over 1000 μm.

c. Since it has high spatial resolution and does not use an electromagnetic lens for detector, the resolution is determined by the diameter of the incident electron beam; several nm to several tens of nm. In addition, it can achieve a resolution nearly equivalent to the beam spot diameter, which is high magnification. SEM with a resolution of 0.5 nm has also been developed, and magnification ranges from low (several times) to high (800,000 times).

d. Because it is possible to depict very small potential changes on the sample surface, it is possible to inspect operating conditions and defects of transistors and integrated circuits.

e. No color is detected. Sometimes colored SEM images are produced, but all of these are created by image processing and are colored artificially.

Recently FE-SEM (Field Emission SEM), a high resolution SEM, has been developed and utilized. FE-SEM uses a cold field emission electron source and has low acceleration voltage/high resolution, so observation at magnifications of several hundreds of thousands is possible.

The SEM defect inspection tools described above can detect defects below 100 nm in size, which is necessary for the latest wafer inspections, and is used to detect micro- defects that have a significant influence on yield. By measuring the electric charge on the sample surface, it is possible to recognize electrical defects, like poor quality of via contacts or disconnection/shorts in the Damascene structure. Because measurement is conducted in a vacuum, it does take more time to obtain a result however, development of specific defect inspection instruments that will work faster is ongoing.

SPM (Scanning Probe Microscope)

SPM is a new analysis method that is used to locally measure the physical quantity of specific substances by scanning the surface of a sample with a probe that is sharpened to the atomic level. The STM (Scanning Tunneling Microscope) and AFM (Atomic Force Microscope) are instruments that can observe the structure of a substance's surface at the atomic level in a vacuum, at atmospheric pressure and in solution. These techniques can observe topography (surface geometry or roughness) at the atomic scale. The size of an atomic nucleus is about 1 fm ($=10^{-15}$ m), and the dimension of an atom including the electron cloud is about 1 Å ($=10^{-10}$ m). These microscopes measure the geometry of the circumference of the electron cloud. However they can only measure features smaller than 100 μm, so it is difficult to measure roughness of the inclined surface.

a) Scanning Tunneling Microscope (STM): STM was proposed by Gerd Binnig and Heinrich Rohrer of IBM in Zurich, Switzerland, in 1982. They were awarded the Nobel Prize in Physics in 1986. The position of a conducting probe tip is controlled so that the distance between the tip and the sample surface is kept constant by applying a bias voltage when the tip is brought very near to the sample surface in about 1 nm (Figure 4.81). Because a piezoelectric element (piezoelectric actuator) is used to control the probe tip movement, the voltage applied to the piezoelectric element itself indicates the surface topography of the sample surface.

The structure of the STM is extremely simple, and consists of a probe that scans the sample surface and a piezoelectric element that drives the probe in the X, Y and Z directions. The probe tip is usually made of tungsten, platinum or gold and is formed by applying a direct or an alternating electric current for electrolytic grinding. The geometry of the probe tip largely determines the resolution. When the tip of a probe gets within several tens of Å of a sample surface, by stretching the piezoelectric element in the Z direction, a tunnel current (tunnel current is set at $0.1 \sim 10$ nA) runs between the two. The amount of tunnel current fluctuates exponentially depending on the distance between probe tip and sample.

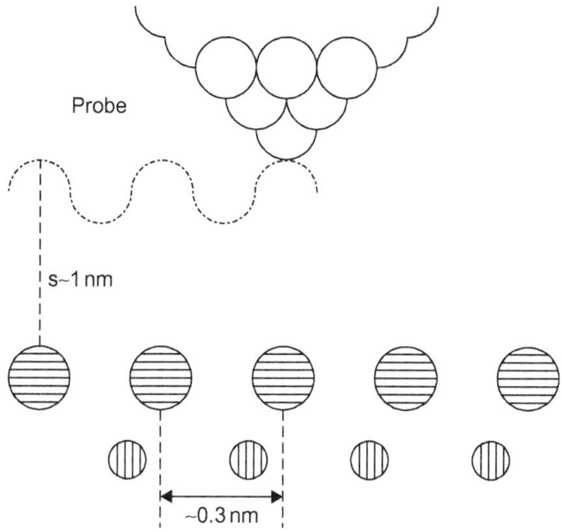

Figure 4.81 Principle of STM measurement

[Fowler Nordheim theory: given by tunnel current $Jt = (\beta V/2\pi\lambda Z)\exp(-2Z/\lambda)$, Z: distance between the surface and probe tip, V: applied voltage.]

For example, when the distance between the two changes by 1 Å, the tunnel current fluctuates by ten times. When the probe tip scans the asperties of the surface, the tunnel current is maintained at a certain level via a feedback circuit; hence, the geometry of the sample surface can be observed from changes in the probe's position, at atomic order resolution (horizontal direction ~3 Å, vertical direction ~0.01 Å = 10^{-12} m = 1 pm). Manipulation of the surface at the atomic scale by STM (atom craft) therefore becomes possible (Figure 4.82).

A disadvantage of STM is that it can only access the structure of conductive samples (metals or semi-conductors) or ultra-thin material (below 1 nm in thickness), as it measures tunnel current as physical quantity. During observation, any external vibration must be fully controlled, and the influence of sonic noises in the atmosphere has to be considered the natural frequency of the measuring component the probe tip has to be as high as possible. Because the probe tip has to be controlled within 10 pm in the Z direction, the coarse and fine mechanisms for the tip movement must be combined.

b) Atomic Force Microscope (AFM): AFM can observe the surface of any substance—insulator, semi-conductor, organic matter, and organisms at the atomic/molecular level—which is not possible with STM. This

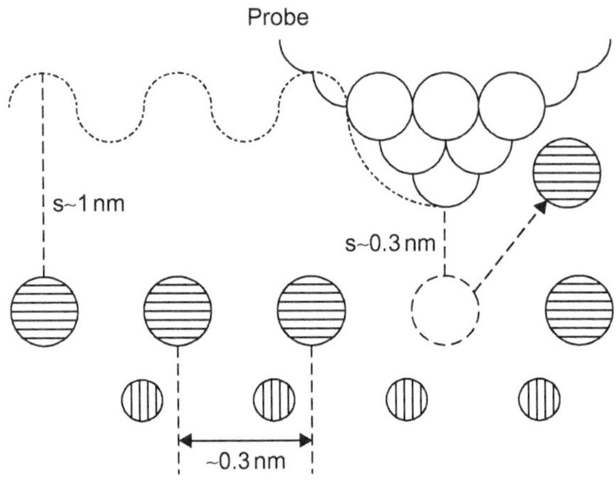

Figure 4.82 Atom craft by STM

means that measurement of not only force from the surface but also friction force of any material can be conducted and it is possible to sense indirectly moisture or contamination thickness adhered to the surface.

Atomic force is the general term for the Coulombic force, van der Waals force and Pauli force between atoms, molecules or ions. When the distance between two atoms is less than 0.8 nm, their electron clouds interfere with each other, and, as a result, fluctuations of the electron cloud occur. At this time electron dipole occurs due to separation of the electric charge, and attractive force between two atoms takes place. This force is known as van der Waals (also called London force, electrostatic attractive force, and dispersion force).

When the two atoms are less than 0.3 nm apart, their electron clouds overlap, and repel each other by Pauli force (exchange repulsion force). Mutual action like chemical bonding force or static reciprocal action force is said to be several tens to several hundreds times larger than atomic force (Figure 4.83).

The measuring principle of AFM maintains a measuring force at a certain level by sensing the amount of deflection of a cantilever with an optical lever in the form of a leaf spring having a probe with small tip curvature ($\sim$50 nm) (fixed force mode = noncontact mode) (Figure 4.84). The positional changes in the cantilever are controlled by utilizing the deformation caused by the piezoelectric effect of a piezoelectric actuator. Because the measurement can be done without the probe touching the sample, there is no chance that the sample will be damaged. AFM utilizes the relationship between the distance and the force between atoms. Besides the aforementioned there are some other measurement modes.

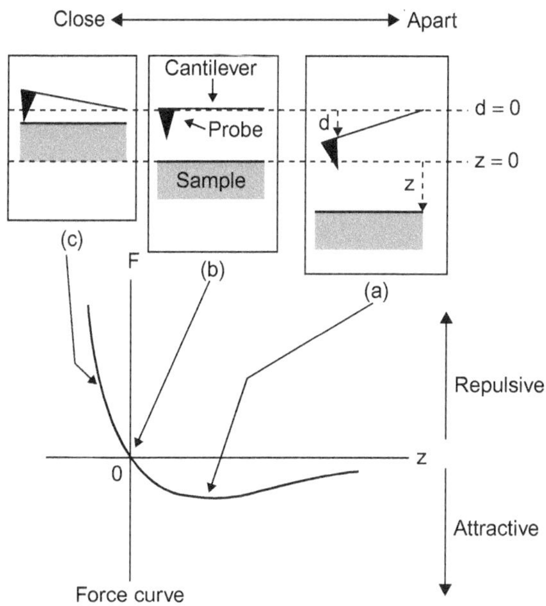

Figure 4.83 Distance and force curve

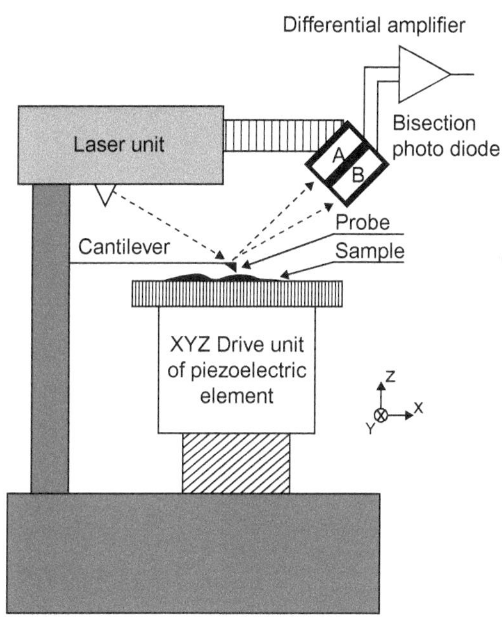

Figure 4.84 Construction of AFM

Contact mode measurement (fixed displacement mode) is when a cantilever is attracted near to the sample due to the atomic force between the probe tip and the sample surface and deforms in contact. By moving the probe tip or the sample up and down to fix the displacements of the cantilever, scanning is done on the sample in a similar way to tracing it. At this time controlled signal is regarded as the height variation of asperities of the sample surface. Contact mode measurement is easy to be implemented, but due to the strong contact and friction forces, a soft sample may be damaged.

In the case of dynamic mode, a resonating probe tip moves the sample surface up and down. Similar to the non-contact mode, the dynamic mode measures the surface condition by detecting changes in the amplitude of oscillation of a cantilever when the tip intermittently contacts the sample surface. This method can be used for observing organisms or fragile samples. It is often used when there is no need to conduct precise measurements or when a result has to be obtained quickly. It is also called Tapping Mode, a registered trademark of Veeco Co. in the US.

AFM measurements require long measuring times, but shortening this time leads to inaccuracy. In addition, AFM can observe surface geometry but cannot identify types of atoms and molecules, so must be combined with other methods.

At first only images of a surface topography were obtained by utilizing atomic forces, but now various modes can be used: LFM (Lateral Force Microscope) senses the amount of twist of a cantilever with horizontal force applied (friction force) while scanning perpendicular to the longitudinal direction of the cantilever; MFM (Magnetic Force Microscope) can simultaneously observe surface geometry and obtain a magnetic image by using a magnetic and conductive cantilever; EFM (Electric Force Microscope) observes electric force; and KFM (Kelvin probe Force Microscope) method to sense an electric current that runs when bias voltage is applied in the contact mode. Recently, in addition to these modes, a hybrid AFM instrument has been developed, which combines a confocal laser scanning microscope and Raman spectroscopy.

As one example of hybrid AFMs, Figure 4.85 shows a LEXT OLS3500 hybrid microscope from Olympus. This combines a confocal laser scanning microscope and SPM. This needs less preparation time before measurement, thanks to the easy and quick positioning of the probe tip which used to be difficult in conventional AFM.

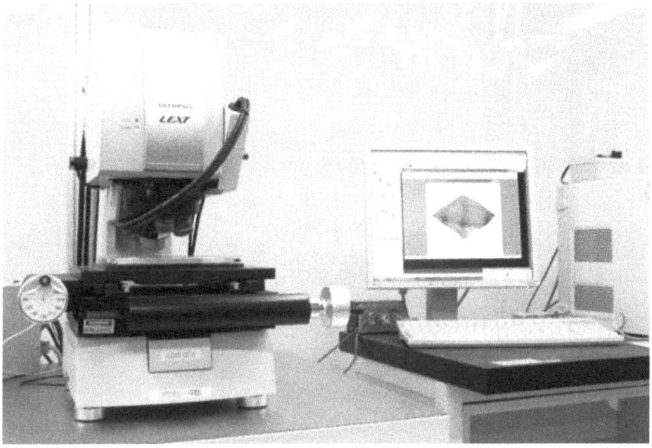

Figure 4.85 Olympus LEXT® OLS3500

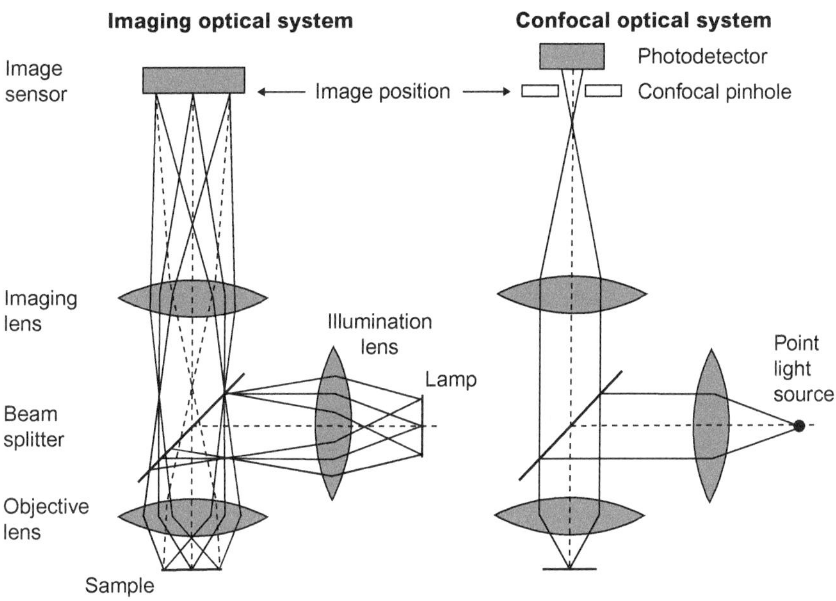

Figure 4.86 Principles of optical and confocal microscope

Confocal Laser Scanning Microscope (CLSM or LSCM)

This technique uses a laser as the measuring probe; hence, measurement at atmospheric pressure, without contacting the sample, is possible. The optical system of a general optical microscope and the confocal optical system are shown in Figure 4.86. In the confocal system, it is possible to detect only light from a focused position by placing a pinhole at the image position which is conjugate to the focal position of an objective lens. In contrast to a general optical microscope, light from a point light source is used by condensing it onto one point of a sample with an objective lens. Laser beams are generally used because they supply strong, coherent light. Light reflected by the sample surface goes back along the same optical path, is separated by a beam splitter and condensed at the pinhole. Light reflecting from all positions except the focal position is mostly cut out by the pinhole, meaning information can only be gained from the focal position.

Information is resolved along the optical axis, so three-dimensional measurement is possible. The pinhole diameter should be less than one third of the expansion of the beam spot due to diffraction. Because decreasing the pinhole diameter causes the amount of detected light to decrease, the final size is determined by the sensitivity of the detector.

The main feature of a CLSM is that it can capture a clear image by scanning the focal position, even if a sample is thick, which gives it different focal distances. Images captured at each scanning point, and these are combined and reconstructed by a computer to get the entire image. By moving the stage of the microscope up and down, information from different focal planes can be collected, so by overlapping two-dimensional images from different focal planes, a three-dimensional image can be created.

Measurement with a CLSM can be broadly classified into two types, namely, line width measurement using high resolution image and three-dimensional measurement by height image. For three-dimensional measurements in particular, the driving accuracy of the Z axis driving mechanism which moves the objective lens and sample, is extremely important. Its resolution needs to be 10 nm at maximum magnification. A high precision linear guide to move the objective lens in several mm at nm resolution is combined with a lead screw and is driven by a pulse motor to read the position by a linear encoder. The confocal optical system and high precision Z axis driving mechanism, give current CLSMs a Z direction resolution of several tens of nm at maximum magnification.

Measuring Surface Roughness Using White Light Interferometry

This is a measuring instrument that uses optical interference as the measurement principle. One variant of this is PSI (Phase Shift Interferometry), which uses monochromatic light and a second is VSI (Vertical Scan Interferometry), which uses white light. PSI calculates the phase difference between referential and measuring lights directly by measuring the intensity change of interfering light at a certain phase difference between the two. With this method the phase difference in each pixel of measured view, can be obtained, and by multiplying it by the wavelength of the measuring light, the deviation from the referential surface image can be obtained. Because of the reference surface is manufactured with very high accuracy, the influence of the reference surface deviation can be ignored and as a result the geometry of the measured surface can be attained. In the case of monochromatic interferometry, theoretical resolution of 1 A (angstrom) can be realized. However, assuming that there is no phase difference of over one-fourth of light wavelength between adjacent pixels, step height of over 160 nm cannot be measured when using a band pass filter with wavelength of 632 nm.

In order to compensate for the small dynamic range of PSI, VSI is used, which measures the change of contrast instead of the phase difference of an interference fringe. The principle of VSI is that the contrast of an interference fringe is at a maximum where the reference light path and measuring light path coincide. By focusing an objective lens at the position where the contrast is the maximum by moving the lens in the Z direction, information on height can be obtained. Because the resolution in the Z direction dominates the total resolution, measurement resolution is around 1 nm, which is larger than that of PSI.

Wyko NT3300S (Figure 4.87) is a Veeco Co. instrument using the white light interference method for measuring surface roughness. It averages

Figure 4.87 Veeco® NT3300

several measured images and has achieved a vertical resolution of 1 Å (measurement range 160 nm) in PSI mode and 1 nm (measurement range 2 mm) in VSI mode.

To use an interferometry, the surface to be measured should be mirror finished, or high spatial resolution cannot be expected, as this depends on magnification and NA. High speed measurement is possible, and the wide vertical measurement range, high resolution and easy handling make this instrument a very useful tool for measuring wafer surface roughness.

4.8.3 Summary

As the miniaturization of devices progresses, the detection of extremely small defects at high device yield becomes very difficult. In order to overcome problems that are difficult to manage with conventional inspection and measuring technology, the pursuit of high speed, high precision evaluation methods for defects generated in the CMP process has been insatiable. Measurement of nanoparticle detection, that was said to be impossible because of the limit of the wavelength of light, has been made possible with various technologies like oblique illumination dark field microscopy, and the use of short wavelength radiation, like UV. It is amazing that such difficulties have been overcome while maintaining high throughput.

Furthermore, the various inspection instruments introduced have been developed as part of a gigantic system to improve the overall throughput of the semiconductor device process, by establishing a system that fits into the environment of the entire plant. They utilize the system network, and include automatic defect review and database control systems.

The semiconductor industry requires timely mass production, so a database and automatic classification of defects to manage the occurrence of killer defects, that are fatal to device function, is extremely valuable. Non-killer defects, that do not affect the post-process or device quality, occur and halting a machine to deal with these defects causes significant losses from the mass production viewpoint. It is therefore meaningful to judge whether a defect is a killer or not. Managing yield by establishing a database for judging killer defects is recently becoming more important.

Combining in-line and off-line inspection will result in an enormous amount of data, and establishing a database of defects and analyzing these data will need quick and efficient feedback.

4.9 PLANARIZATION SIMULATION TECHNIQUE (PREDICTION/MANAGEMENT/EVALUATION TECHNIQUE)

4.9.1 In the Beginning

Important technological developments continue in the semiconductor industry, and the need for planarization and micro-roughness reduction of wafer materials increases. The size of wafers used in mainstream semiconductor applications has increased to 300 mm (12″), and further increases to 400 mm (16″) are envisaged, in order to improve the productivity and efficiency of the semiconductor production process.

When fabricating high-precision wafers, lapping and ultra-precision polishing (by chemical mechanical polishing (CMP)) are performed to improve the flatness and the micro-roughness of the surface of the as-cut wafers, which are in turn obtained by multi-wire sawing using free abrasives. Polishing equipment such as double-sided four-way polishing systems and double-sided polishing systems with small-circular-motion mechanisms are widely used. In polishing, a slurry is supplied to the system, and multiple wafers perform orbital and rotational motions so that both sides of the wafers are polished simultaneously.

In an actual CMP process, the polishing rate and accuracy are affected not only by the flow of the slurry and the characteristics of the polishing pads, but also by the mechanical action between the wafer and the platen, the chemical action due to the chemical components of the slurry and the interactions between them. In particular, the mechanical action depends on the relative sliding distance between the wafer and the platen; this dependence is known as Preston's law. While improvements made to the polishing process have resulted in enhanced polishing rates and wafer flatness, it is necessary to ensure that the relative sliding distance for a wafer is increased and maintained constant over the entire wafer surface. The same holds for wear of the platen; we should consider flatness management of the platen to realize planar wafers.

Note that the platen usually has grid-like or radial grooves. The main function of these grooves is to improve the circulation and discharge of the polishing slurry in the working area. However, since the grooves influence the relative sliding distance, which in turn affects the mechanical action between the wafers, they must be positioned such that the relative sliding distance on the wafer surface remains unchanged.

In this section, theoretical formulae for the relative sliding distance of the wafer and the platen are presented, and a series of simulation results are discussed. The proposed formulae can be used for predicting the relative sliding distance required to maintain the flatness of the wafer and platen surfaces during polishing and for confirming whether the interaction between the mechanical and chemical actions in CMP is stable.

Further, we discuss the method for optimizing the groove arrangement on the platen from the viewpoint of keeping the relative sliding distance on the wafer surface constant. In particular, we discuss the simulation results and show how a genetic algorithm (GA), which mathematically expresses biological evolution by repeated selection, crossover, and mutation, can be employed to maintain a constant relative sliding distance. In other words, we show the method of applying GA to our problem and the optimized distribution of grooves in the radial direction of the platen.

4.9.2 Simulation of Relative Sliding Distance on the Wafer and Platen Surfaces in the Double-Sided Four-Way Polishing System

4.9.2.1 Mathematical Expressions for Wafer-Side Sliding Distance

We will use the schematic in Figure 4.88 to calculate the wafer-side relative sliding distance between the wafer and the platen in the four-way system. Although the platen usually has grid-like or radial grooves, as discussed earlier, we show here the case of a platen without grooves. R_{lapi} is the inner radius of the platen; R_{lapo}, the outer radius of the platen; R_{car}, the carrier radius; R_{wafer}, the wafer radius; and x_o, the wafer offset from the carrier center (pitch circle radius of holes for holding wafer in carrier). ω_{lap}, ω_{wafer} and ω_{car} are the angular velocities of platen rotation, carrier rotation, and carrier revolution, respectively. We assume the counterclockwise and clockwise directions to be positive and negative, respectively. Further, Figure 4.88(a) is assumed to show the initial processing stage (time $t = 0$), and hence, point A is on the periphery of the wafer, on the X-axis. In Figure 4.88(b), we can see a baseline $O_{wafer\_t}A$, which connects the center of the wafer O_{wafer_t} and point A, and point N (r, ω_0, t) lying at r ($0 < r < R_{wafer}$) and making an arbitrary central angle ω_0 ($0 < \omega_0 < 2\pi$) with the baseline. We will now discuss the relative sliding velocity at point N (r, ω_0, t) at time t.

Assuming N (x_N (r, ω_0, t), y_N (r, ω_0, t)) as the coordinates of point N (r, ω_0, t) in the X-Y plane, we have:

$$\begin{aligned} x_N(r,\theta_0,t) &= (R_{lapi}+R_{car})\cos\omega_{car}t \\ &+ x_o\cos[(\omega_{car}+\omega_{wafer})t] \\ &+ r\cos[(\omega_{car}+\omega_{wafer})t+\theta_0] \end{aligned} \tag{4.7}$$

$$y_N(r, \theta_0, t) = (R_{lapi} + R_{car})\sin\omega_{car}t + x_o \sin[(\omega_{car} + \omega_{wafer})t] + r\sin[(\omega_{car} + \omega_{wafer})t + \theta_0] \tag{4.8}$$

Thus, at point N (r, ω_0, t), the rotation velocity V_{lap} (r, θ_0, t), carrier revolution velocity V_{car} (r, θ_0, t), and wafer movement velocity associated with the carrier rotation V_{wafer} (r, θ_0, t) are given by:

$$V_{lap}(r, \theta_0, t) = \sqrt{x_N(r, \theta_0, t)^2 + y_N(r, \theta_0, t)^2} \cdot \omega_{lap} \tag{4.9}$$

$$V_{car}(r, \theta_0, t) = \sqrt{x_N(r, \theta_0, t)^2 + y_N(r, \theta_0, t)^2} \cdot \omega_{car} \tag{4.10}$$

$$V_{wafer}(r, \theta_0, t) = \sqrt{(x_o + r\cos\theta_0)^2 + (r\sin\theta_0)^2} \cdot \omega_{wafer} \tag{4.11}$$

By resolving V_{lap} (r, ω_0, t), V_{car} (r, ω_0, t) and V_{wafer} (r, ω_0, t) into the corresponding X-axis and Y-axis components, (indicated by subscripts x and y) we have:

$$Vx_{lap}(r, \theta_0, t) = V_{lap}(r, \theta_0, t) \cdot \cos[\Theta(r, \theta_0, t) + \pi/2] \tag{4.12}$$

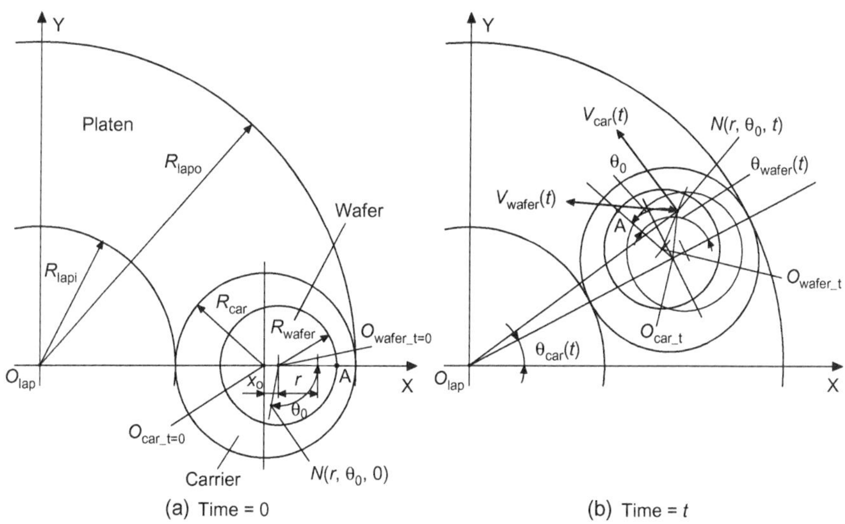

Figure 4.88 Schematic drawing of four-way wafer system for calculating sliding distance at all wafer surfaces

$$Vy_{lap}(r, \theta_0, t) = V_{lap}(r, \theta_0, t) \cdot \sin[\Theta(r, \theta_0, t) + \pi/2] \qquad (4.13)$$

$$Vx_{car}(r, \theta_0, t) = V_{car}(r, \theta_0, t) \cdot \cos[\Theta(r, \theta_0, t) + \pi/2] \qquad (4.14)$$

$$Vy_{car}(r, \theta_0, t) = V_{car}(r, \theta_0, t) \cdot \sin[\Theta(r, \theta_0, t) + \pi/2] \qquad (4.15)$$

$$Vx_{wafer}, \theta_0, t) = V_{wafer}(r, \theta_0, t) \cdot \cos[\Theta(r, \theta_0) + (\omega_{car} + \omega_{wafer})t + \pi/2] \qquad (4.16)$$

$$Vy_{wafer}(r, \theta_0, t) = V_{wafer}(r, \theta_0, t) \cdot \sin[\Theta(r, \theta_0) + (\omega_{car} + \omega_{wafer})t + \pi/2] \qquad (4.17)$$

where we assume:

$$\Theta(r, \theta_0, t) = \tan^{-1} \frac{y_N(r, \theta_0, t)}{x_N(r, \theta_0, t)} \qquad (4.18)$$

$$\Theta(r, \theta_0) = \tan^{-1} \frac{r \sin \theta_0}{x_o + r \cos \theta_0} \qquad (4.19)$$

Therefore, the relative sliding velocity $V_{relative}(r, \omega_0, t)$ of point N (r, ω_0, t) on the wafer against the platen is given by:

$$V_{relative}(r, \theta_0, t) = \sqrt{\begin{array}{l} \{Vx_{lap}(r, \theta_0, t) - [Vx_{car}(r, \theta_0, t) + Vx_{work}(r, \theta_0, t)]\}^2 \\ + \{Vy_{lap}(r, \theta_0, t) - [Vy_{car}(r, \theta_0, t) + Vy_{work}(r, \theta_0, t)]\}^2 \end{array}} \qquad (4.20)$$

and the relative sliding distance $L_{relative}(r, \omega_0)$ of point $N(r, \omega_0, t)$ between time $t = 0$ and T is given by:

$$L_{relative}(r, \theta_0) = \int_0^T V_{relative}(r, \theta_0, t)dt \qquad (4.21)$$

4.9.2.2 Mathematical Expressions for Platen-Side Sliding Distance

We will use the schematic shown in Figure 4.89 for calculating the platen-side relative sliding distance between the platen and the wafer in the four-way system. Further, as in the case of the wafer, we will assume that the platen has no grooves. Figure 4.89(a) is assumed to show the initial processing stage (time $t = 0$), and thus, point A is on the periphery of the wafer, on the X-axis.

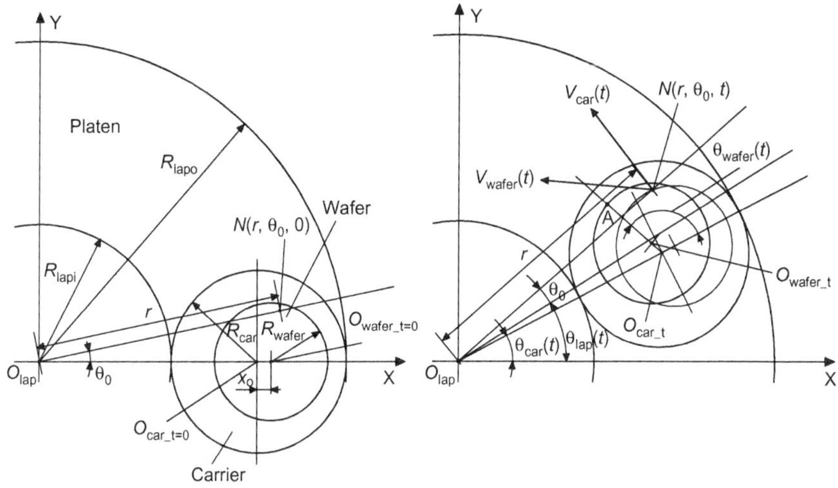

Figure 4.89 Schematic drawing of four way wafer system for calculating sliding distance at all platen surfaces

In the initial stage of processing (time $t = 0$) (Figure 4.89(a)), we assume an arbitrary point N $(r, \omega_0, 0)$, which lies on the platen at r $(R_{lapi} < r < R_{lapo})$ and makes an arbitrary central angle ω_0 $(0 < \omega_0 < 2\pi)$, and discuss the relative sliding velocity at this point at time t.

Assuming N $(x_N (r, \omega_0, t), y_N (r, \omega_0, t))$ to be the coordinates of point N (r, ω_0, t) in the X-Y plane, we express $x_N (r, \omega_0, t)$ and $y_N (r, \omega_0, t)$ as:

$$\begin{aligned} x_N (r, \theta_0, t) &= r \cos(\theta_{lap}(t) + \theta_0) \\ &= r \cos(\omega_{lap} t + \theta_0) \end{aligned} \quad (4.22)$$

$$\begin{aligned} y_N (r, \theta_0, t) &= r \sin(\theta_{lap}(t) + \theta_0) \\ &= r \sin(\omega_{lap} t + \theta_0) \end{aligned} \quad (4.23)$$

On the other hand, assuming $O_{car\_t}(x_{car}(t), y_{car}(t))$ as the coordinates of the carrier center O_{car_t} at time t, we obtain $x_{car}(t)$ and $y_{car}(t)$ as:

$$\begin{aligned} x_{car}(t) &= (R_{lapi} + R_{car}) \cos(\theta_{car}(t)) \\ &= (R_{lapi} + R_{car}) \cos(\omega_{car} t) \end{aligned} \quad (4.24)$$

$$\begin{aligned} y_{car}(t) &= (R_{lapi} + R_{car}) \sin(\theta_{car}(t)) \\ &= (R_{lapi} + R_{car}) \sin(\omega_{car} t) \end{aligned} \quad (4.25)$$

In addition, we assume O_{wafer_t} ($x_{wafer}(t)$, $y_{wafer}(t)$) as the coordinates of the wafer center O_{wafer_t} at time t and express $x_{wafer}(t)$ and $y_{wafer}(t)$ as:

$$\begin{aligned} x_{wafer}(t) &= (R_{lapi} + R_{car}) \cos(\theta_{car}(t)) + x_o \cos(\theta_{car}(t) + \theta_{wafer}(t)) \\ &= (R_{lapi} + R_{car}) \cos(\omega_{car} t) + x_o \cos(\omega_{car} + \omega_{wafer})t \end{aligned} \tag{4.26}$$

$$\begin{aligned} y_{wafer}(t) &= (R_{lapi} + R_{car}) \sin(\theta_{car}(t)) + x_o \sin(\theta_{car}(t) + \theta_{wafer}(t)) \\ &= (R_{lapi} + R_{car}) \sin(\omega_{car} t) + x_o \sin(\omega_{car} + \omega_{wafer})t \end{aligned} \tag{4.27}$$

Thus, at point N (r, ω_0, t), the rotation velocity V_{lap} (r, ω_0, t) of the platen, carrier revolution velocity V_{car} (r, ω_0, t), and wafer movement velocity V_{wafer} (r, ω_0, t), which depends on the carrier rotation, are given by:

$$V_{lap}(r, \theta_0, t) = r \cdot \omega_{lap} \tag{4.28}$$

$$V_{car}(r, \theta_0, t) = \sqrt{x_N(r, \theta_0, t)^2 + y_N(r, \theta_0, t)^2} \cdot \omega_{car} \tag{4.29}$$

$$V_{wafer}(r, \theta_0, t) = \sqrt{(x_N(r, \theta_0, t) - x_{car}(t))^2 + (y_N(r, \theta_0, t) - y_{car}(t))^2} \cdot \omega_{wafer} \tag{4.30}$$

Resolving V_{lap} (r, θ_0, t), V_{car} (r, θ_0, t), and V_{wafer} (r, θ_0, t) into the corresponding X-axis (Vx_{lap}, Vx_{car}, Vx_{wafer}) and Y-axis (Vy_{lap}, Vy_{car}, Vy_{wafer}) components denoted by subscripts x or y in Figure 4.89, we have:

$$Vx_{lap}(r, \theta_0, t) = V_{lap}(r, \theta_0, t) \cdot \cos(\omega_{lap} t + \theta_0 + \pi/2) \tag{4.31}$$

$$Vy_{lap}(r, \theta_0, t) = V_{lap}(r, \theta_0, t) \cdot \sin(\omega_{lap} t + \theta_0 + \pi/2) \tag{4.32}$$

$$Vx_{car}(r, \theta_0, t) = V_{car}(r, \theta_0, t) \cdot \cos(\omega_{lap} t + \theta_0 + \pi/2) \tag{4.33}$$

$$Vy_{car}(r, \theta_0, t) = V_{car}(r, \theta_0, t) \cdot \sin(\omega_{lap} t + \theta_0 + \pi/2) \tag{4.34}$$

$$Vx_{wafer}(r, \theta_0, t) = V_{wafer}(r, \theta_0, t) \cdot \cos[\Theta(r, \theta_0, t) + \pi/2] \tag{4.35}$$

$$Vy_{wafer}(r, \theta_0, t) = V_{wafer}(r, \theta_0, t) \cdot \sin[\Theta(r, \theta_0, t) + \pi/2] \tag{4.36}$$

where we assume that:

$$\Theta(r, \theta_0, t) = \tan^{-1}\frac{y_N(r, \theta_0, t) - y_{car}(t)}{x_N(r, \theta_0, t) - x_{car}(t)} \tag{4.37}$$

Therefore, the relative sliding velocity $V_{relative}(r, \omega_0, t)$ of point N (r, ω_0, t) on the platen against the wafer is given by:

$$V_{relative}(r, \theta_0, t) = \sqrt{\begin{array}{l}\{Vx_{lap}(r, \theta_0, t) - [Vx_{car}(r, \theta_0, t) + Vx_{work}(r, \theta_0, t)]\}^2 \\ + \{Vy_{lap}(r, \theta_0, t) - [Vy_{car}(r, \theta_0, t) + Vy_{work}(r, \theta_0, t)]\}^2\end{array}} \tag{4.38}$$

Here, the condition:

$$\sqrt{(x_N(r, \theta_0, t) - x_{wafer}(t))^2 + (y_N(r, \theta_0, t) - y_{wafer}(t))^2} < R_{wafer} \tag{4.39}$$

must be satisfied, and hence, we replace $V_{relative}(r, \omega_0, t)$ with zero when the abovementioned expression is not satisfied at time t.

By using the numerical formulae discussed above, we can determine the relative sliding distance related to the conditioning characteristics for the platen equipped with a conditioner (dresser). Since a conditioner usually has a cavity, we can expand the numerical formulae, as shown below.

We assume that the outer radius of the conditioner is equal to R_{wafer}, as mentioned earlier, and the inner radius of the cavity is R_{wafer}; then, we replace Eqn. (4.39) with the following equation and determine the relative sliding distance for the platen when a conditioner that has a cavity is used:

$$R_{wafer} < \sqrt{(x_N(r, \theta_0, t) - x_{wafer}(t))^2 + (y_N(r, \theta_0, t) - y_{wafer}(t))^2} < R_{wafer} \tag{4.40}$$

Thus, the relative sliding distance $L_{relative}(r, \omega_0)$ between time $t = 0$ and T for point $N(r, \omega_0, t)$ is given by:

$$L_{relative}(r, \theta_0) = \int_0^T V_{relative}(r, \theta_0, t)dt \tag{4.41}$$

4.9.2.3 Simulation of Relative Sliding Distance between Wafer and Platen

Figure 4.90 shows the calculated relative sliding distance over the whole surface of (a) the wafer and (b) the platen for the polishing conditions adopted in the four-way system (shown in Table 4.12). For the calculation, we divide the wafer and platen surfaces into 100 small sections in the radial

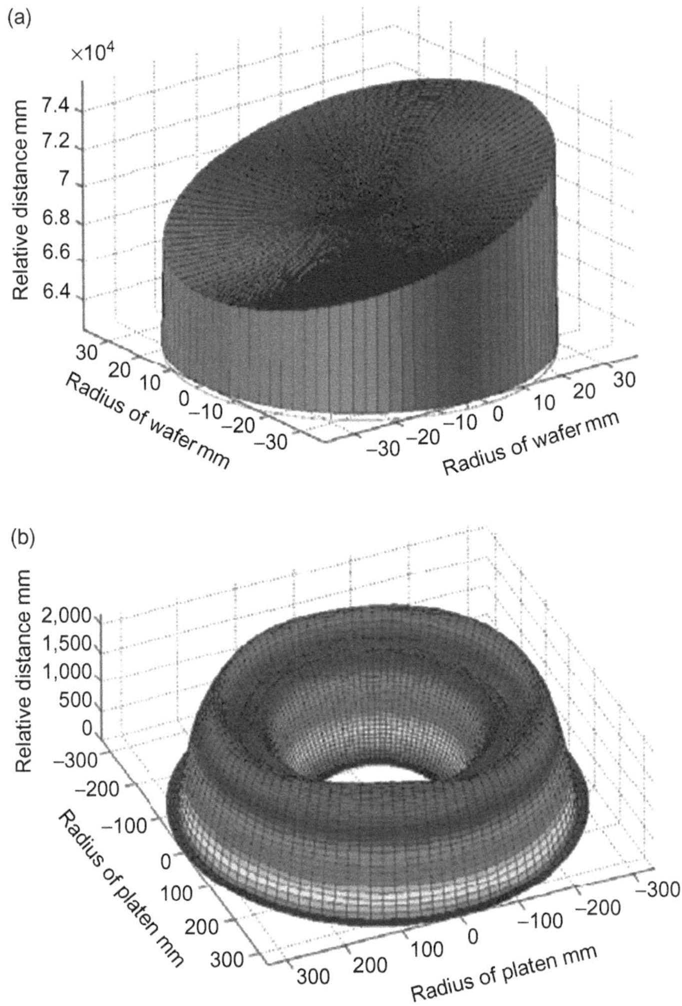

Figure 4.90 Calculated results of relative sliding distance. (a) Wafer (b) Platen

and circumferential directions, respectively. In other words, we first calculate the relative sliding velocity, which varies with time, between the wafer and the platen for each of the small sections, and then multiply the calculated velocity by the time spent in each section. Figure 4.91 shows the ratio of the relative sliding distances over (a) the wafer and (b) platen surfaces. Note that in (a), the relative sliding distance of the wafer center is used as the reference (=1), and in (b), the maximum relative sliding distance of the platen is used as the reference (=1). The result shows that the relative

Table 4.12 Calculation Conditions

Inner radius of lapping plate R_{lapi}: 211.67/2 mm
Outer radius of lapping plate R_{lapo}: 668.87/2 mm
Radius of wafer R_{wafer}: 75/2 mm
Offset x_o: 60 mm
Rotation angular velocity of platen ω_{lap}: $-3.33 \times 2\pi/60$ rad/s
Revolution angular velocity of carrier ω_{car}: $3.3 \times 2\pi/60$ rad/s
Rotation angular velocity of carrier ω_{wafer}: $2.5 \times 2\pi/60$ rad/s
Calculated time T: $2\pi/\omega_{car} \times 2\pi/\omega_{wafer}$

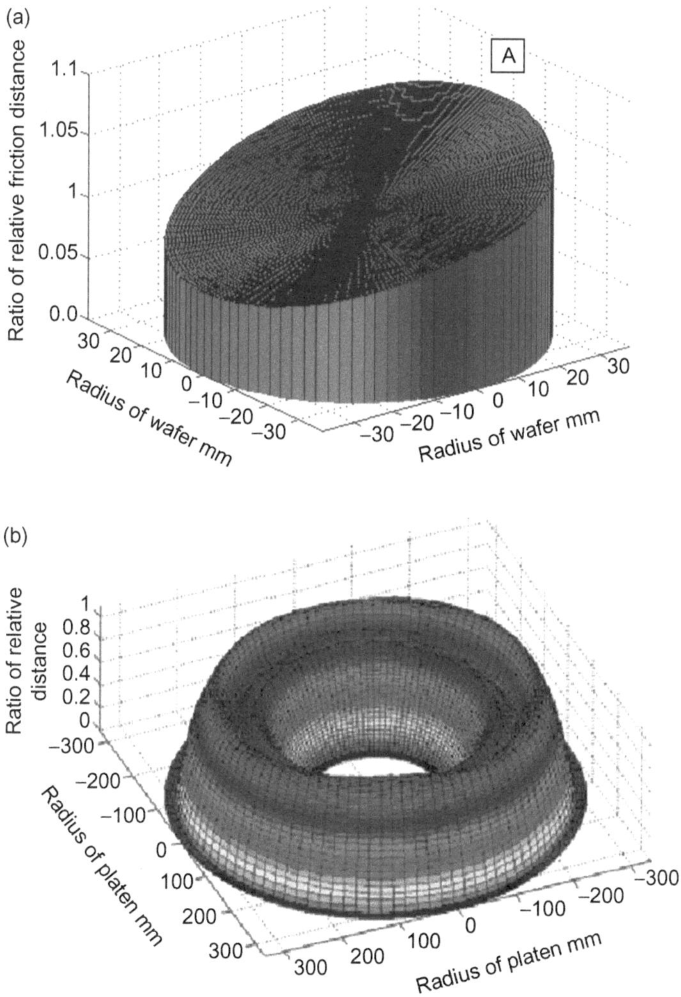

Figure 4.91 Calculated results of relative sliding distance ratio. (a) Wafer (b) Platen

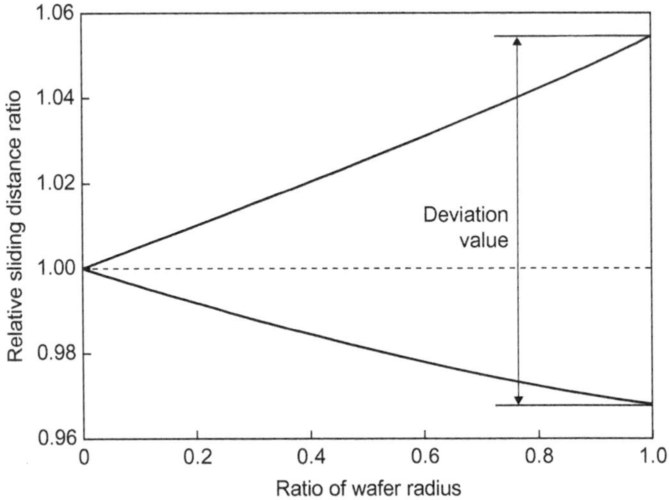

Figure 4.92 Calculated results of relative sliding distance at two angle directions which show the maximum deviation

sliding distance ratio in the four-way polishing process varies markedly between the radial and central angle directions of the wafer. It is also shown that for the platen, the relative sliding distance is large at the center of the radial direction.

Here, we focus on the relative sliding distance of the wafer side. On the basis of the results shown in Figure 4.91(a), we propose two cases with the maximum and minimum relative sliding distance ratios in the radial direction of the wafer (Figure 4.92). Under the polishing conditions assumed, the relative sliding distance ratio deviates by around 8.7% over the entire surface of the wafer, and hence, it is clear that this deviation affects the wafer flatness during polishing, as discussed earlier.

4.9.3 Use of GA to Determine the Optimum Proportion of Grooves in the Radial Direction of the Platen

From the discussion in the preceding section, it is clear that in the four-way system, the relative sliding distance between the wafer and the platen differs with the location on the wafer surface. Further, we can say that it is necessary to optimize the distribution of grooves on the platen so that the relative sliding distance remains constant for different locations. Therefore, in this section, we propose a method involving the use of a GA (genetic algorithm) and discuss the effectiveness of the method for optimizing the groove design and determining the rate of change of the proportion of grooves in the radial direction of the platen.

The GA to be used for the optimization calculation helps identify the best solution through a mathematical examination of the evolution of living beings. As shown in the GA flowchart in Figure 4.93, generation transition occurs successively through selection, crossover, and mutation in the genome of a generation in which there are N chromosomes. Here, it is assumed that a chromosome is made of M genes.

When applying the GA for optimizing the proportion of grooves in the radial direction of the platen, we define the numerical expression for the gene as follows. We regard the platen as a collective entity of M rings, as shown in Figure 4.94, and express the surface area ratio for the area without grooves in the ring ($R_{lap,m-1} < R < R_{lap,m}$) on the basis of the value of the mth gene belonging to the nth chromosome of the kth generation. Here, $R_{lap,m}$ is given by:

$$R_{lap,m} = \frac{m}{M}(R_{lapo} - R_{lapi}) + R_{lapi} \tag{4.42}$$

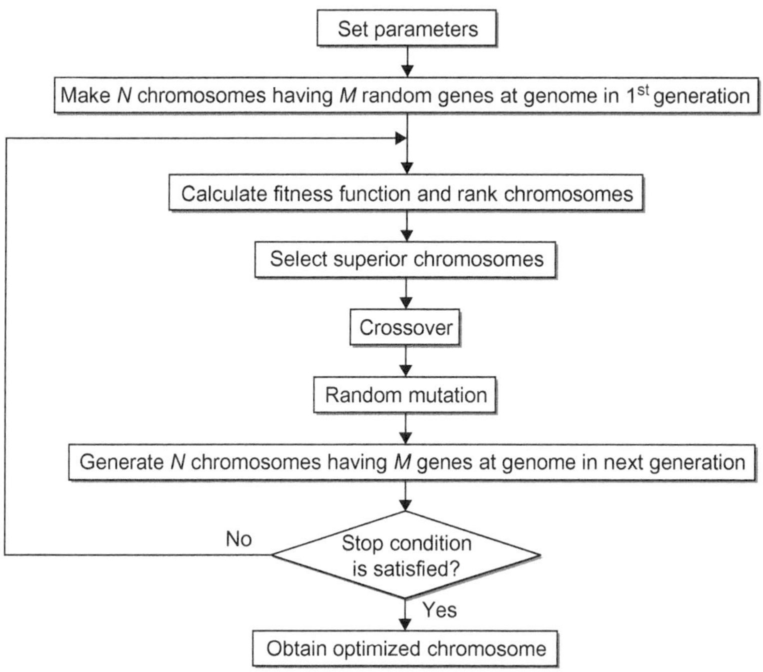

Figure 4.93 Flow chart of genetic algorithm applying to the optimization of platen groove rates in radius direction for four-way wafer system

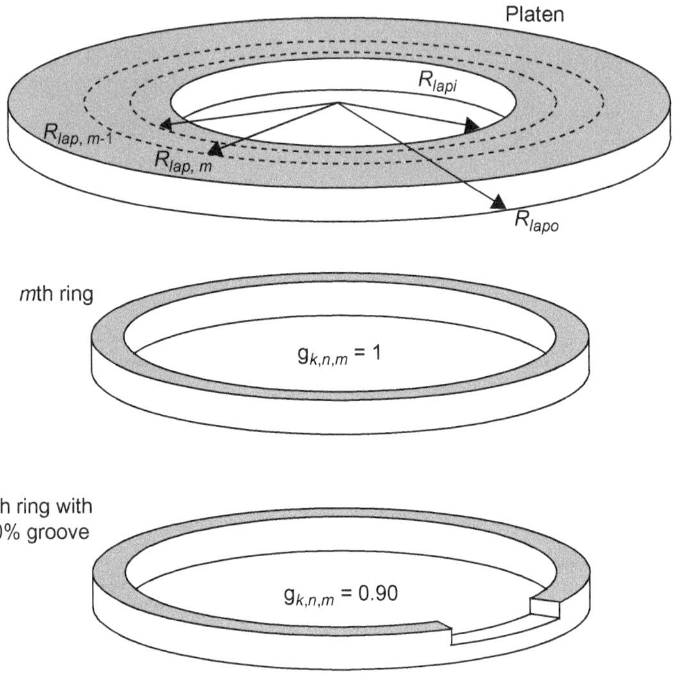

Figure 4.94 Relationship between the gene value and the configuration of ring surface area in genetic algorithm operation

Let us assume that the numerical value of the gene is between 0.8 and 1. If this value is 0.9 in the mth ring, as shown in the figure, we can say that grooves are formed in the ring and cover 10% of the surface area in the circumferential direction; if the numerical value is 1.0, no grooves are formed.

Figure 4.95 shows a genome in the kth generation as an example. We assign numerical values ranging from 0.8 and 1 to all the genes, as discussed above, and assume the resolution for the numerical values to be 0.001. In other words, we allow for the possibility of groove formation in each ring with a precision of 0.1% in the circumferential direction. In addition, we assume a minimum genetic value of 0.8 so that the grooves do not cover more than 20% of the surface area; this is because a large increase in the groove-covered area may result in a decrease in the polishing efficiency, although it is known that addition of a large number of grooves would cause a decrease in the sliding distance and help control the relative sliding distance.

Thus, when we consider the platen as an aggregate of divided platen rings, as shown in Figure 4.94, it is necessary to confirm the polishing

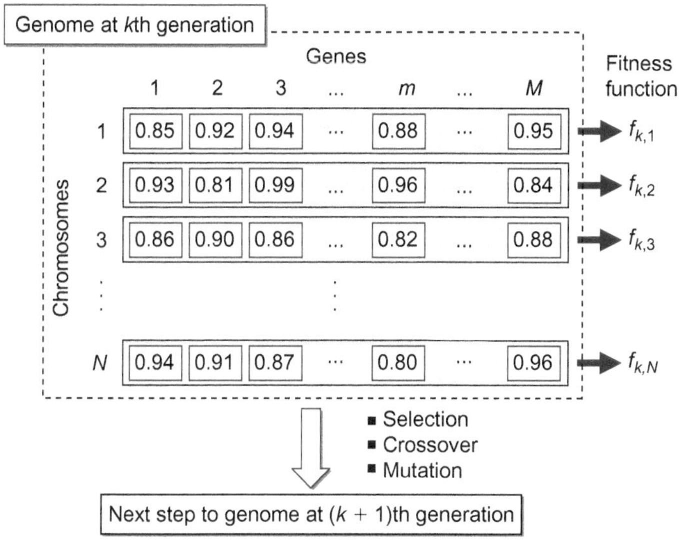

Figure 4.95 Expression method of the genome at genetic algorithm operation

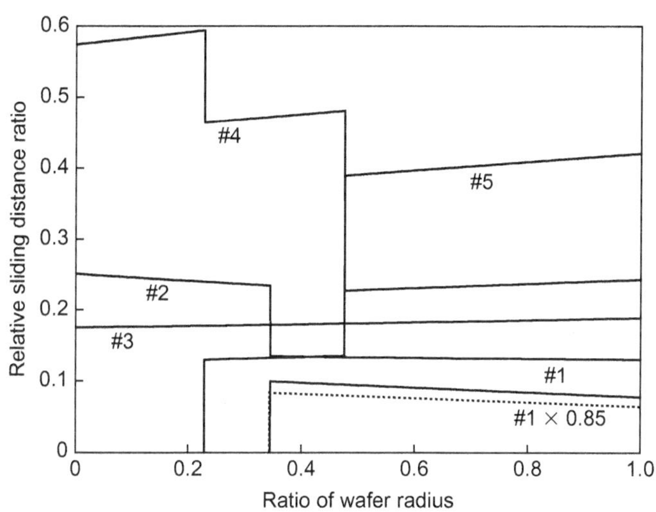

Figure 4.96 Calculation result of relative sliding distance by each ring of platen

rate (the degree of contribution to polishing) in each ring. For example, if we assume M = 5, the platen would consist of five rings. Figure 4.96 shows the calculated ratio at which rings #1–5 influence the relative sliding distance corresponding to each wafer radius point under the polishing conditions listed in Table 4.13. Although the figure shows only the case

Table 4.13 GA operating conditions

Number of chromosomes at one generation N: 50
Number of genes at one chromosome M: 40 (=Number of divided plate)
Crossover probability P_c: 0.25 (25%)
Mutation probability P_m: 0.01 (1%)
Number of generation: 5000

of $\omega_0 = 0$, the influence of each ring over the entire wafer surface can be determined by adopting the same procedure used for the former case. In addition, calculations are performed after dividing the wafer into 2000 small sections in the radial direction, but only the ratios shown in Figure 4.96 are used for the calculations. Here, if the numerical value for a gene, which gives the proportion of surface area of ring #1, is 0.85, as shown in the figure, the polishing rate for ring #1 may decrease to 85%.

Although the polishing rate is assumed to depend on the circumferential length of each ring, the position of the grooves in the circumferential and radial directions must be considered for evaluating the polishing rate realized when grooves are actually formed on the platen. Meanwhile, under the polishing conditions adopted in the general four-way system or in this section, the rotational angular velocity of the platen, revolution angular velocity of the carrier, and rotational angular velocity of the carrier are so set that they are not integral multiples of one another. This implies that under these polishing conditions, each motion is not synchronized, and when polishing is continued for a long time, the wafer crosses the same point in each circumferential direction. Therefore, the sliding distance averaged over a long processing time can be given by the first approximate solution without considering the groove position in the circumferential direction, as mentioned in this section.

Thus, various conditions are chosen for the GA, as listed in Table 4.13 and in the following section, by taking into account the change in the proportion of grooves in the radial direction for optimizing the shape of the platen.

a. Table 4.12 shows the conditions for the basic four-way system.

b. Equation (4.43) gives the fitness of each chromosome and the maximum deviation of the relative sliding distance over the entire surface of the wafer polished under the no-groove condition with respect to the relative sliding distance at the wafer center (Figure 4.91(a)). In other words, a small numerical value for a gene indicates a strong chromosome, with a small deviation of the relative sliding distance. Here, $L_{relative,GA,k,n}$ (r, ω_0) is the relative sliding distance of a wafer calculated from the nth chromosome of the kth generation obtained by GA.

$$f_{k,n} = \frac{\max\left(L_{relative,GA,k,n}(r,\theta_0)\right) - \min\left(L_{relative,GA,k,n}(r,\theta_0)\right)}{L_{relative}(r,\theta_0)|_{r=0}} \quad (4.43)$$

c. We first selected one-half (N/2) of the high-fitness chromosomes from among N chromosomes and used the roulette rule method to select the remaining half.

d. We assumed a crossover probability pc and decided the crossover position by generating random numbers; in this case, we assumed that all the crossover combinations are independent of one another.

e. Assuming a mutation probability of pm, we randomly generated mutations from the N × M genes in a genome. Here, we randomly generated numerical values in the range 0.8–1.0 for the mutated gene.

Figure 4.97 shows an example of the optimized proportion of grooves in the radial direction of the platen determined by the GA on the basis of the aforementioned problem setting. The figure shows the minimum and average of the fitness of the chromosomes in each generation, and indicates how the fitness of N chromosomes varies with the generation when the GA is implemented. It is seen that when applying the GA, the fitness decreases as the generation changes; moreover, the

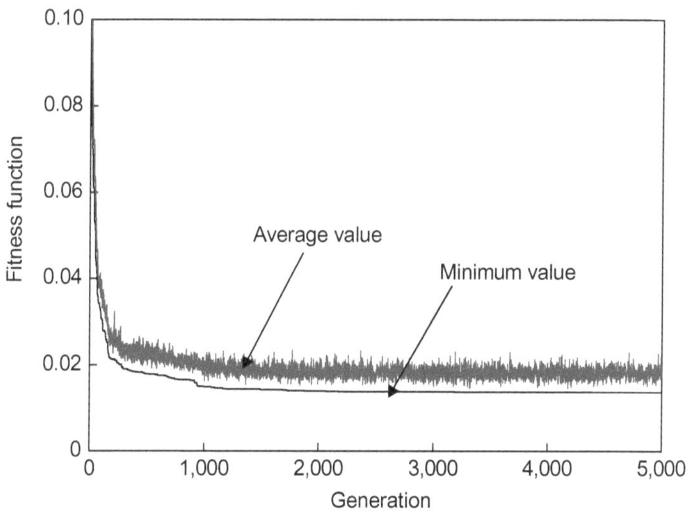

Figure 4.97 Change in fitness function when GA generation was operated

method of determining the proportion of grooves in the radial direction of the platen while ensuring a small deviation of the relative sliding distance is shown.

Figures 4.98 and 4.99 show the relative sliding distance characteristics, which are the best in the 5000th generation, the last generation in the calculation shown in Figure 4.97. Similar to Figure 4.91(a), Figure 4.98 shows the

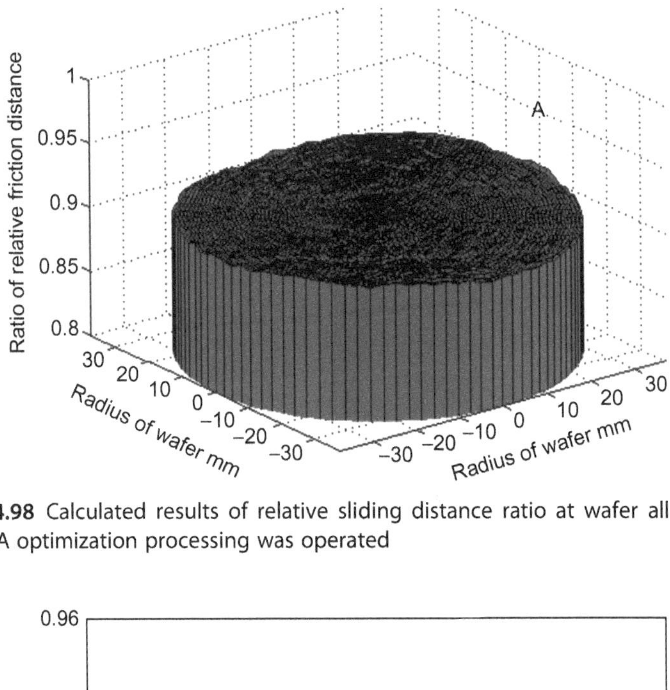

Figure 4.98 Calculated results of relative sliding distance ratio at wafer all surfaces when GA optimization processing was operated

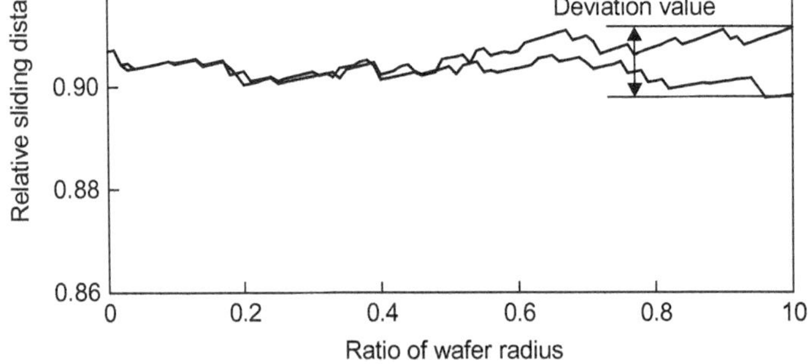

Figure 4.99 Calculated results of relative sliding distance ratio at two angle directions which show the maximum deviation when GA optimization processing was operated

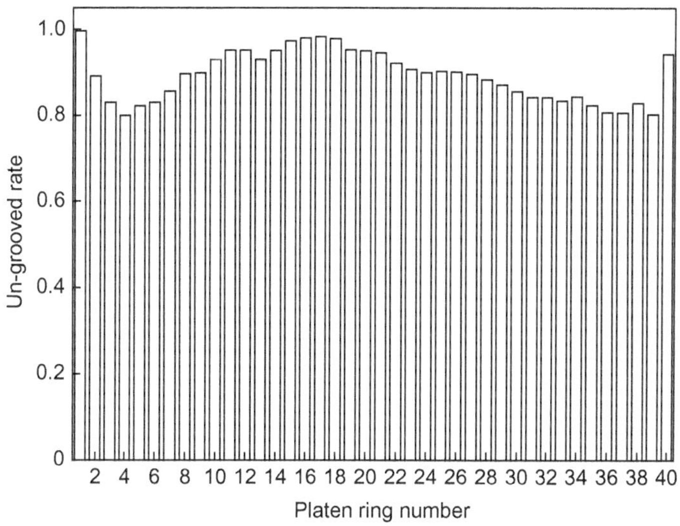

Figure 4.100 Relationship between platen ring number and un-grooved rate when GA optimization processing was operated

sliding distance calculated over the entire wafer surface, and Figure 4.99 shows (similar to Figure 4.92) the deviation between the two cases with the maximum and minimum relative sliding distance ratios in the radial direction of the wafer. The use of the GA helps reduce the deviation to about 1.6% over the entire wafer surface; this value is much smaller than that obtained in the no-groove case (deviation: 8.7%, as shown in Figures 4.91(a) and 4.92). Figure 4.100 illustrates the proportion of grooves in the radial direction of the platen, represented by the genetic value of a chromosome instead of the relative sliding distance characteristics, and gives the optimized proportion of grooves in the platen, as predicted by the GA.

4.9.4 Summary

In this section, we have presented theoretical formulae to determine the relative sliding distance between the wafer and the platen in the four-way polishing system. We have also discussed how the GA can be used to determine the optimum proportion of grooves in the radial direction of the platen, so that the relative sliding distance is constant over the entire wafer surface.

REFERENCES

1. T. Doi, Details of Semiconductor CMP Technology (in Japanese), Kogyo-chosakai Publishing Co., Tokyo, Japan, 2001.
2. B. Davari, C.W. Koburger, R. Shulz, J.D. Warnock, T. Furukawa, M. Jost, et al., IEDM Tech. Digest 61, 1989.
3. B. Davari, C.W. Koburger, T. Furukawa, Y. Taur, W. Noble, A. Megdanis, et al., IEDM Tech. Digest 88, 1988.
4. C.W. Kaanta, et al., Proc. of VLSI Multilevel Interconnects Conf., 1991, 144.
5. S. Yokogawa, et al., Proc. IITC, 2002, p. 127.
6. E.T. Ogawa et al., Proc. IRPS 2002, p. 312.
7. T. Doi, M. Kinoshita, H. Hiyama (Eds.), A library of CMP Planarization Technology & Application (in Japanese), Global Net Corp., 2006.
8. <http://public.itrs.net/Common/2005ITRS/Home2005.htm/>
9. M. Tada, et al., Symp. VLSI Technol. (Kyoto, JPN), Dig., 2005, 18.
10. M Abe, et al., Proc. IEEE Int. Electron. Device Mtg., 2005, pp. 83.
11. Y. Seike, D. Denards, K. Miyachi, T. Doi et al., Development and analysis of a high-pressure micro jet pad conditioning system, J. J. Appl. Phys., 44.
12. F.W. Preston, The theory and design of plate glass polishing machines, J. Society of Glass Tech., 1927, 214−256.
13. T. Doi, et al., Planarization processing technology for silicon-deviced wafers, Fall-meeting of JSPE, 1990.
14. T.K. Doi, T. Kasai, et al., A new proposal for a cylindrical linear contact polishing (CMP) system, Polish Ceramic Soc., 2000, 375−381.
15. T.K. Doi, et al., New ultra-precision CMP technique applying direct air-back method and non-foaming plastic pad, Proc. of ISSM'99/IEEE, 1999, 225−228.
16. A. Kaller, Naturwiss 43, 1956, p. 156.
17. J. Regh, G. Silvey, Electrochem. Soc. Fall Meeting (Phila.), 1996, 14.
18. E. Mendel, Polishing of Silicon, SCP & Solid State Technol. 10(8), 1967, 27−39.
19. A. Kaller, Monatsschr. Feinmech. u Optik 79(5), 1962, 135.
20. L.M. Cook, Chemical process in glass polishing, J. Non-Crystalline Solids 120, 1990, 152−171.
21. N.J. Brown, Some speculations on the mechanisms of abrasives grinding and polishing, Precision Eng. 9(3), 1987, 129−138.
22. M. Pourbaix, Atlas of Electro-Chemical Equilibrium, Pergamon Press-CEBELCOR, 1966.
23. Y. Karasawa, T.K. Doi, et al., Production technology of high perfomance slurry without CMP dispersing agent, Proc. of ISSM'99/IEEE, 1999, 437−440.
24. T.K. Doi et al, Development of both sides simultaneous polishing system for 300 mm silicon wafers, aiming at the production of nanotopography-free surfaces, Proceedings of the 31th International Symposium on Advanced Sci and Tech. of Silicon Materials 2000, pp. 248–257.
25. T.K. Doi, et al., Global Planarization Technique/CMP by High-Precision and Its Characteristics, Proc. of ISSM'95/IEEE, 1995, 214−217 and J. Sorooshian: Principles of Dielectric Chemical Mechanical Planarization – Tribological, Thermal and Kinetic Characterization of Dielectric and Tungsten Chemical Mechanical Planarization Process, VDM Publishing House Ltd, USA, 2009.
26. T. Doi, Details of Semiconductor CMP Technology (in Japanese), Kogyo-chosakai: Publishing Co., Tokyo, Japan, 2001.
27. M. Oshida, S. Haba, K. Yoshida, M. Kinoishita, New observation technique and applied research of surface roughness for silicon polishing pad, Proceedings of JSPE Spring Meeting, 2008, pp. 197–198.

28. Planarization and CMP Technical Committee: Proceedings of 49th and 50th Meetings of Planarization and CMP Technical Committee, 40/50 2006, p. 31.
29. M. Uneda, K. Okabe, N. Moriya, T. Kobayashi, K. Shibuya, K. Ishikawa, et al., Study of effect of dressing method on pad surface characteristics in polishing technique, Proceedings of JSPE Autumn Meeting, 2008, pp. 849–850.
30. M. Uneda, N. Moriya, T. Kobayashi, K. Shibuya, K. Ishikawa, H. Suwabe, Study of effect of dressing method on pad surface characteristics in polishing technique – application of evaluation parameters to commercially available apparatus, Proceedings of JSPE Spring Meeting, 2009, pp. 1033–1034.
31. Y. Sakai, Sensitive detection of trace pairs of gold nanoparticles for biosensing, KAST 4, 2004.
32. H. Yamada, Project Display of Radar Image Using Laser and Fluorescent Screen, J. Inst. TV Engnr. Jpn. 42(10), 1988, 1104–1111.
33. T. Doi, M Kinoshita, H. Hiyama (Eds.), A library of CMP Planarization Technology & Application (in Japanese), Global Net Corp., 2006.
34. E.Y. Zhao, et al., Chemical mechanical cleaning for post-CMP application: defects and metals results, Materials Reserch Society Symp. Proc. 477, 1997, 137.
35. I.J. Malik, et al., Post-CMP cleaning of W and SiO_2: A model study, Materials Res. Society Symp. Proc. 386, 1995, 109.
36. H. Aoki, et al., Post-Cu-CMP cleaning using oxalic acid, Extended Abstracts. of SSDM 282, 1998.
37. H. Aoki, et al., A degradation-free Cu/HSQ Damascene technology using metal mask patterning and post-CMP cleaning by electrolytic ionized water, Proc. of Interl. Electron Devices Meeting Technol. Symp. 777, 1997.
38. C. Jin, L. Ting, K.Taylor, T. Seha, J. Luttmer, in Proceedings 2nd International Dielectric for VLSI/ULSI Multilayer Interconnect Conference (DUMIC), 21 1996.
39. Planarization and CMP Technical Committee (Ed.), A Library of CMP Planarization Technology and Application, JSPE, Global Net Corp., 2006 in Japanese.
40. B. Pinto, J. Saito, W. Shen, L. Cheung, A. Wang, New Inspection Technology for 45 nm Wafers, www.kla-tencor.com/ymsmagazine, Winter 2007, pp. 28–32.
41. M. Yoshise, Evaluation technique of wafer geometry, Proceedings of the 119th Meeting of the 145th Committee on Processing and Characterization of Crystals of JSPS, 2009, pp. 1–5 (in Japanese).
42. S. Ramaswamy, N. Yathapu, G. Lembach, M. Guse, M. Linnane, Leveraging blanket wafer film inspection to efficiently characterize root cause for lithographic micro-masking patterning defects, Proceedings of IEEE/SEMI Advanced Semiconductor Manufacturing Conference, 2010, pp. 165–168.
43. A. Wocko, S. Radovanovic, P. Dighe, Study of Relationship between 300 mm SI Wafer Surface and Annealing Temperatures for Advanced Semiconductor-based Applications, Proc. of IEEE/SEMI Advanced Semiconductor Manufacturing Conference, 2010, pp. 154–157.
44. S. Halder, R. Vos, M. Wada, D. Tsvetanova, M. Clase, P. Mertens, et al., Evaluation of post ion-implantation resist strip with the background signal of a light scattering tool, Jpn. J. Appl. Phys. 49, 2010, pp. 056504-1-056504-5.
45. W. Shen, High volume, high yield production, Advanced Substrate News, Winter 2009–2010, 7.
46. K. Turner, S. Veeraraghavan, J. Sinha, Predicting distortions and overlay errors due to wafer deformation during chucking on lithography scanners, J. Micro/Nanolith. MEMS MOEMS 8(4), 2009, pp. 043015-1-043015-8.
47. D. Feiler, S. Radovanovic, P. Dighe, A. Kitnan, G. Simpson, G. Schwager, et al., A proven methodology for detecting photo-resist residue and for qualifying

photo-resist material by measuring fluorescence using SP2 bare wafer inspection and SURFmonitor, Proc. of SPIE, 7520 2009, pp. 75201M-1-75201M-8.
48. B. Bhushan (Ed.), Handbook of Nanotechnology, Springer-Verlag, 2004.
49. H. Kuwano (Ed.), Fundamentals and applications of MEMS/NEMS Engineering, Techno System, 2009 in Japanese.
50. <http://www.azonano.com/>
51. <http://en.wikipedia.org/wiki/Atomic_force_microscopy/>
52. <http://en.wikipedia.org/wiki/Confocal_laser_scanning_microscopy/>
53. <http://www.olympus.co.jp/jp/>
54. International Technology Roadmap For Semiconductors 2010 Update Overview.
55. T. Doi, K. Ichikawa, Nanotopography in planarization CMP, and a new double-sided simultaneous polishing machine with micro-motion mechanism aimed at nanotopography reduction, Jpn. Soc. Abrasive Technol. 48(8), 2004, 421.
56. F.W Preston, The theory and design of plate glass polishing machines, J. Glass Technol. 11(44), 1927, 214.

CHAPTER 5

Promising Future Processing Technology

5.1 ELECTROLYTIC CMP (E-CMP) AND APPLIED TECHNIQUES

Porous low dielectric constant (low-k) films are being developed for manufacturing faster copper multilayer interconnects to be used in LSIs with 65 nm nodes or smaller. A porous low-k film has a low density because cavities of several nanometers in size are incorporated into the material. Hence, such a film is easily peeled off and damaged by the mechanical stress applied during the chemical mechanical polishing (CMP) process, so decreasing the mechanical pressure applied becomes necessary. In recent LSIs, however, the thickness of the upper global routing copper wiring layer is greater than 1 μm, often as much as 5–10 times greater than the thickness of the lower, high-density copper layers. In other words, a reverse scaling phenomenon occurs in the upper global routing copper wiring layers. A copper film of such substantial thickness requires a longer CMP time, thus decreasing the throughput. While the removal rate may be improved if the CMP pressure is raised, the low-k film of the lower, high-density wiring layer may be damaged. Under the present conditions, very few CMP slurries can give a satisfactory removal rate, combined with accuracy to the level of several nanometers. As a result, a two-step CMP process, employing two types of slurry, one of high-speed and one of high-precision, is sometimes adopted. However, the use of several types of CMP slurry is not advisable, and it is desirable to devise new planarization techniques. Given these factors, electrolytic polishing is seen as a promising technique for next-generation CMP, since it can offer low mechanical pressure and high speeds. This section discusses the compound actions of CMP – the current planarization technique used for device wafers – with the electrolytic action imparted by electrolytic processing (E-CMP). In addition, we describe electrolytic polishing using ultrapure water (ECP-DI), which is an advanced, eco-friendly type

Advances in CMP/Polishing Technologies
DOI: 10.1016/B978-1-4377-7859-5.00005-3

of E-CMP. Ebara Corporation and Osaka University jointly developed the ECP-DI technique.

5.1.1 Characteristics of Non-Contact Electrolytic Polishing and Advanced Electrolytic CMP

Electrolytic polishing or electrolytic CMP (E-CMP) is a technique that combines the removal action of conventional CMP with that of electrodissolution. In conventional CMP, the slurry has both the capability of chemically reforming and mechanically scratching the surface. Now, we include in E-CMP the process of electrochemical dissolution, which can reduce the extent of mechanical polishing and result in a more chemically refined copper surface. The severity of the damage caused can be minimized, and the surface micro-roughness can be reduced. As an added benefit, the electrochemical removal action does not affect the insulating low-k material during the planarization of device wafers. Therefore, the removal rate should decrease and the precision in detecting the process endpoint should improve.

The conventional E-polishing methods can be broadly classified into two groups: the non-contact type, in which no mechanical action is exerted at the wafer surface, and the contact type of electrochemical mechanical polishing (electrolytic CMP, or E-CMP), in which an initially uneven wafer surface is planarized by chemical mechanical (e.g. with a polishing pad and slurry) and electrolytic action.

With the non-contact type, even a low-k film of $k < 2.0$ can be polished at a removal rate of 600 nm/min without any damage, because no mechanical action is involved; however, because the planarization capability is low, CMP is usually also needed. Selectivity against the underlying barrier-metal layer may be good, but corrosion by etching action may occur because copper dishing is vigorous and, in particular, the resistance of isolated metal wires increases. Furthermore, process optimization may be more difficult than in the case of CMP because the micro-roughness of the copper surface may increase depending on the value of the electric current. Experience shows that good surface uniformity can be obtained for the wafer (i.e. the within-wafer non-uniformity (WIWNU) can be reduced) when electrolytic polishing is carried out by applying a reverse voltage to the electrolytic solution using a copper plating device. The plating solution is the commonly used copper sulfate, and the polishing conditions should be as follows: removal rate, 500–700 nm/min; voltage, 2–3 V; electric current 10–15 A. However, the copper surface becomes increasingly roughened as the non-contact type of polishing is extended, as determined by scanning electron microscopy (SEM) observations of the polished copper wafer. Visual observation of the wafer does not reveal a mirror surface. It is therefore necessary to use E-CMP, which requires mechanical contact with the wafer, to achieve planarization at the nanometer scale. Realization of a mirror surface by non-contact electrolytic polishing will require optimization of the polishing conditions and the electrolytic solution composition.

5.1.2 Electrolytic CMP (E-CMP) and its Basic Polishing Performance

Electrolytic CMP (E-CMP) is a planarization technique that involves both electrolytic and chemical mechanical polishing of the wafer surface (using pads and slurries, for example). A representative method is Ecmp™, which IBM and Applied Materials are developing jointly. The mechanism of planarization by E-CMP involves polishing the surface to remove the protective film on the convex portion of the copper surface; this film is formed by the protective-film-forming agent contained in the electrolytic solution. To do this, a resin pad made of a polymeric material such as polyurethane is first used and local electrolytic polishing is subsequently conducted. For a filmed concave surface, however, electrolytic polishing is slow, and the planarization proceeds only through a mechanism such as CMP.

ASM/NuTool is developing ECMD™, which combines E-CMP and plating. In other words, instead of using the Ecmp™ protective-film-formation agent, ECMD™ controls the amount of additive in the plating liquid (suppressor and accelerator) that adheres to the copper surface, by rubbing the surface using a polishing pad, as shown in Figure 5.1. In this technique, bottom-up growth occurs selectively on the concave surface of the copper. As a result, as shown in the SEM photograph in Figure 5.2, it is possible to produce a copper film that is thin in the flat (convex) area, while the concave

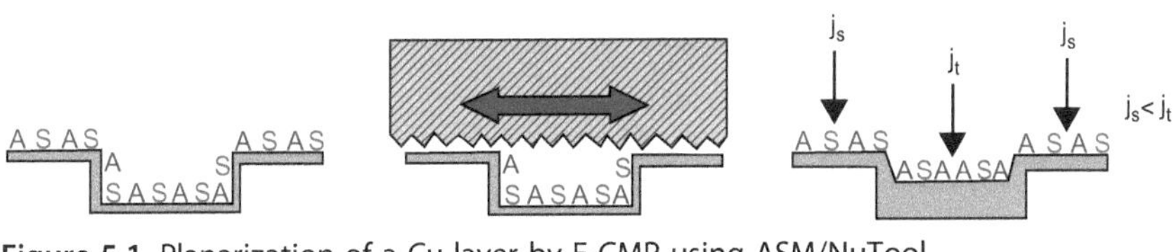

Figure 5.1 Planarization of a Cu layer by E-CMP using ASM/NuTool

Figure 5.2 The SEM photograph

trench is completely filled with copper. This can reduce process costs by minimizing the subsequent polishing time by CMP.

The E-CMP process does present a different set of problems. First, this technique requires the metal electrode to be in direct contact with the copper film on the wafer surface. Therefore, this phenomenon may produce scratches as a result of wafer rotation during electrolytic polishing. Damage to a low-k film cannot be avoided without making improvements to the contact method or optimizing the shape and composition of the electrode. In addition, as the conventional method generally positions the contact electrode at the wafer periphery, the removal rate differs between the wafer center and the peripheral area owing to a voltage drop, and this difference increases as the copper thickness decreases. Thus, it is difficult to ensure uniformity over the wafer.

It is therefore necessary to devise a new type of pad for E-CMP. Assuming the polishing of copper films and aiming to improve the current application of E-CMP, we will now describe a cell pad consisting of an insulating layer with through holes and a conductive sheet. Further, we will discuss the E-CMP technique, in which current is applied by ensuring direct contact of the entire copper film surface with the conductive sheet.

Figure 5.3 shows a schematic diagram of the electrolytic cell pad, which consists of two layers; an outer conductive layer, and an insulating layer with many holes penetrating through it. In addition, a conductive stainless steel sheet is placed below the lower part of the two-layer structure to block the through holes and prevent contamination of the platen by copper plating. We can connect negative pole of the power supply directly to the platen and use the stainless steel sheet as the cathode if the electrolytic cell pad is adhered to the platen using conductive adhesive tape.

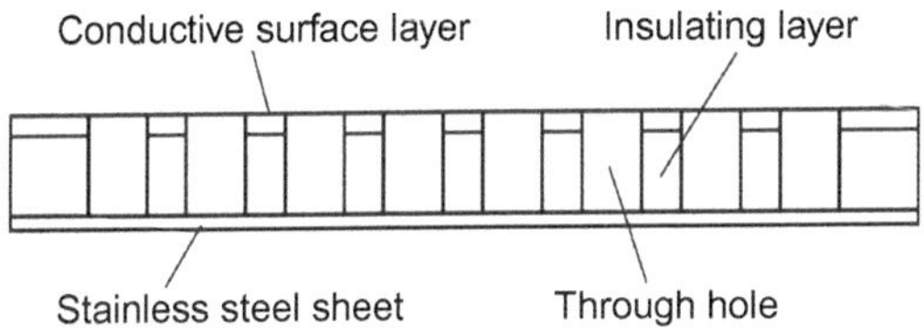

Figure 5.3 Basic structure pattern diagram of electrolytic cell pad

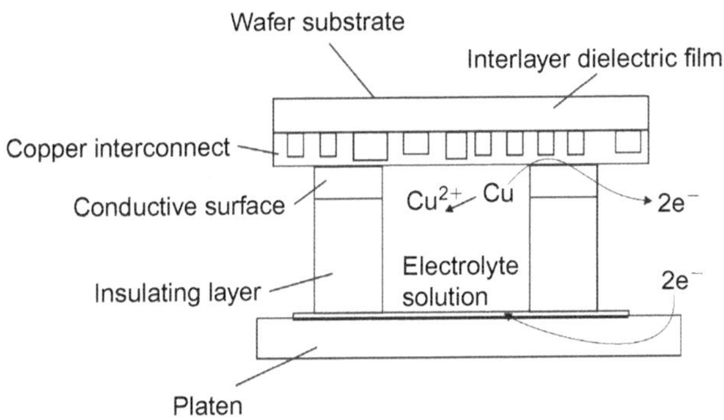

Figure 5.4 Schematic diagram of E-CMP with electrolytic cell pad

Figure 5.4 shows E-CMP using an electrolytic cell pad schematically. When the through holes are filled with the electrolyte, an electrolytic cell is formed between the copper wiring film and the stainless steel sheet. After connecting the positive electrode of the power supply to the outer surface of the electrolytic cell pad and the negative electrode to the platen, polarization occurs, and the copper wiring film acts as the anode, while the stainless steel sheet functions as the cathode. As a result, the electrochemical reaction shown in Eqn. (5.1) occurs, and the copper wiring film dissolves and is removed.

$$\text{Anode}: \mathrm{Cu} \rightarrow \mathrm{Cu}^{2+} + 2\mathrm{e}^- \tag{5.1}$$

$$\text{Cathode}: 2/\mathrm{nM}^{n+} + 2\mathrm{e}^- \rightarrow 2/\mathrm{nM} \tag{5.2}$$

where, M^{n+} is a positive ion in the electrolyte.

In the electrolytic cell, copper is oxidized at the interface between the anode and electrolyte, and the generated electrons move through the outer conductive layer to the cathode while reducing the positive ions in the electrolyte and consuming the electrolyte. The positive ions vary according to the selected electrolyte; for example, hydrogen ions are evolved in the case of an acidic aqueous solution such as phosphoric acid, or copper ions in the case of copper sulfate.

For general CMP devices of the rotary-platen type, relative motion occurs between the platen, which rotates and fixes the electrolytic cell pad, and the polishing head, which rotates and supports the copper base plate. The wafer is polished uniformly because the copper wiring film, which is in contact with the electrolytic cell, is continuously sliding. In addition, the electrolyte solution is churned by this relative movement and is thus automatically replenished, and the entire surface of the copper wiring film touches the outer conductive layer; this implies that the electrolysis current density is hardly affected by the IR drop (voltage drop between the working and opposite electrodes).

As discussed earlier, the general, rotary-platen-type CMP devices consist of a platen with an attached pad and its drive mechanism, the polishing head and its drive mechanism and the slurry feed mechanism. A power unit is also added to the basic configuration of the CMP system. The analysis of the test results for E-CMP operation is discussed below. Figure 5.5 shows the proposed configuration of the E-CMP system. The structure of the platen and the polishing head is similar to that of the conventional rotary-platen-type CMP system; an electrolyte is supplied instead of the slurry, and an electrolytic cell pad (outer diameter ϕ300 mm) is used in place of a polishing pad. The power unit consists of

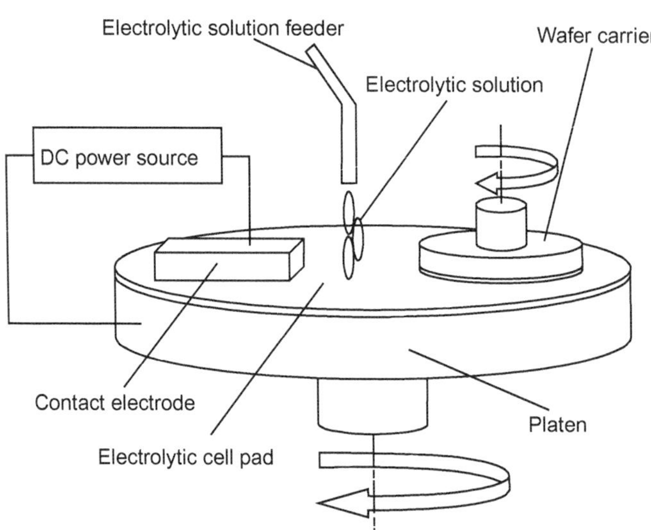

Figure 5.5 Proposed pattern diagram of E-CMP system

a direct current power supply and contact electrodes, and the anode bias is applied to the copper wiring film through the contact electrode and the outer conductive layer. The sample to be polished is a copper blanket wafer of dimensions 40 mm × 40 mm (where a copper film of thickness 1–30 μm is deposited on the Si substrate by electroplating). Copper sulfate pentahydrate is used as the electrolyte.

The removal rate is plotted against the polishing pressure at varying platen rotation speeds, for an electrolyte concentration of 20 g/L and power supply voltage of 5 V in Figure 5.6. The removal rate is approximately constant, and independent of the polishing pressure. As the platen rotation speed increases, the contact electrical resistance will decrease when the copper film is physically lifted off the surface of the wafer upon hydroplaning. Another possibility is that the amount of electrolyte in the electrolytic cell may be reduced by the centrifugal force, and the contact area between the copper film and the electrolytic cell may decrease. The latter mechanism would have greater influence at a pad rotation speed of 40 rpm.

Figure 5.7 shows the relationship between the removal rate and the power supply voltage with the electrolyte concentration as a parameter, for certain values of polishing pressure and platen rotation speed. The removal rate increases linearly with the power supply voltage, and it shows that electrolysis does not occur unless the power supply voltage exceeds 1 V.

On the basis of these results – the pressure dependence and power supply voltage dependence of the removal rate during E-CMP of a copper wiring film – we can assume that electrochemical polishing is the

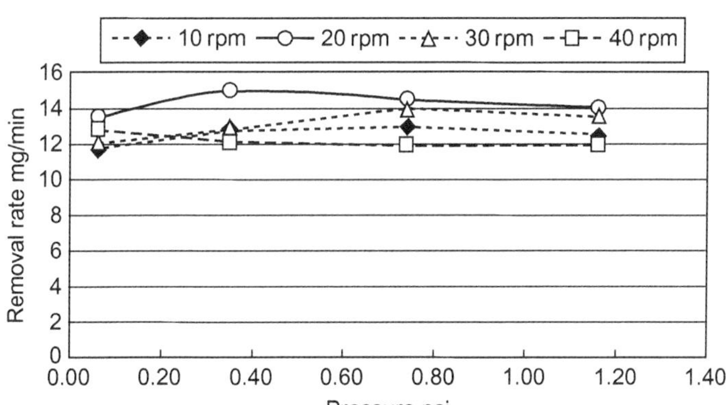

Figure 5.6 Relationship between polishing pressure and removal rate

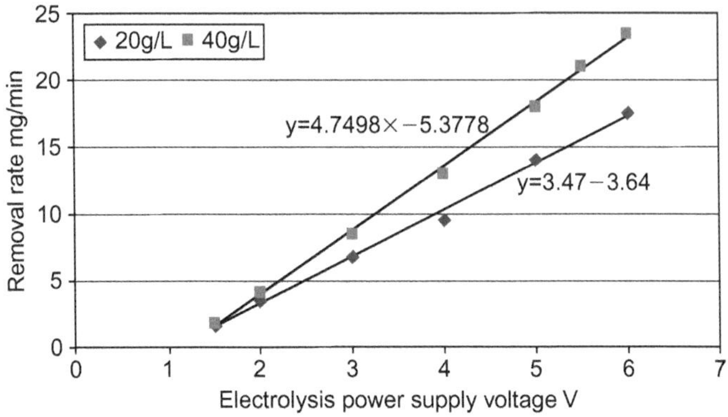

Figure 5.7 The relationship between the removal rate and the power supply voltage with the electrolyte concentration

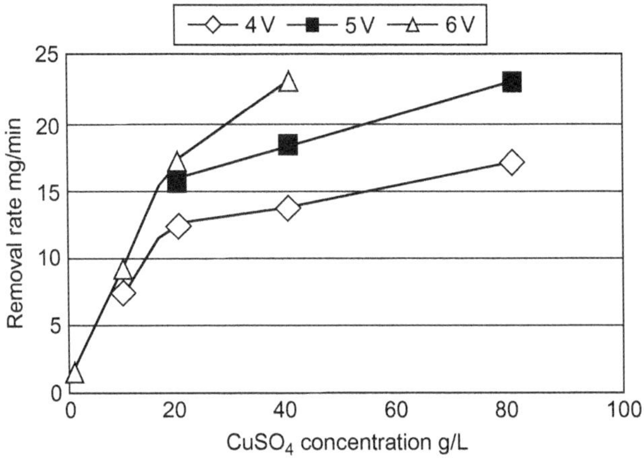

Figure 5.8 The typical relationship between the removal rate and the electrolyte concentration (Pressure: 0.36 psi, Rotational platen speed: 20 rpm)

only active mechanism taking place, and the chemical and mechanical polishing actions found in conventional CMP can be ignored. Figure 5.8 depicts a typical relationship between the removal rate and the electrolyte concentration, which shows that the removal rate reaches a plateau above a certain electrolyte concentration.

Figure 5.9 shows a typical result – on the film thickness – for polishing a substrate (initial copper film thickness: 1 μm) with dimensions of 40 mm × 40 mm, under the following conditions: polishing time, 20 s; polishing pressure, 0.35 psi (24 g/cm^2); platen rotation, 20 rpm; electrolyte

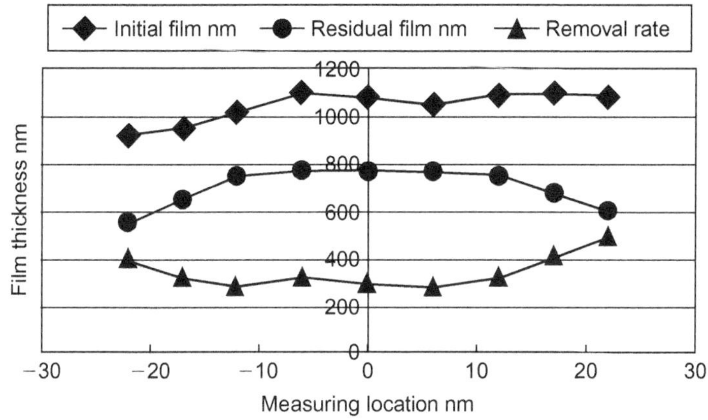

Figure 5.9 Typical effect on film thickness of polishing a substrate (pressure: 0.36 psi, rotational platen speed: 20 rpm)

density, 40 g/L; and power supply voltage, 4 V. The central area of the substrate is polished flat. The electrical contact between the electrolytic cell pad and the substrate is good. This result suggests that the influence of the IR drop in the remaining film is small.

Judging from the preceding test result, the copper dissolution reaction involves two mechanisms: the electrochemical reaction (see below) and its accompanying chemical reaction. For control of the thickness of the removed film, it is necessary to consider the chemical reaction.

Anode

$$Cu \rightarrow Cu^{2+} + 2e^{-} \tag{5.3}$$

$$2H_2O \rightarrow O_2 + 4H^{-} + 4e^{-} \tag{5.4}$$

Cathode

$$3Cu^{2+} + 6e^{-} \rightarrow 3Cu \tag{5.5}$$

Chemical reaction

$$2Cu + O_2 \rightarrow 2CuO \tag{5.6}$$

$$2CuO + 4H^{+} \rightarrow 2Cu^{2} + 2H_2O \tag{5.7}$$

Low-friction polishing using an electrolytic cell pad and application of E-CMP to the rotary-platen-type polishing system would be required to obtain high polishing efficiency and uniform planarization over the surface. However, in the Cu/low-k dual-Damascene wiring process, it is necessary to remove the uneven copper wiring film in order to form interconnects with minimized dishing.

Kondo et al. have developed a conductive carbon pad for E-CMP, and full-scale tests are underway on a 300 mm orbital-type CMP system. The three layers that comprise the 400 mm (16 inch) carbon pad are (from the surface side) an anodic carbon layer, an insulating layer, and a cathodic electrode sheet whose thickness is approximately 5 mm. As shown in Figure 5.10, more than 100 electrolytic cells in which E-CMP occurs are formed in the pad. The middle insulating layer also works as a cushion to improve the WIWNU of the polishing profile. A soft and smooth carbon material, which is in direct contact with the copper wafer, is selected to avoid any damage to the copper surface. The power supply is connected to the peripheries of the carbon pad and the electrode sheet.

In the conductive carbon pad for E-CMP developed by Kondo et al., the electrolyte is supplied at a rate of 300 mL/min from under the platen to fill the electrolytic cells formed in the carbon pad, and is then drained out from the periphery of the carbon pad through the groove (channel) in the insulating layer between the anode and the cathode. Proper choice of electrolyte allows the ionized Cu^{2+} to be drained concomitantly, thus avoiding the generation of scratches.

As is shown by the direct dependence of the removal rate on the supplied current (Figure 5.11), the removal rate improves in proportion to the supplied current, and the WIWNU is improved by using the cushion layer. A higher removal rate is obtained than with conventional CMP in the present case, despite the low rotary speed of the platen and head, and a comparatively large value of the removal-rate-to-electric-current ratio is obtained. With desirable linearity, the removal rate is easily controlled by the supplied current.

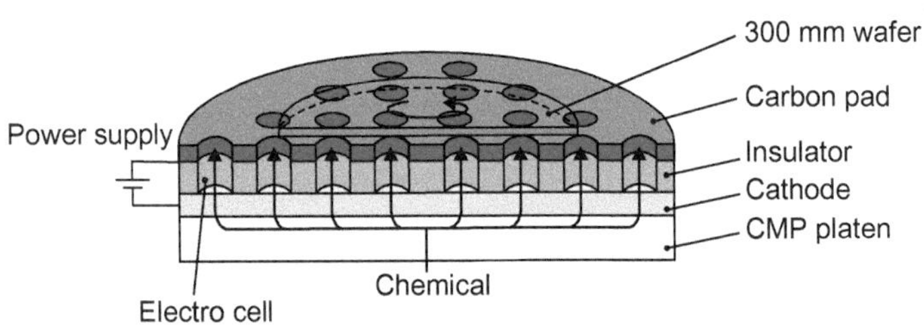

Figure 5.10 Structure of conductive carbon pan in E-CMP

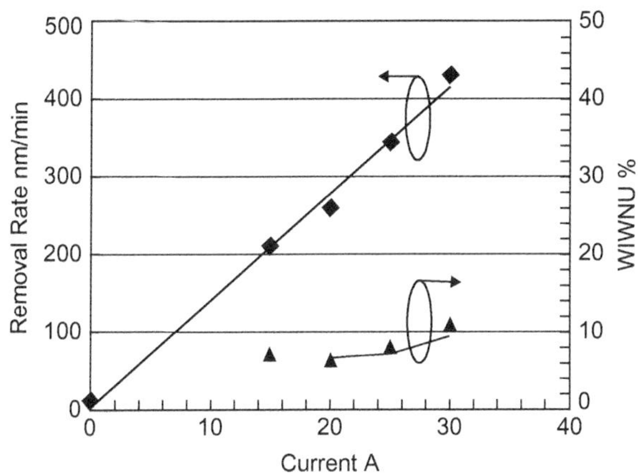

Figure 5.11 Dependence of copper removal rate on supplied electrical current

Figure 5.12 An external view of the E-CMP system developed by AMAT

Kondo et al. polished porous low-k (k = 1.8) films using the E-CMP technique and compared the polishing damage with that in conventional CMP. Their results confirmed that E-CMP suppresses the peeling off of the porous low-k film more effectively than the mechanical and stressful CMP process.

E-CMP, as discussed above, is expected to be a promising polishing technique for porous low-k/Cu wiring. Figure 5.12 shows an external view of the E-CMP system developed by AMAT, as well as the E-CMP polishing area.

5.1.3 Electrolytic Polishing Using Ultrapure Water (ECP-DI: Electrochemical Polishing in DI Water)

Electrolytic polishing using only ultrapure water (ECP-DI) exploits the electrochemical interactions between the OH^- ions in ultrapure water and polishes the atoms on the material surface. Because ECP-DI is electrochemical in nature, mechanical surface damage is avoided. Further, the use of only ultrapure water, rather than slurries or electrolyte solutions, reduces the extra burden of cleaning after polishing, potentially lowering process costs and environmental load.

This new polishing method was jointly developed by Ebara Corporation and Osaka University. A catalytic material is placed between the electrode and the polished material. The dissociation reaction of the ultrapure water molecules is accelerated by catalysis, enabling the current density to increase, as shown in Figure 5.13. Thus, catalytically generated OH^- ions are supplied by means of an external field to the surface of the polished material. The catalyst is required in order to (1) decrease the activation energy of the dissociation reaction of the water molecules and (2) to avoid contamination of the ultrapure water.

Figure 5.14 shows the average electric-field dependence of the electrolysis current density for the electrolysis of ultrapure water effected by an ion exchanger (a typical catalytic material) positioned between Pt (platinum) electrodes. According to Ebara Corporation and Osaka University, the use of a catalytic material affords an electrolysis current

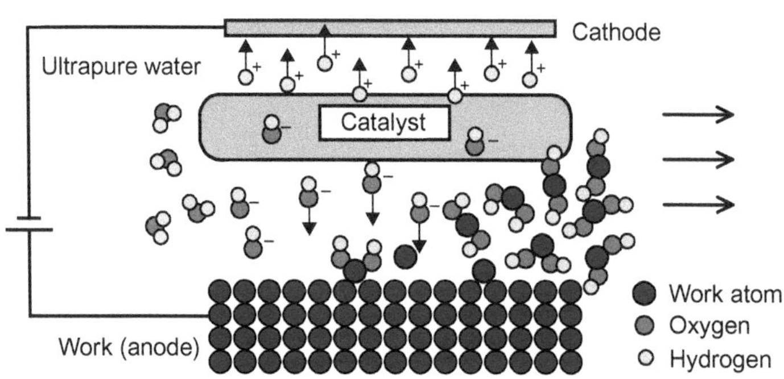

Figure 5.13 Electrochemical polishing in DI water (ECP-DI) using a catalyst

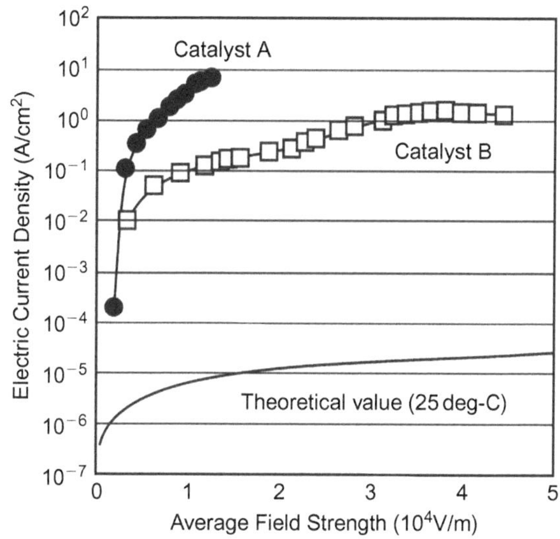

Figure 5.14 Relationship between average field strength and electric current density

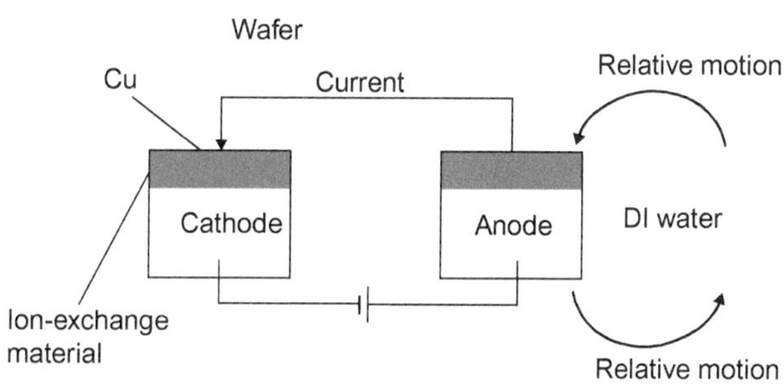

Figure 5.15 Overview of electrode structure

density that is about 1,000,000 times the theoretical value at normal temperature and pressure.

The basic electrode structure plays an important role; the structure comprises a polishing electrode, which polishes copper, and a power-feeding electrode, as shown in Figure 5.15. Both electrodes are placed on the copper side of the wafer to be polished. The ion exchanger is placed on the surface of the main body of each electrode, so that the ion exchanger is in contact with the copper surface. The ultrapure water is supplied to the interface between the wafer and ion exchanger, and is also supplied to the interface between the main body of the electrode and the ion exchanger. The ion density of the ultrapure water is increased by the ion exchanger, and the electric current follows the path:

Feeding electrode → Cu wafer surface → Polishing electrode

as copper is polished on the polishing electrode side by the abovementioned mechanism. The relative movement of the electrode and the wafer is intended to ensure polishing uniformity over the wafer surface.

When polishing over the entire wafer surface, multiple components are placed to expand the total area of the polishing electrode.

The removal rate was examined in detail as a function of the pressure on the polishing surface, the relative speed between the wafer and electrode, the current density, and the electrode area (polishing area). It is hardly influenced by the pressure on the polished surface and the relative speed between the wafer and electrode; these results show that polishing is possible at low pressures and that the removal rate remains constant even if the randomness of the relative movement is intensified to improve uniformity over the surface. In addition, the removal rate increases almost linearly with the current density and electrode area, indicating that the polishing efficiency is approximately constant.

Thus, ECP-DI is expected to be a promising next-generation polishing technique, since a removal rate of 800–1600 nm/min (equal to or almost twice that realized with conventional CMP) is obtained if the current density and the electrode area (polishing area) are optimized.

5.2 ELID GRINDING OF SAPPHIRE – EXPERIMENTAL APPROACH

Single crystal sapphire possesses a unique combination of excellent optical, mechanical, physical and chemical properties which make it invaluable in a wide range of applications such as optics, electronics and high temperature sensors. In many of these applications, super surface finish and high accuracy are required. However, the fine grinding of sapphire is quite challenging because of its low fracture toughness and thermal conductivity, making it sensitive to cracking. Another problem with conventional grinding is the rapid grit wear due to the high hardness of sapphire. In this research, ELID (electrolytic in-process dressing) grinding was performed on sapphire, and the processing parameters and surface finish were evaluated. Experimental studies have been carried out to obtain optimum conditions for ELID grinding of sapphire. The effect of grinding parameters such as wheel speed, feed rate and depth of cut on the surface roughness has been investigated. A mathematical model will be developed for estimating the surface roughness on the basis of experimental results.

5.2.1 Introduction

Sapphire, which is a single-crystal form of α-alumina, is widely used in a range of high-technology applications such as optics, electronics and high temperature sensors because it combines excellent optical, mechanical, physical and chemical properties. In many of these applications, super surface finish and high accuracy are required. Nowadays, grinding is the major manufacturing processe. A metal-bonded super-abrasive wheel is widely used to obtain a mirror surface finish on the advanced ceramic materials. In the grinding processe, low efficiency, high abrasive wear rate, and long wheel dressing time are the main problems. In order to reduce the surface roughness and subsurface damage, grinding wheels with smaller diamond grains are needed, but such wheels are susceptible to loading and glazing, especially when fine grains are used. Periodic dressing is required to minimize the above problems in conventional grinding, which makes the process less efficient. The electrolytic in-process dressing (ELID) technique is a relatively new technology, which can continuously dress the wheel during the grinding process by using electrolysis.

In this section, the ELID technique is applied in order to minimize the problems in conventional grinding of sapphire. Most of the experiments investigated the influence of wheel speed, feed rate, and depth of cut on the ground surface roughness. A mathematical model is developed for estimating the surface roughness on the basis of experimental results.

5.2.2 Experimental Set-Up

The ELID grinding experiments were conducted on a Thompson Creep Feed Grinder equipped with Fanuc and ELID system with a feed accuracy of 1 μm. As shown in Figure 5.16, a copper electrode, covering 1/6 of the perimeter of the grinding wheel, was used for electrolytic dressing. The metal-bonded diamond grinding wheel was connected to the positive pole by contact with a carbon brush. The wheel had a diameter of 305 mm and a width of 10 mm. The gap between the grinding wheel and copper electrode was adjusted to between 0.1 and 0.2 mm.

Figure 5.16 ELID system applied to the grinding wheel

Table 5.1 Experimental System Specifications

Grinding Wheel	Metal-bonded diamond #2000 Diameter: 305 mm, Width: 10 mm
Electrolyte	TRIM229 (1:20), pH: 8.5 ~ 9 Conductivity: 4.26 mS (5%)
Pre-dressing	On (3 μs), off (3 μs)
Electrode	Copper
Workpiece Material	Sapphire (diameter: 76 mm)
Measuring Tool	HOMMEL-TESTER T 1000
ELID Power	60 V, 9A
Pulse Width	6 μs

A coolant, TRIM229, was diluted with distilled water at a ratio of 1:20 and used as electrolyte and coolant for the experiment. TRIM229 is a synthetic coolant concentrate with refractometer factor 1.14 and typical pH operating range of 8.5–9.0. The conductivity of TRIM229 at concentration 5% is about 4.26 mS. A direct current pulse generator was used as the power supply to generate a square pulse wave for the process. The surface finish was measured by the HOMMEL-TESTER T 1000. The method of measuring the surface roughness was the stylus method. The specifications and conditions of the experimental system are listed in Table 5.1.

5.2.3 Chip Thickness Model in ELID Grinding

Material removal in grinding is mainly a chip formation process, in which a work piece surface profile is generated by the grooves left by grain paths. Many attempts have been made to predict the performance of grinding processes by using chip thickness models. The equation for estimating the maximum chip thickness or grain penetration depth in grinding can be expressed as follows:

$$h_{max} = \left[\frac{4}{Cr}\left(\frac{v_w}{v_s}\right)\left(\frac{a_e}{d_e}\right)^{1/2}\right]^{1/2} \tag{5.8}$$

where h_{max} is the maximum undeformed chip thickness, C is the number of active particles or grains per unit area of the wheel periphery, r is the ratio of chip width to average undeformed chip thickness, v_s is the wheel peripheral speed, v_w is the work speed, a_e is the wheel depth of cut, and d_e is the equivalent wheel diameter. The value of r is reported to be in the range of 10–20, and a value of 10 was selected by Mayer. In the case of surface grinding, the equivalent wheel diameter is the wheel diameter. From the above expression, it is clear that the maximum chip thickness depends on both kinematic and wheel parameters.

The number of active particles or grains per unit area, C, can be expressed as follows:

$$C = 4f/\left[d_g^2(4\pi/3v)^{2/3}\right] \tag{5.9}$$

where is the equivalent spherical diameter of diamond particles, is the volume fraction of diamond in the grinding wheel, and is the fraction of diamond particles that actively cut in grinding.

$$d_g(\text{mm}) = 15.2\ M^{-1} \tag{5.10}$$

where M is the grit mesh size.

During the ELID grinding process, due to the formation of an oxide layer, the active grits on the wheel are bonded in the metal oxide matrix, which has a lower strength than the metal bond. The reduction of bond strength reduces the grits' depth of cut, which is the reason for the better surface finish. In ELID grinding, the chip thickness model can be modified by adding a constant k into the existing model, then h_{max} can be expressed as:

$$h_{max} = k\left[\frac{4}{Cr}\left(\frac{v_w}{v_s}\right)\left(\frac{a_e}{d_e}\right)^{1/2}\right]^{1/2} \tag{5.11}$$

where k is the ELID constant that ranges from 0 to 1, and is a function of the dressing current, voltage, current duty ratio, and the coolant properties. Increasing the applied voltage and current duty ratio will increase the formation rate and thickness of the oxide layer on the wheel surface, which will consequently result in a lower value of k and chip thickness. Therefore, in ELID grinding, a thicker oxide layer is preferred when a super surface finish is needed. In addition, k also depends on the properties of the work piece, such as its modulus of elasticity, because the grinding is a wheel/work piece interaction process, in which the properties of the work piece influence the performance of the oxide layer.

The random geometry and probabilistic distribution of the cutting edges on the wheel surface make the grinding process stochastic in nature. Therefore, the random distribution of cutting points on the wheel surface leads to variation in the size of the undeformed chips. The chip thickness can be described by the Rayleigh probability function as:

$$f(h) = \begin{cases} (h/\sigma^2)e^{-h^2/2\sigma^2} & h \geq 0 \\ 0 & h < 0 \end{cases} \tag{5.12}$$

where σ is a parameter which completely defines the probability density function and is a function of work piece material, grinding wheel topography, dynamic conditions, and kinematic conditions.

The expected value is expressed as:

$$E(h) = \sqrt{\frac{\pi}{2}}\sigma \tag{5.13}$$

5.2.4 Surface Roughness Model

The surface roughness, Ra, is the average deviation of the profile from the mean line over the length of assessment. Ra is determined by the following formula:

$$Ra = \frac{1}{l}\int_0^l \left|y - y_{cl}\right| dl \tag{5.14}$$

where y_{cl} is the location of center line, chosen such that the areas above and below the line are equal.

Under the assumptions that back transferring of materials, grain overlapping, etc., do not take place, the surface roughness, Ra, can be calculated from the grooves generated during the grain/work piece interaction by using the probability density function. In Hecker and Liang's model, where the geometrical analysis of the grooves left on the surface was carried out by assuming the groove shape to be triangular from the projection of the assumed conical shape of grains, the expected surface roughness can be mathematically expressed as:

$$\mathrm{E(Ra)} = 0.46\sigma \tag{5.15}$$

Or it can be expressed as a function of expected chip thickness value:

$$\mathrm{E(Ra)} = 0.37E(h) \tag{5.16}$$

In Sanjay and Rao's model, the surface roughness is obtained by assuming the profile of groove generated by an individual grain to be an arc of a circle, in which the expected surface roughness can be expressed as:

$$\mathrm{E(Ra)} = 0.532\sigma \tag{5.17}$$

Or it can be expressed in terms of expected chip thickness value:

$$\mathrm{E(Ra)} = 0.423E(h) \tag{5.18}$$

The assumption that the transverse shape of the grooves is triangular or an arc of circle form two limit conditions, both of which may not be true. As we know, the triangle is the polygon possessing the least number of line segments, and the arc of a circle is a polygon in which the number of line segments is infinite. The real condition must be between these two limit conditions, so it is more realistic to take the average value of these two models, which can be expressed as:

$$E(R\mathrm{a}) = \frac{0.37 + 0.423}{2} E(h) = 0.397(h) \tag{5.19}$$

Substituting the chip thickness model into the above equation, we get:

$$\mathrm{E(Ra)} = 0.397\mathrm{k}\left[\frac{4}{\mathrm{Cr}}\left(\frac{\mathrm{v_w}}{\mathrm{v_s}}\right)\left(\frac{\mathrm{a_e}}{\mathrm{d_e}}\right)^{1/2}\right]^{1/2} \tag{5.20}$$

This shows a proportional relationship between the surface roughness and expected chip thickness value. The variable C needs to be analyzed in more detail to estimate the surface roughness. Also, more study needs to be done in order to determine the ELID constant k. However, although the variables C and k are unknown in our study, the surface roughness model can still be used to evaluate the effect of kinematic parameters on the surface roughness for a specific grinding wheel, work piece material, and ELID conditions.

5.2.5 Experimental Results and Discussion

A 33 factorial design was used to evaluate the ELID grinding of sapphire. The effect of kinematic parameters such as wheel speed, feed rate and depth of cut on the surface roughness was investigated. The depths of cut used were 1, 2 and 3 μm per pass. The rpms used were 600, 800 and 1000, which have the corresponding wheel speeds 9.6 m/s, 12.8 m/s and 16 m/s, respectively. The feed rates applied were 0.5 m/min, 1 m/min, and 2 m/min. ELID grinding was conducted under $E_0 = 60$ V, $I_{initial} = 9$ A, $\tau on = \tau off = 3$ μs, gap: 0.1 ~ 0.2 mm, and pre-dressing time: 30 min. For each set of parameters, 5 repeated grinding tests were run, and 15 roughness measurements were taken in total.

The results of these grinding trials show that the achieved surface roughness can be as good as 50 nm when the super-abrasive wheel and ELID technique are used. The effects of wheel speed, feed rate, and depth of cut on surface finish in ELID grinding of sapphire are shown in Figures 5.17 to 5.21. They show that wheel speeds of 9.8 m/s to 16 m/s have a significant effect on the surface finish of sapphire. Feed rates from 0.5 m/min to 2 m/min also have an effect but this is not as significant as wheel speed.

The ANOVA analysis of the experimental results is shown in Table 5.2.

The Model F-value of 318.35 implies the model is significant. There is only a 0.01% chance that a 'Model F-Value' this large could occur due to noise. Values of "Prob > F" less than 0.05 indicate model terms are significant.

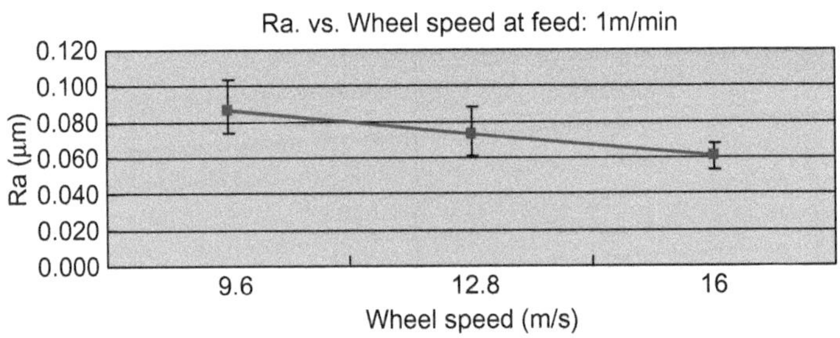

Figure 5.17 Effect of wheel speed on Ra

The regression equation was determined as follows:

$$\begin{aligned}\mathrm{Ra} = {} & 0.11344 \\ & -4.00347\mathrm{E}\text{-}003 \times V_s\ (\mathrm{m/s}) \\ & + 2.89312\mathrm{E}\text{-}003 \times V_w\ (\mathrm{m/min}) \\ & + 6.84815\mathrm{E}\text{-}003 \times \mathrm{DOC}\ (\mu\mathrm{m})\end{aligned} \tag{5.21}$$

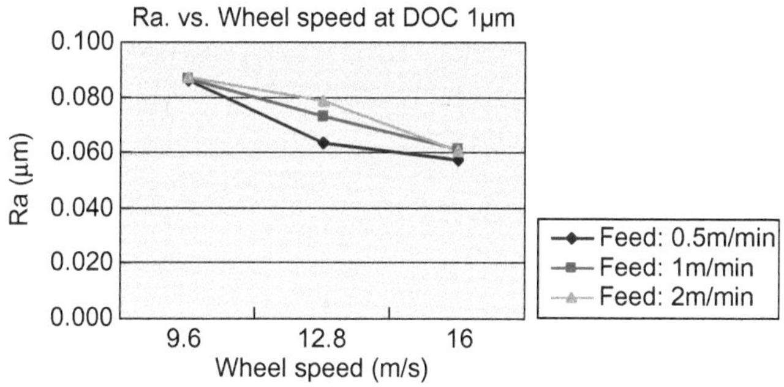

Figure 5.18 Effect of wheel speed on Ra

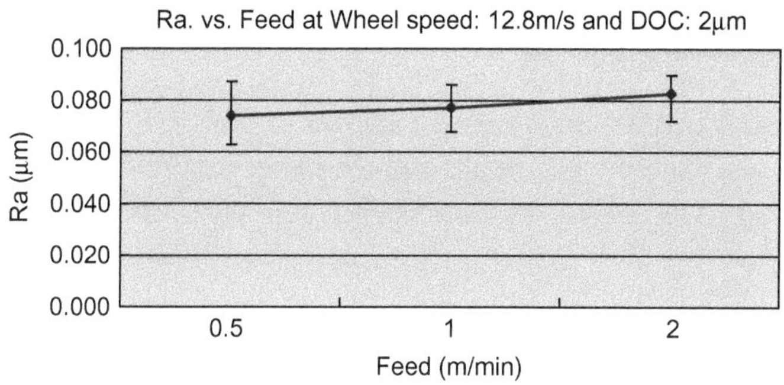

Figure 5.19 Effect of feed rate on Ra

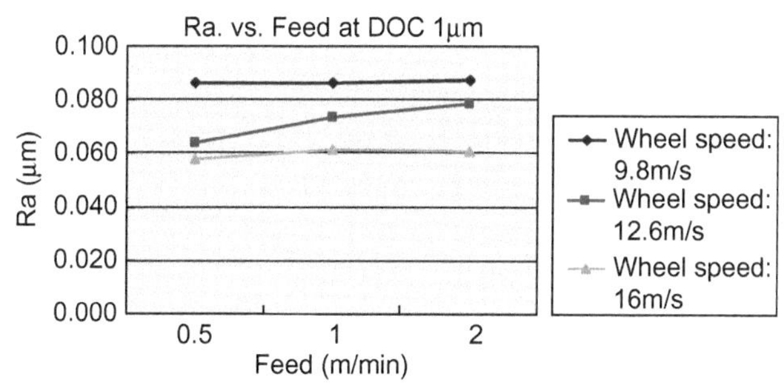

Figure 5.20 Effect of feed rate on Ra

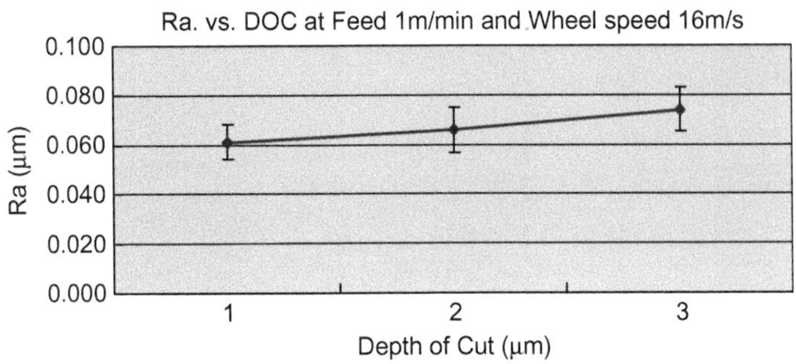

Figure 5.21 Effect of depth of cut on Ra

Table 5.2 ANOVA for 3^3 Factorial Design

Source	Sum of Squares	DOF	Mean Square	F Value	Prob > F
Model	0.058	3	0.019	318.35	
A: wheel speed	*0.044*	1	*0.044*	*726.01*	*<0.0001*
B: feed rate	*1.318E-3*	1	*1.318E-3*	*21.60*	*<0.0001*
C: depth of cut	*0.013*	1	*0.013*	*207.45*	*<0.0001*
Residual	0.024	401	6.104E-5		
Lack of Fit	*2.358E-3*	23	*1.025E-4*	*1.75*	*0.0182*
Pure Error	*0.022*	378	*5.851E-5*		
Cor Total	0.083	404			

The estimated response surface is shown in Figures 5.22 to 5.24.

From the experimental results and the regression model, it is clear that the depth of cut, wheel speed, and feed rate affect the surface finish in the ELID grinding of sapphire, which is consistent with the theoretical surface roughness model. Increasing wheel speed, reducing the feed rate, or reducing the depth of cut reduces the chip thickness, and thus improves the quality of surface finish.

5.2.6 Conclusions and Recommendations

Based on the theoretical analysis and experimental results, the following conclusions can be derived regarding the ELID grinding of sapphire:

1. The formation of an oxide layer on the wheel surface reduces the bond strength, and thus reduces the grit depth of cut, which results in a better surface finish.
2. In this study, the surface roughness (Ra) of the ground sapphire is as good as 50 nm when super-abrasive wheel and ELID techniques are used.
3. This study shows that increasing the wheel speed (from 9.8 m/s to 16 m/s), reducing the feed rate (from 2 m/min to 0.5 m/min), or

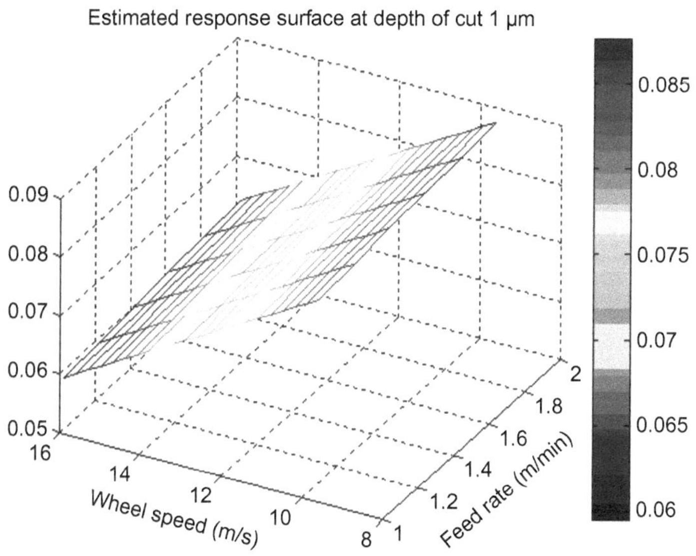

Figure 5.22 Estimated response surface for depth of cut 1 μm

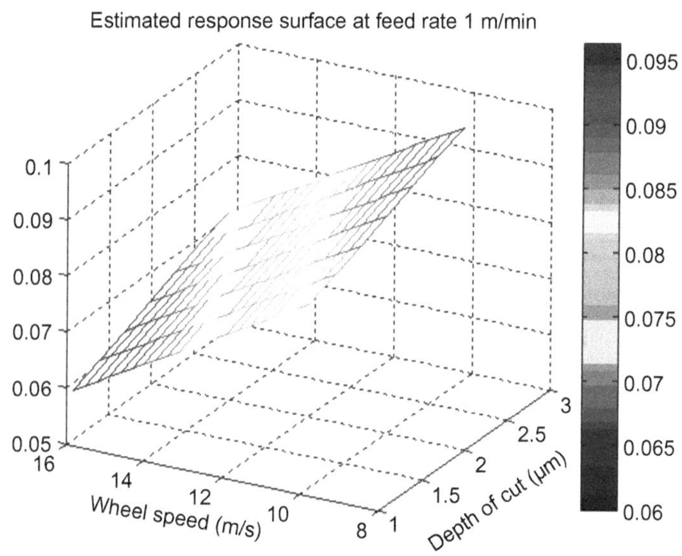

Figure 5.23 Estimated response surface for feed rate 1 m/min

reducing the depth of cut (3 μm to 1 μm) increases the quality of surface finish in ELID grinding sapphire.

4. In this study, the wheel speed tends to have a larger effect than the feed rate.

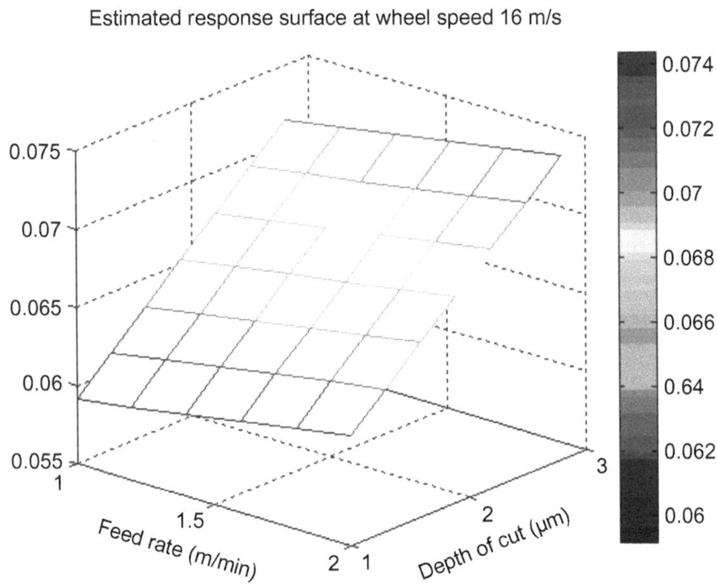

Figure 5.24 Estimated response surface for wheel speed 16 m/s

5.3 ELID GRINDING OF SAPPHIRE WITH ACOUSTIC EMISSION MONITORING

Single crystal sapphire is of significant interest due to its combination of excellent optical, electrical, and mechanical properties. However, fine grinding of sapphire is quite challenging because of its high hardness and low fracture toughness, making it sensitive to cracking. Wheel loading is a common problem in conventional grinding of hard and brittle materials. ELID grinding shows great promise for achieving a mirror surface finish relatively efficiently. The ELID grinding of sapphire was investigated using acoustic emission. The effects of processing parameters and acoustic emission signals on surface finish were evaluated. Correlations were found between the dressing current intensity, surface finish and acoustic emission signals. A smoother surface was obtained using a higher dressing current at the cost of a higher wheel wear rate. The wheel wear mechanism in ELID grinding of sapphire was dominated by bond fracture because the bond strength is reduced by electrolysis. Results indicate that the acoustic emission technique has potential to be used for monitoring the ELID grinding process, detecting the condition of the grinding wheel, and investigating the mechanisms of ELID grinding.

5.3.1 Introduction

A new technique, electrolytic in-process dressing (ELID) grinding, shows great promise in overcoming the problems of conventional grinding of hard and brittle materials. This technology provides dressing of the metal-bonded wheels during the grinding process, while maintaining sharp abrasives from the super-abrasive wheels. ELID grinding was introduced by Murata et al. to grind ceramics with metal-bonded diamond wheels of grit size less than #400. It was further improved by Ohmori and Nakagawa in 1990 for fine grinding. Over the past 10 years, ELID has been used to grind various hard and brittle materials, for which it is very efficient. It has been reported that ductile-mode grinding with smoother surface finish and lower subsurface damage was obtained on silicon wafers by using super-abrasives and the ELID technique. The improved surface finish was explained by the formation of a beneficial oxide layer on the grinding wheel surface, which acts as a spring damper, reducing the grit depth of cut, and absorbing the vibrations in the ELID grinding process. Force was commonly used by previous researchers to monitor the ELID grinding process and to investigate its mechanisms. A significant reduction of grinding force has been reported with the application of ELID. Lim investigated the influence of dressing parameters on ELID grinding, and pointed out that the grinding force is reduced with the increase of dressing current duty ratio. Furthermore, he also found that the tangential force is unstable due to the microfracture of the oxide layer from the wheel surface.

Acoustic emission (AE) monitoring of the grinding process was investigated by Dornfeld and Cai. They discovered that the AE RMS (root mean square) is most strongly influenced by the chip thickness determined by the depth of cut, and that it is more sensitive than force measurement to detect the wheel/work piece contact and spark-out. They also showed that the AE energy increases with the increase of wheel loading and wear. This was further confirmed by Inasaki. Stephenson studied the AE RMS and AE raw signals in monitoring ELID grinding of BK7 glass and Zerodur. Higher AE energy was found at gentle dressing parameters in ELID grinding. The main advantage of AE monitoring during the grinding process is that the frequency of AE signals ranges from 20 kHz to 1 MHz, which is above most structural natural frequencies, such as frequencies of machine vibrations and other mechanical noises. It is known that the AE RMS is a relatively reliable parameter in AE monitoring of the grinding process. However, its main limitation is unpredictable changes in signal linearity and amplitude with time. These fluctuations may be caused by minor and unpredictable changes in grinding conditions. In order to overcome this problem, an innovative grit mapping technique was proposed by Gomes de Oliveira and Dornfeld to give information about the grinding wheel topography and the dressing operation. They used "Fast AE RMS analysis" to characterize and map each abrasive-tool contact.

In this study, the acoustic emission technique was used to monitor ELID grinding of sapphire. The influence of grinding conditions and dressing parameters on surface finish and acoustic emission signals was investigated. AE RMS was used as the parameter to analyze the AE signals generated in the ELID grinding process of sapphire.

5.3.2 Experimental Set-Up

The experiment was performed on a Thompson Creep Feed Grinder equipped with Fanuc CNC, an ELID system, a Dynamic Balance System, and an acoustic emission system. A copper electrode, covering 1/6 of the perimeter of the grinding wheel, was used for electrolytic dressing. A cast iron bond diamond grinding wheel was connected to the positive pole through the contact of a carbon brush. The wheel had a diameter of 305 mm, a width of 11 mm, and grit size no. #4000 (average grain size 4 μm). The gap between the wheel and electrode was maintained at 0.2 mm. A special coolant, TRIM C270, was diluted with distilled water to a ratio of 1:20 and used as an electrolyte and coolant for the experiment. A direct current pulse generator was used as the power supply to generate a square pulse wave for the ELID process. The surface finish was measured by the HOMMEL-TESTER T 1000. The stylus method was used to measure the surface roughness. AE testing was performed with the aid of the MISTRAS 2001 system, which was developed by the Physical Acoustic Corporation. A piezoelectric type transducer, R15, was used in the study. The sensor was attached to the surface of the work piece holder. The acoustic emission signals detected by the sensor were amplified to usable voltage levels by a 2/4/6 type preamplifier. The 2/4/6 type preamplifier was set at a gain of 100(40 dB). The AE signals detected by the sensor were first amplified by the preamplifier and then fed to the AEDSP-32/16 board where the processing of the raw AE signals was done and stored for later analysis. A frequency of 4 MHz sample rate was selected for signal acquisition. The experimental configuration is shown in Figure 5.25.

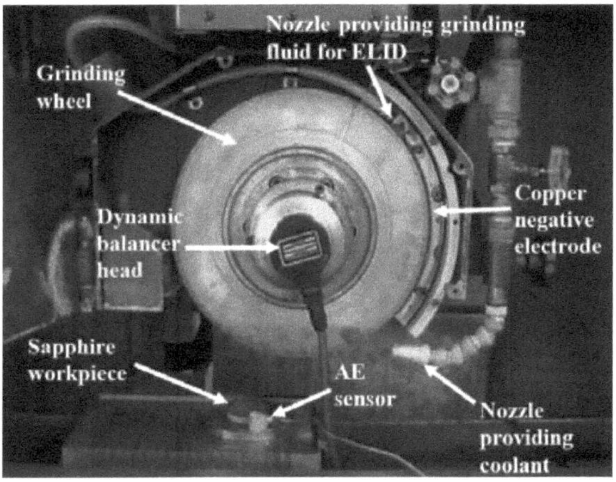

Figure 5.25 Configuration of experimental set-up

5.3.3 Results and Discussion

A series of ELID grinding tests were conducted to investigate the effect of grinding conditions on surface finish of the ground sapphire work piece and AE signals under the following conditions:

- Wheel speed: 16 m/s (1000 rpm)
- Feed rate: 0.5 m/min
- Depth of cut: 1 μm, 3 μm, 5 μm, 7 μm, 9 μm, 11 μm, 13 μm, 15 μm
- Initial ELID current: 5 A, 7 A, 9 A

5.3.3.1 Pre-Dressing and Oxide Layer Thickness

ELID pre-dressing of the grinding wheel is a preliminary stage, which is required before starting ELID grinding. During this stage, the abrasive grains are made to protrude by removing the metal bond material from the wheel surface through electrolysis. At the same time, a beneficial oxide layer forms on the wheel surface. The diameter of the grinding wheel increases after pre-dressing due to the formation of this oxide layer. In the case of a cast iron bonded grinding wheel, the cast iron bond is first ionized into Fe^{2+}. The ionized Fe then reacts with OH^- to form hydroxides, which further transform into oxide (Fe_2O_3), forming an insulating oxide layer on the grinding wheel surface. This insulating layer causes a decrease in current flow and an increase in voltage, known as "anode effect." The oxide layer growth rate decreases with the decrease in the current flow, and the electrical resistance in the electrolytic circuit

Figure 5.26 Yellow-brown oxide layer formed on cast iron bonded wheel surface

increases with the increase of oxide layer thickness. The oxide layer formed on the cast iron bonded wheel surface is yellow-brown, as shown in Figure 5.26.

The thickness of the oxide layer during ELID pre-dressing was investigated at three different pre-dressing currents. The initial dressing currents were 5 A, 7 A, and 9 A, respectively. The gap between the grinding wheel and the negative electrode was maintained at 0.2 mm. The applied voltage was fixed at 60 V. The cast iron bonded grinding wheel was pre-dressed for 15 minutes in all experiments. The oxide layer thickness is defined as the value of the wheel radius with the oxide layer minus that without the oxide layer. The oxide layer thickness was measured by reading the X coordinates of the CNC grinder before and after a period of electrolytic dressing when the grinding wheel surface touched the same work piece surface. The AE system was used to detect the contact between the wheel surface and the work piece surface. The results show that the oxide layer thickness generated in ELID pre-dressing increases with increased dressing current, as shown in Figure 5.27. According to Faraday's Law, the dissolution rate of a metal anode in electrolysis is proportional to the applied current. Clearly, the current is the determining parameter in the metal removal and oxide layer formation processes. After dressing, the cast iron bond on the wheel surface is replaced by its softer oxide (Fe_2O_3).

During grinding, the insulating layer and abrasive grains are scraped off and removed, resulting in a decrease in the wheel's electrical resistance. As a result, the electrical current through the wheel and the

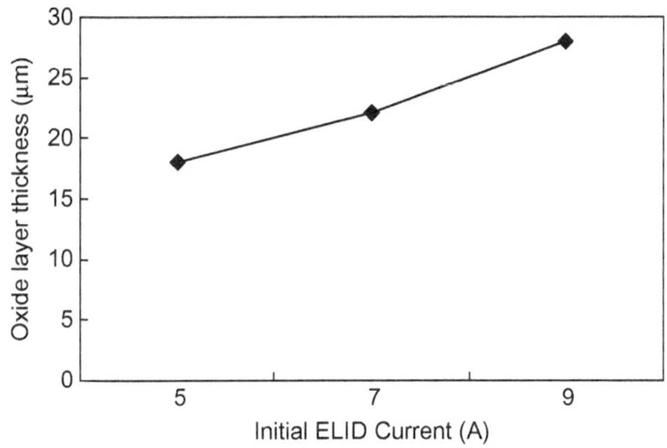

Figure 5.27 Effect of electrolytic dressing current on oxide layer thickness

electrode increases and electrolysis restarts. This dressing cycle repeats during the ELID grinding process. In addition to continuously exposing new, sharp diamond grains, ELID can also improve the surface finish quality of the ground work piece.

5.3.3.2 Effect of Electrolytic Dressing Current on Surface Finish

A series of ELID grinding trials were conducted under the above grinding and pre-dressing conditions. The average surface roughness (Ra) was measured after each grinding test. The influence of initial ELID current on the average surface roughness of the ground sapphire work piece is shown in Figure 5.28. As shown in this figure, increasing the ELID current slightly improves the surface finish quality of the ground sapphire work piece. Because of the soft oxide layer that forms on the wheel surface, the active grits are bonded in a metal oxide matrix that has a lower strength than the original metal bond. The soft oxide layer acts as a spring damper, which plays a role in absorbing the vibration and reducing the grit depth of cut during ELID grinding. The chip thickness model in ELID grinding can be modified by adding a constant k to the existing model and h_m can be expressed as:

$$h_m = k\left[\frac{4}{Cr}\left(\frac{v_w}{v_s}\right)\left(\frac{a_e}{d_e}\right)^{1/2}\right]^{1/2} \tag{5.22}$$

where k is the ELID constant, which ranges from 0 to 1, and is a function of dressing current, voltage, current duty ratio, and coolant properties. The experimental results show that increasing the electrolytic

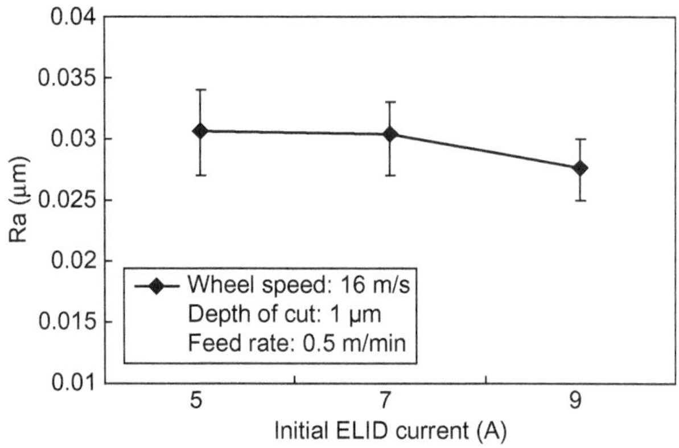

Figure 5.28 Effect of electrolytic dressing current on the average surface roughness

dressing current increases the formation rate and thickness of the oxide layer on the wheel surface, which consequently results in a lower value of k and chip thickness. ELID grinding with a thicker oxide layer produces a smoother surface finish. It is known that ductile-mode grinding can be achieved when the maximum chip thickness or grit depth of cut is less than the critical depth of cut. The calculated critical depth of cut for sapphire is approximately 30 nm. The application of ELID in grinding of sapphire provides a method to control the grit depth of cut. By controlling the dressing parameters, it is possible to achieve ductile-mode grinding by using a super-abrasive grinding wheel and the ELID technique, which could eliminate the requirement for finishing processes such as lapping and polishing.

5.3.3.3 Effect of Electrolytic Dressing Current on AE Signals

From the above experimental results, we know the influence of dressing current on the oxide layer thickness and its role on surface finish. This section investigates the effect of dressing current on AE signals. The ELID grinding trials were monitored by the AE system. Figure 5.29 plots the influence of initial ELID currents on the average AE RMS. The average AE RMS was slightly affected by the dressing current, and found to be lower at higher dressing current. These results indicate that AE energy decreases with an increase in dressing current. This is because the oxide layer formed on the wheel surface reduces the bond strength and the grit depth of cut during ELID grinding. The increase in dressing current

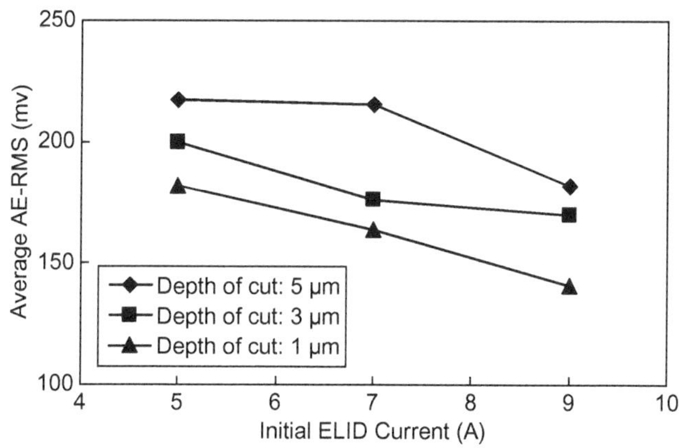

Figure 5.29 Effect of electrolytic dressing current on the average AE RMS

increases the oxide layer thickness and decreases the chip thickness. Since the AE RMS is a function of chip thickness, the increase in ELID current decreases the average AE RMS.

5.3.3.4 Effect of Depth of Cut on Surface Finish and AE Signals

This section investigates the influence of depth of cut on surface finish of the ground sapphire work piece and the AE RMS during ELID grinding. Figure 5.30 plots the influence of depth of cut on surface finish of sapphire at a wheel speed of 16 m/s, a feed rate of 0.5 m/min, and an initial ELID current of 5 A over a range of depth of cuts ($a = 1-15\ \mu m$). The average AE RMS during grinding is shown in Figure 5.31. The surface roughness and average AE RMS generated during ELID grinding were found to be strongly affected by depth of cut. They both increase with an increase in depth of cut because they are proportional to the chip thickness determined by depth of cut. Furthermore, both figures show that there is a transition point between $a = 7\ \mu m$ and $a = 9\ \mu m$. Increasing the depth of cut from 1 μm to 7 μm slightly increases the surface roughness of the ground sapphire work piece. However, the value of Ra increases more quickly with the increase of grinding depth of cut from $a = 7\ \mu m$ to $a = 15\ \mu m$. The AE RMS shows the same trend.

It has been identified by Lim and Fathima that the dominant wear mechanism, when the bond strength is reduced during ELID grinding, is bond fracture which is characterized as grain pull-out and oxide layer breakage from the wheel surface. The transition point may be caused by the combined effects of the oxide layer breakage and the subsequent

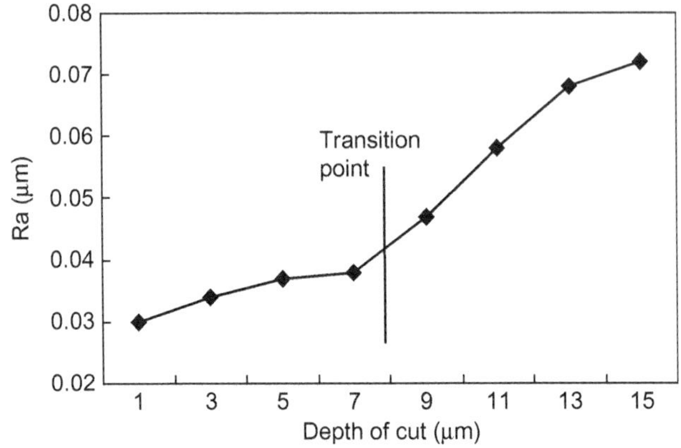

Figure 5.30 Influence of depth of cut on surface finish

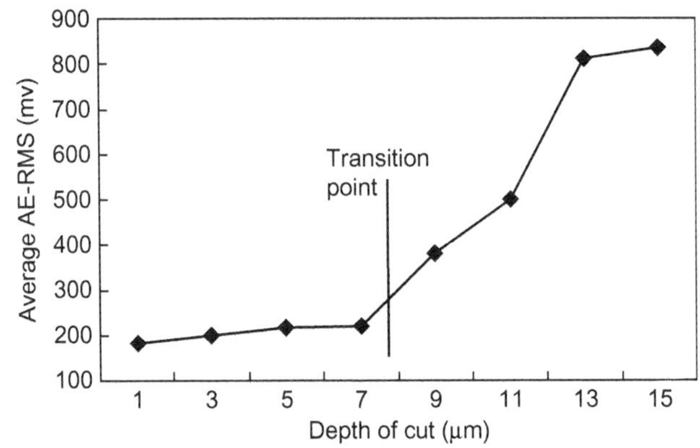

Figure 5.31 Influence of depth of cut on average AE RMS

transition of grinding mechanism from ductile-mode grinding to brittle-mode grinding due to the high depth of cut and oxide layer breakage. Ductile-mode grinding can be realized in ELID grinding under two conditions: a small grinding depth of cut and a relatively thick oxide layer. The smoother surface finish obtained at low depth of cut may be due to ductile-mode grinding. It can be seen from Figures 5.30 and 5.31 that both the surface roughness and AE RMS increase more quickly at grinding depths of cut higher than 7 μm, because the oxide layer breakage at high depth of cut occurs, resulting in a transition of the grinding mechanism from ductile-mode grinding to brittle-mode grinding.

The oxide layer breakage can be further verified by examining the electrical current in the ELID grinding process. After the ELID grinding starts, the oxide layer and the abrasives bonded in the metal oxide matrix begin to wear due to the interactions between the oxide layer and the sapphire work piece. This wear consequently results in an increase in the electrical current through the wheel and the electrode. The wear rate of the oxide layer depends on the grinding conditions. When the grinding depth of cut is low, the wear rate of the oxide layer and abrasive grains are small due to the gentle grinding force. The breakage of the oxide layer will take place at a relatively high depth of cut due to the large grinding force. As shown in Figure 5.32, the current flow increases slowly when the grinding depth of cut is less than 7 μm. However, the electrical current increases rapidly at a grinding depth of cut higher than 7 μm, due to the breakage of the oxide layer from the wheel surface.

A key issue in ELID grinding is sustaining the balance between the removal rate of the bond metal by electrolysis and the wear rate of diamond abrasives. In the grinding process, there are two major forces acting in opposition on the grit; the grit holding force exerted by the bond material of the grinding wheel, and the grinding force. The grinding force increases with an increase in depth of cut. When this force is larger than the grit holding force, the grits will be pulled out. Compared to the abrasives in conventional grinding, the abrasives in ELID grinding are more susceptible to pull-out due to the reduction in bond strength by electrolysis. Due to the formation

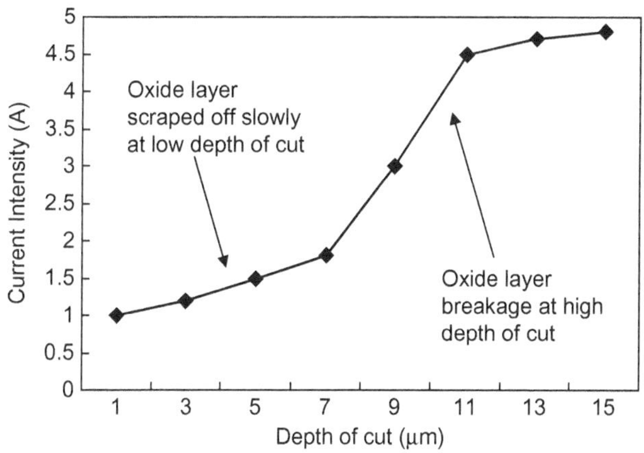

Figure 5.32 Dressing current change in ELID grinding with the increase of depth of cut

of an oxide layer in ELID grinding, the active grits on the wheel surface are bonded in a metal oxide matrix, which has a lower strength than the original metal bond. ELID grinding at a large depth of cut will increase the chance of grain pull-out and oxide layer breakage from the wheel surface. This phenomenon can be detected by AE signals, and is characterized as a rapid increase in the AE RMS value. The oxide layer breakage can also be noticed by examining the electrical current in ELID grinding. A rapid increase in the electrical current indicates the oxide layer breakage. Oxide layer breakage indicates that the wear rate of the oxide layer and abrasives is much higher than the electrolytic dressing rate. Therefore, in order to sustain a relatively stable ELID grinding process, it is better not to perform ELID grinding at a depth of cut that could cause oxide layer breakage.

5.3.4 Conclusions and Recommendations

Based on the experimental results discussed above, several conclusions and recommendations can be obtained regarding ELID grinding of sapphire:

1. In ELID grinding, the soft oxide layer acts as a spring damper, which plays a role in absorbing the vibration and reducing the grit depth of cut. Increasing the dressing current increases the oxide layer formed on the wheel surface. The surface finish also improves as the dressing current increases.
2. The AE RMS was found to be affected by the electrolytic dressing current. The AE RMS value was observed to be higher at lower ELID currents.
3. The average AE RMS increases with an increase in depth of cut, because the average AE RMS is proportional to the chip thickness determined by depth of cut.
4. Oxide layer breakage occurs at high depths of cut in ELID grinding. This causes a rapid increase in the AE RMS value, which indicates the transition of the grinding mechanisms.

5.4 NOVEL BELL-JAR SHAPED, SEALED, ATMOSPHERE CONTROLLED CMP MACHINE, AND PRECISION PROCESSING OF VARIOUS FUNCTIONAL MATERIALS

5.4.1 Introduction

There are a number of materials that could be used for the fabrication of electronic components. In order to bring out the maximum potential of either silicon, oxide, metal, sapphire or ceramics, it is essential that they are processed to high precision, high quality and high efficiency. Since new functional materials for fabricating new components and systems will continue to be developed, versatile processing technology becomes indispensable. In the ultra-precision processing methods that turn functional materials into high performance electronic components, loose abrasive processing is ideal for minimizing the amount of processing. Lapping is applied to rough processing, and polishing or CMP (Chemical Mechanical Polishing) to the finishing process. This finishing process is considered critical in processing high performance components.

Polishing, on which CMP is based, is conducted while the area to be processed is exposed to the atmosphere. However, CMP, which is a combination of chemical and mechanical actions, needs be conducted in a controlled environment in order to secure stable processing characteristics. Taking into account the advent of novel functional materials, and given that the processing environment is controlled, it has become essential to investigate the effect of processing conditions on CMP, which has driven us to develop a novel CMP technology.

This section introduces a new bell-jar type CMP machine, which is under development as part of our study of planarization CMP technology. This machine is designed such that the processing unit where CMP is performed is hermetically sealed within a bell-jar (chamber) to enable control of the ambient atmosphere around the processing area. In these studies, CMP was performed under various ambient atmospheres using several work pieces for opto-mechatronics including difficult materials such as SiC and sapphire.

5.4.2 New Bell-Jar Type CMP Machine and Experimental Methods

5.4.2.1 A Closed Bell-Jar Type CMP Machine, Designed to Control Processing Atmosphere

The need to develop this new polishing machine has arisen from the sensitivity of the CMP process to ambient conditions, such as temperature and presence of chemicals. To put it differently, conventional open type processing methods should be reviewed in order to be able to achieve stable processing characteristics, and particularly in the case of CMP, the ambient atmosphere around the processing unit should be controlled.

A prototype compact polishing machine was devised, featuring the ability to change and control the ambient atmosphere of the processing unit. This machine is termed a "bell-jar type atmosphere controlled polishing machine."

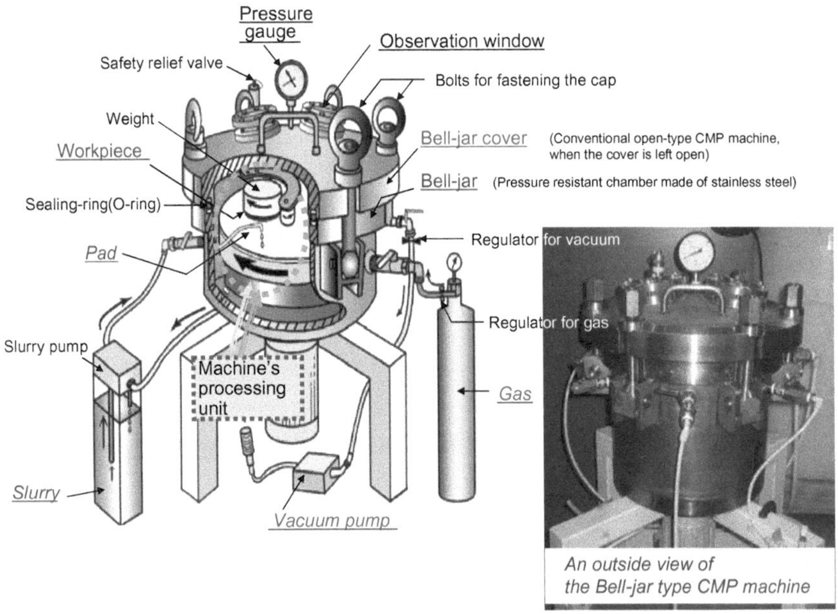

Figure 5.33 Exterior view and structure of bell-jar CMP machine

A photograph (main body excluding controller) and the basic specifications of this new machine are shown in Figure 5.33. The inside of the bell-jar (stainless steel pressure-proof chamber) can be pressurized by charging gases in the range 0 to 1000 kPa. The chamber can also be depressurized down to −100 kPA with a vacuum pump. A ring-type polishing method is adopted, as in conventional polishing machines. The pressure-proof bell-jar is equipped with a manometer, gas supply inlet, exhaust outlet, inspection window (quartz glass), and slurry supply and discharge pipes. The dead weight method is employed for applying the process pressure to the work piece. The rotation rate of the pad and platen can be set at any speed between 10 and 100 rpm by the controller. A constant amount of slurry can be supplied onto the pad (ϕ200 mm) by a circulating pump.

5.4.2.2 Experimental Method and Processing Conditions

Materials of different stability, including silicon, metal copper film (electroplated Cu layer), silicon dioxide film (P-TEOS) and sapphire (Al_2O_3) were investigated.

The pressure inside the bell-jar was adjusted to normal (regular CMP), depressurized (vacuum) or high pressure. Air and oxygen were

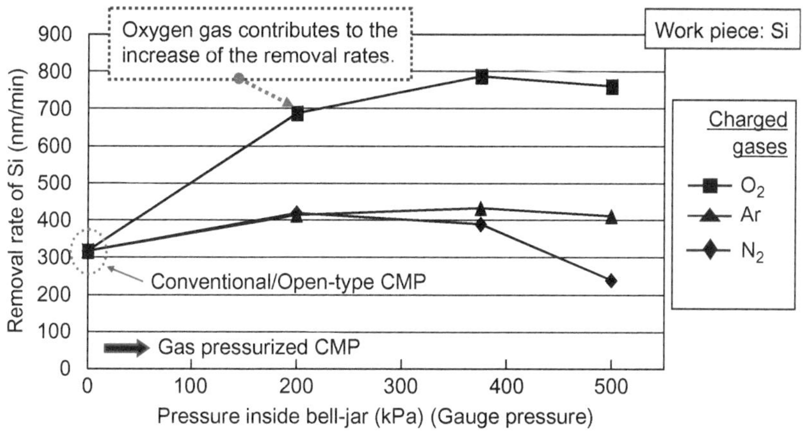

Figure 5.34 Relationships between Si removal rates and ambient pressures applied with various gases

used, along with the inert gases like nitrogen and argon, as the ambient gases inside the bell-jar to consider the oxidation of work pieces during the CMP process. By changing gas composition and pressure in the bell-jar, the processing characteristics of each work piece have been obtained.

Colloidal silica (Compol-80/Fujimi Incorporated) for silicon and sapphire, fumed silica (SS-25/Cabot) for silicon dioxide film, and alumina slurry (EP-C5001/Cabot) for Cu were selected as slurries. Usually, hydrogen peroxide (H_2O_2) is added to the slurry for Cu, but not in this case, since the effect of dissolved oxygen was being investigated. A foamed polyurethane pad with an elastic film in the back (IC1400/perforate type) was used.

5.4.3 Experiment Results – Characteristics of Polishing by Bell-Jar Type CMP Machine

5.4.3.1 Mechanism and Basic Polishing Characteristics

Figure 5.34 shows the effect of gas type on the processing characteristics of silicon wafers. When the bell-jar is filled with air or oxygen, the processing rate increases with internal pressure, but processing rates remain almost unchanged in the presence of inert gases, or decrease in the presence of nitrogen. These results suggest that oxygen is one factor that promotes processing rates. Under vacuum conditions the processing rates also increased, which is likely attributable to the fact that heat generated by friction could be dispersed less easily, increasing the effect of mechanical action on the wafer surface.

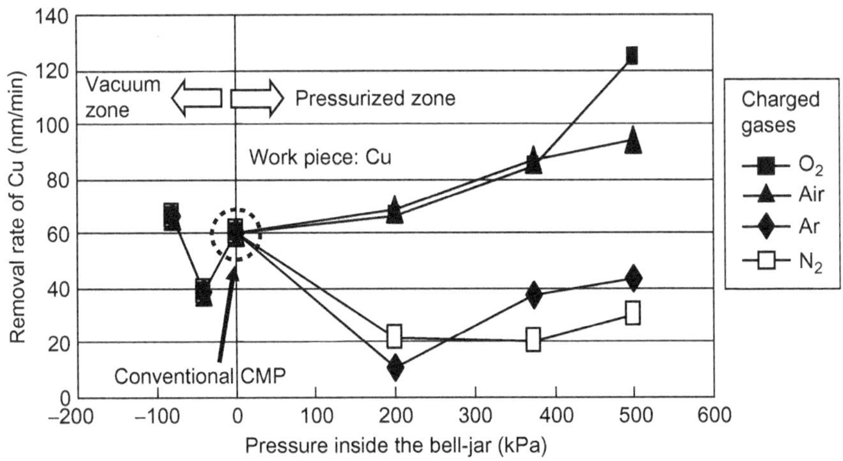

(Slurry: alumina (pH4) (Without H_2O_2), Polishing pressure: 500 g/cm^2, Revolution number: 90 rpm)

Figure 5.35 Relationships between the removal rates of Cu and the ambient pressures charged with various gases

The processing characteristics of easily oxidizable copper wafers showed bigger changes in processing rates than those of Si wafers (see Figure 5.35). When oxygen or air was used, the processing rate increased along with the pressure inside the bell-jar, as in the case of silicon. However, when inert gases were used, processing rates dropped sharply. This suggests that inert gases do not form an oxide layer over the Cu surface, but rather serve to inhibit the acceleration of processing. These results support the processing model of Cu CMP, which can be summarized as: "a Cu oxide layer is first formed on the Cu surface, and then mechanically removed by the abrasives and pad."

Figure 5.36 is a summary of the processing rates obtained for silicon dioxide film (SiO_2), silicon (Si), Cu film, and sapphire at 500 kPa inside the bell-jar, filled with air, oxygen, nitrogen and argon, and at −90 kPa (vacuum). The ratio, α, of processing rate with bell-jar type CMP to rate with conventional CMP is shown in the diagram, which indicates the degree to which processing characteristics vary with work piece type and atmosphere present. Oxygen at high pressure can increase this ratio by 1.5 times for all work pieces.

On the other hand, processing rates do not increase when the bell-jar is filled with inert gases. When the ambient atmosphere inside the bell-jar

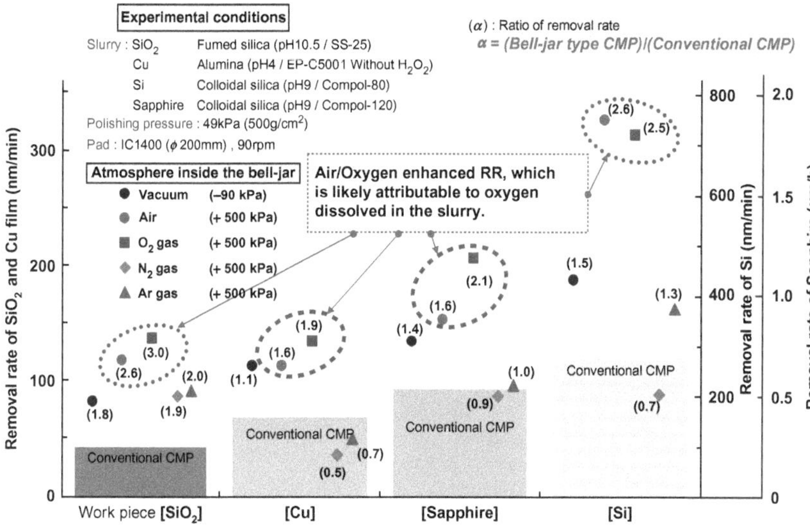

Figure 5.36 Relationship of removal rates between bell-jar CMP with different atmospheres and conventional CMP

is air or oxygen, the possibility of gases working directly on the work pieces is small, due to the presence of slurry. Assuming that gases have an effect on the work pieces through the slurry, the acceleration of the processing rate is dependent on the oxygen dissolved in the slurry. Consequently, it is considered that when high pressure oxygen is charged into the bell-jar, the level of oxygen dissolved in the slurry increases, which promotes hydration reactions, increasing processing rates of silicon and oxide films. The processing mechanism of sapphire is also based on hydration reactions. In the case of silicon, under an ambient atmosphere of high-pressure oxygen, the oxygen becomes dissolved in the slurry, which increases the formation of groups such as siloxane.

5.4.3.2 Polishing Characteristics of SiC Single Crystals in a Bell-Jar Type CMP Machine

Single crystal SiC wafers are considered difficult to process, and are currently gaining attention as a material suitable for power devices, so they were considered suitable subjects for these high efficiency, high quality polishing experiments.

Figure 5.37 shows the evolution of semiconductor power devices using Si, SiC, GaN and diamond. As shown, Si has a theoretically limited

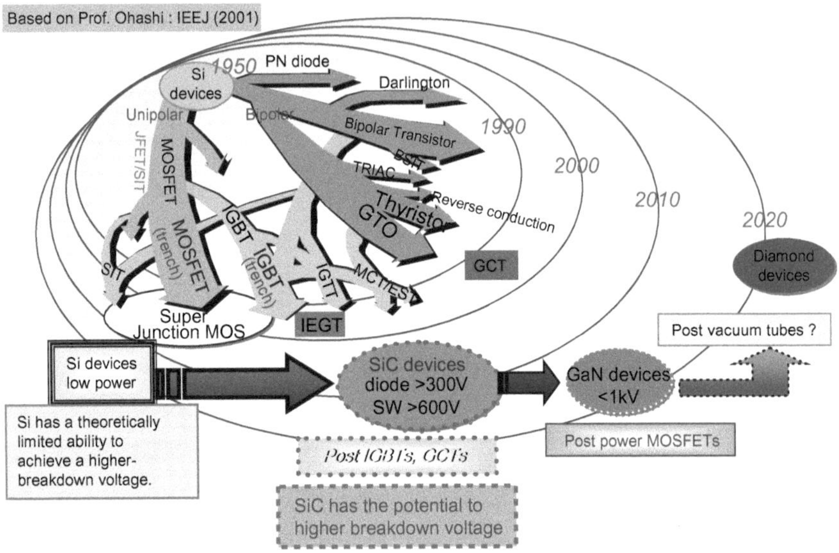

Figure 5.37 Evolution of semiconductor power devices

ability to achieve a high breakdown voltage, but that of SiC could exceed it. Figure 5.38 shows potential fields and market size for SiC power devices as post-silicon devices. Breakdown voltages and current ratings 10 to 100 times those of Si could be realized.

SiC wafers are hence anticipated to become a next generation power device, suitable for cars, trains, airplanes and so on. However, they are very stable chemically and mechanically, which makes polishing extremely difficult. Figure 5.39 shows an example of the traditional CMP characteristics of SiC single crystals; the crystal plane (0001) only is used, and the polishing rate is less than 0.1 μm per hour. This is several hundred times less than that for silicon.

As for the previous experiments, the bell-jar was filled with oxidizing gas (oxygen or air). The processing rates of SiC wafers increased when the pressure of oxidizing gas was increased. In particular, when oxygen was used, the processing rate increased to 2.5 times that of conventional CMP at 500 kPa. As anticipated, this proved that oxygen plays a very important role in the polishing of single crystal SiC wafers.

These results indicate that improved CMP conditions could be determined for SiC wafers. From this viewpoint, we focused on the photocatalytic reactions of titania (TiO_2) particles, shown in Figure 5.40. We added fine

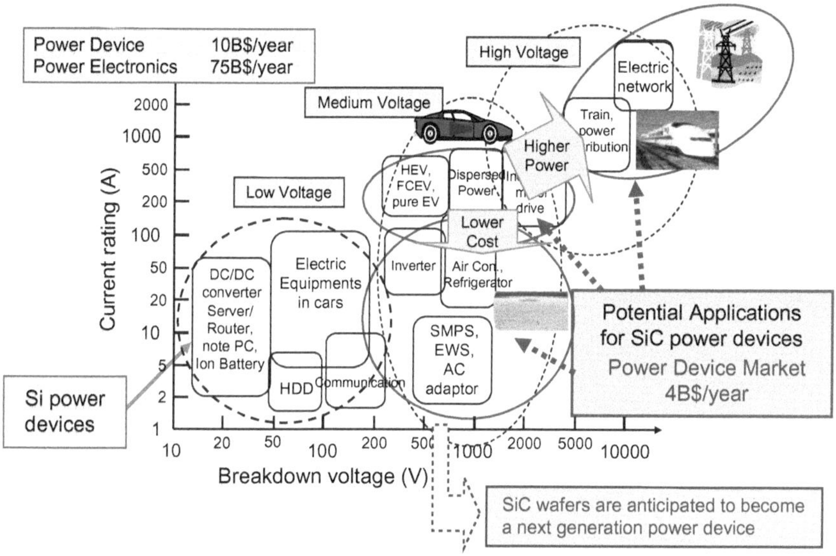

Figure 5.38 Potential fields and market size for SiC power devices

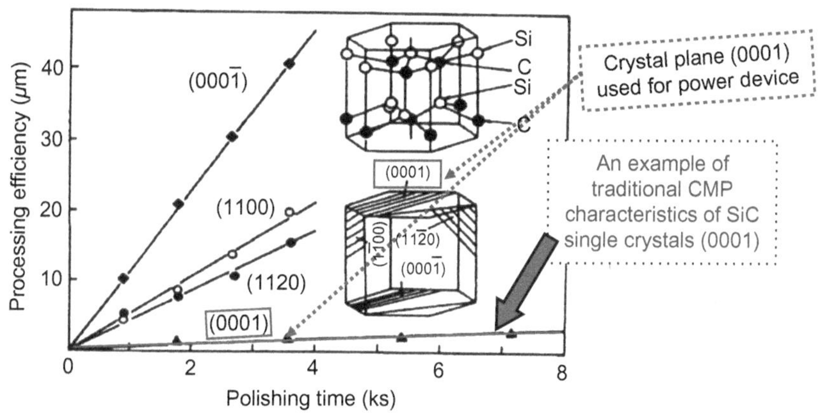

Figure 5.39 CMP/MCP characteristics of SiC single crystal (0001) by Cr_2O_3 abrasives (Ref. Tool engineer, 37, 4 (1993) p.113 by T.Suga)

TiO_2 particles to the slurry (colloidal silica) for polishing SiC wafers. To initiate the photocatalytic reactions, the particles needed to be irradiated by ultraviolet light. In this experiment, the slurry was constantly irradiated by ultraviolet light via the quartz inspection window (see Figure 5.41).

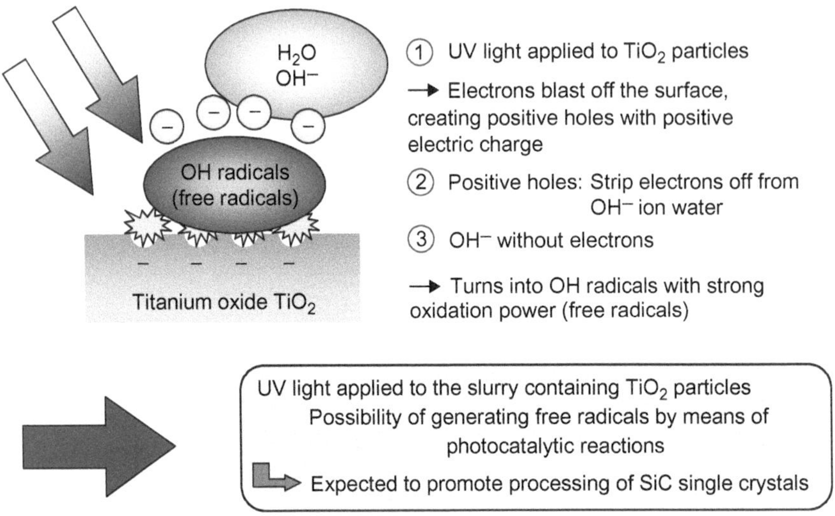

Figure 5.40 Mechanism of photocatalytic reaction

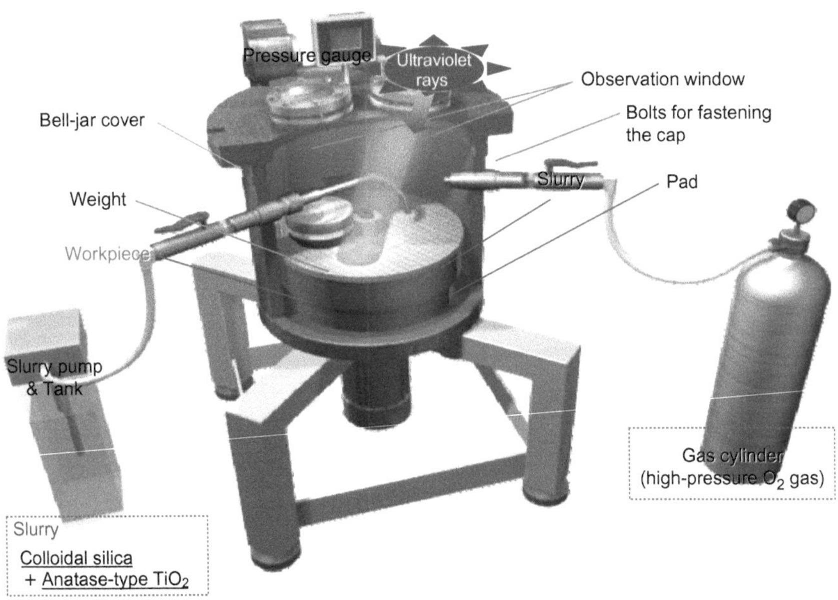

Figure 5.41 The photocatalytic reaction under high-pressure gas (UV is applied to the slurry on the pad surface)

Figure 5.42 shows results of an experiment where anatase-type TiO_2 particles (0.5 μm) were added to the colloidal silica slurry at 0.5 wt%, and single crystal SiC wafers were polished under ultraviolet radiation. The

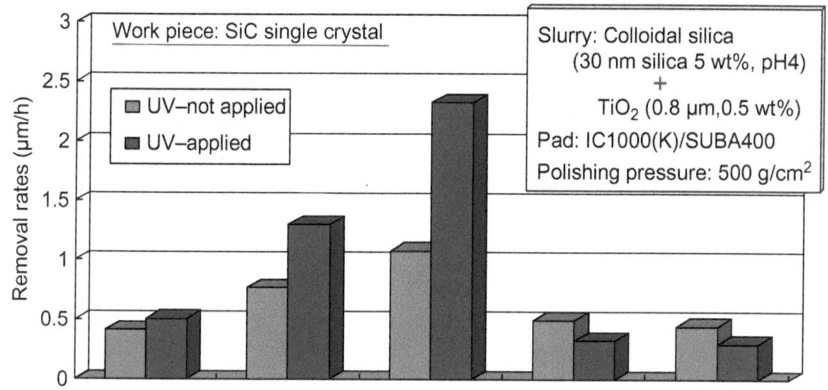

Gas charged inside bell-jar	None	Air	O_2	N_2	Ar
Pressure inside bell-jar	Atmospheric pressure	500 kPa	500 kPa	500 kPa	500 kPa

Figure 5.42 Removal rates of SiC wafers with different gases in a bell-jar type CMP machine (effectiveness of UV application and photocatalytic reaction)

processing rate increased slightly under normal atmospheres, but increased by a factor of two in an oxygen atmosphere, which verifies the effect of ultraviolet radiation. These processing rates are 5 to 6 times better than those of conventional CMP, and suggest that high-pressure oxygen gas enhanced the generation of active oxygen via the photocatalytic reactions of TiO_2. According to Zeta potential measurements, the isoelectric point or pH value shifted toward the acid side by 1.2 in the presence of TiO_2 and ultraviolet radiation, which increased the hydrophilic nature of the slurry, contributing to the increased processing rates. By contrast, use of inert gases and ultraviolet radiation caused a decrease in processing rate. This is presumably due to the reduction actions (effect of active hydrogen) of inert gases.

As indicated in Figure 5.43, the surface roughness (Ra) of the single crystal SiC wafers was excellent - as small as 0.2 nm.

In these experiments it is thought that single crystal SiC wafers have been processed by a mechanism involving active oxygen. This was generated by reaction between hydroxide ions and water, photocatalyzed by TiO_2 particles in the slurry and ultraviolet irradiation. Such strong oxidation broke the Si-C bonds in the crystals. The effect was most pronounced in high pressure oxygen atmospheres.

Closed bell-jar CMP machines under high pressure oxygen atmospheres and using photocatalytic reactions have opened a path for the

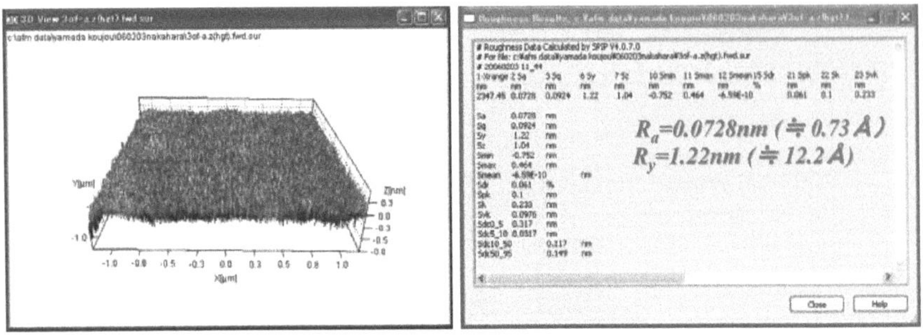

Figure 5.43 Surface roughness of SiC wafers polished by a bell-jar type CMP machine, assisted by photocatalytic reaction using oxidant gases

efficient, high-quality polishing of materials considered extremely difficult to process, such as SiC.

5.4.4 Summary

An innovative "bell-jar type atmosphere controlled CMP" method has been proposed in this section, which allows the control of the atmosphere inside the bell-jar by pressurizing with different gases or depressurizing the bell-jar. Using a prototype bell-jar type CMP machine, CMP characteristics of silicon, SiO_2, sapphire and SiC were obtained to investigate the fundamental processing mechanism of this new machine.

The experiments confirmed that processing rates increase by a factor of 1.5-3 under vacuum, or in an oxygen atmosphere. These increases are not observed when inert gases like argon or nitrogen are used.

The CMP mechanism has been thoroughly examined, which has led to the development of the highly efficient, high quality, photocatalyst assisted method for the CMP of single crystal SiC wafers. This new method dramatically improves the surface quality and processing rate, by four times that of conventional CMP.

Based on these experiments, we have designed and manufactured a prototype of a sealed, bell-jar shaped, both-side simultaneous CMP machine (See Figure 5.44). This innovative machine uses photocatalytic reactions under ultraviolet light inside the sealed bell-jar chamber to integrate high efficiency CMP and both-side simultaneous polishing into one machine.

This new processing method and machine have made it possible to control the processing rate and processing quality, and is truly expected to become a next-generation CMP technology.

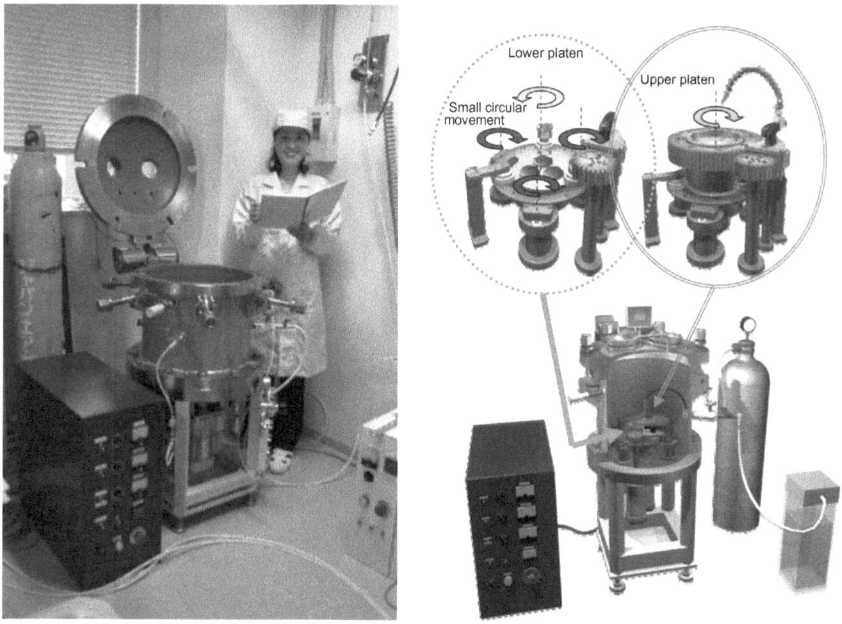

Figure 5.44 Advanced bell-jar shaped, both-side simultaneous CMP machine (exterior view and structure of the machine)

5.5 DRY ETCHING AND PLANARIZATION CMP APPLICATIONS FOR SURFACE MICRO-MACHINING

5.5.1 Introduction

In device fabrication, etching is used between the lapping process and CMP to remove residual particles and damaged layers, as explained in Chapter 3. The cleaning process after CMP can also be regarded as an etching process. Since these etching techniques utilize acid or alkaline chemical solutions, they are known as wet etching. Alternative etching procedures use a gas etchant, and are known as dry etching processes. Historically, wet etching was the first technique to be used.

Wet etching utilizes a chemical reaction between an etchant and a substrate, and has the advantage that quite a large area can be etched at the same time, giving high productivity and cost performance. Wet etching has therefore played an important role in the early stages of the bulk micro-machining of MEMS (Micro Electro Mechanical Systems) for creating cavities, pits, membranes, and cantilevers at dimensions of a few hundred micrometers. Wet etching is isotropic by nature. Of course, the crystal orientation of a substrate yields anisotropic etching. Isotropic etching, however, has a disadvantage for high precision micro-machining, because it makes it hard to control the etched shapes, which limits the structures that can be produced. The etching area tends to spread slightly beyond the design due to over-etching, and the undercut phenomenon also causes problems.

Anisotropic etching can be carried out by dry etching under controlled conditions. Selection of appropriate etching systems and gas etchants, allows control of the reaction mechanisms (chemical, physical, or a combination). Nowadays, a dry etching technique can process large volumes in a short time, but it requires an expensive etcher system, so production costs are high.

In this section, the basic characteristics of dry etching and some examples etched by electron beam (EB) lithography are briefly discussed. A few dry etching applications for surface micro- and nano-machining are also introduced.

5.5.2 Dry Etching Characteristics

Dry etching can etch small patterns with smaller undercut and higher aspect sratio than wet etching. Reactive Ion Etching (RIE), which uses a high pressure plasma as the etchant gas, is widely used for micro-machining.

Figure 5.45 illustrates the RIE process. It works by a combination of chemical reaction and physical action. In the plasma, reactive species are created and react with substrate material. By applying a magnetic field, ionized species attack atoms in the material and activate them. Reaction products are removed as exhaust gases. If an inert gas, such as argon (Ar), is added, its ions are accelerated, collide with the substrate and break atomic bonds on the surface (ion milling). Since those ions collide more on the horizontal surface than on the sidewalls, the etching rate is faster in the vertical direction, which results in anisotropic etching. The voltage of the applied magnetic field is therefore a significant parameter for both chemical reaction and physical action. Although etching is faster in the vertical direction, the sidewall is also etched at a low rate. To achieve better anisotropy, the sidewall must be actively protected.

Deep Reactive Ion Etching (DRIE) is a technique for fabricating high aspect ratio structures. This is a cyclic process of repeated etching and deposition steps (Figure 5.46). Since the deposition is mainly to protect the sidewall against etching, it is called passivation or passivation

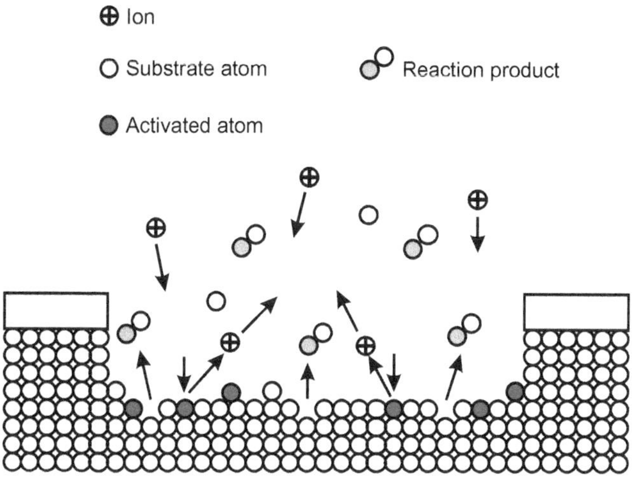

Figure 5.45 Illustration of dry etching mechanisms in Reactive Ion Etching (RIE)

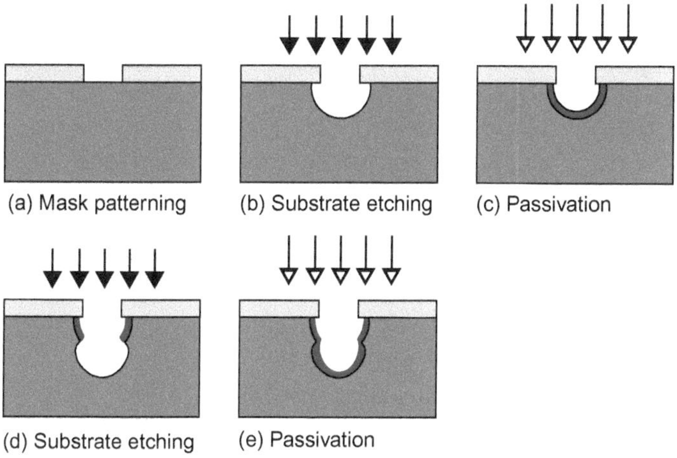

Figure 5.46 Deep Reactive Ion Etching (DRIE)

deposition. The passivation layer on the horizontal surface is removed by Ar^+ ion milling during the etching step. Under correct conditions, it is possible to create very deep structures, with high aspect ratios. Etching that passes through the whole thickness of the substrate (about 0.5 mm, for example) is also possible, by repeating several hundred two-step cycles. The etched sidewall, however, tends to have a periodic wave shape, called "scalloping," due to the cyclic nature of the DRIE process.

5.5.3 Micro-Machining Using Dry Etching

Many micro- and nano-manufacturing techniques have been proposed for manufacturing tiny parts: machining by micro-cutting, micro-electro-discharge manufacturing, micro-fabrication by scanning probe microscope (SPM), molecular beam epitaxy, the lithography technique, X-ray lithography, and electroplating (in German: Lithographie, Galvanoformung, Abformung: LIGA) process, and so on. Of these, lithography is widely used for mass production. Lithography itself has several variants, for example, photo-lithography, electron beam lithography, ultraviolet beam lithography, X-ray lithography, ion beam lithography, nano-imprint lithography, and their combinations. Most of these techniques are normally used to create small pits or high aspect walls. The most popular and high resolution technique is a combination of electron beam (EB) lithography and dry etching technique, used on silicon substrates.

Figure 5.47 is an illustration of the whole EB lithography and dry etching procedure. The substrate is silicon (Si), on which silicon dioxide (SiO_2) is formed as an etching mask layer. Electron beam resist (EB resist) is pasted onto the SiO_2 layer by spin coating. The desired pattern is drawn by electron beam on the EB resist, and this is exposed by irradiating an electron beam at it (Figure 5.47a). After development, the exposed area is removed, and the desired pattern is formed by the difference in solubility of the exposed and unexposed parts of the resist layer (Figure 5.47b). The SiO_2 layer is etched by using the remaining resist as a mask (Figure 5.47c).

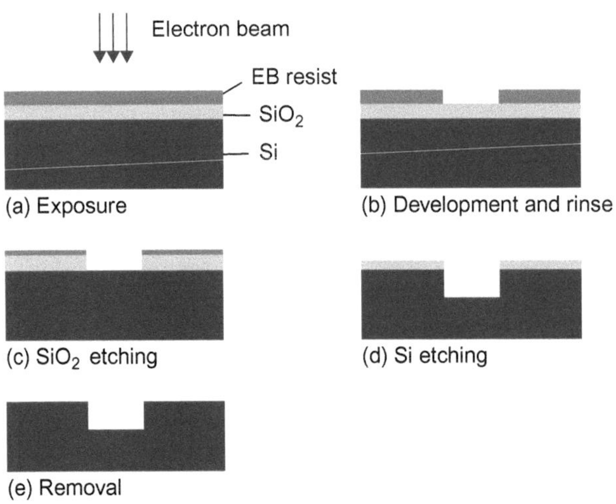

Figure 5.47 The whole procedure of EB exposure and dry etching process

The resulting SiO_2 layer acts as the mask for Si substrate etching. Si substrate is etched with SiO_2 layer mask (Figure 5.47d), and finally, the SiO_2 layer is removed if necessary (Figure 5.47e).

This procedure is well known, but the actual operating conditions used have a considerable impact on the etching results. In the following subsection, an actual example of etching conditions is provided, to illustrate their influence on substrate patterns.

5.5.4 Influence of Dry Etching Conditions

The configuration of a silicon work piece is shown in Table 5.3, along with the thickness of each layer. The chip is 10 mm square, and the thickness of the Si substrate is about 500 μm. SiO_2 forms a layer 200 nm thick above this. The thickness of the spin-coated EB resist (ZEP520A, ZEON Corp.) is about 320 nm.

EB exposure is carried out by a high-resolution electron beam lithography system. The system is based on a scanning electron microscope (SEM) combined with an electron gun as an exposure system (S-4300SE type SEM by Hitachi, Ltd.). Its drawing capability is as follows: minimum line width, 20 nm; range of the drawing field is from 50 μm^2 to 1 mm^2; accuracy of drawing field stitching, about 0.3 μm; electron gun, ZrO/W Schottky emission type, with an acceleration voltage from 0.5 to 30 kV; exposure beam current, maximum 25 nA; beam blanking frequency, 10 MHz. Designed patterns are radial gratings drawn by use of the Vector Works CAD system, and electrical drawing data of the patterns are transferred to the electric gun system.

Development after EB exposure is carried out using developing solution (ZED-N50), and development time is about 3 min. After development, a rinse procedure is carried out for about 2 min 30 s.

Dry etching was carried out using a dry etcher (SAMCO Model RIE-10iP) in this example. The type of dry etching is RIE by utilizing inductive coupled plasma (ICP). This creates a high-temperature plasma by applying a high voltage and a high frequency variable magnetic field

Table 5.3 Configuration of the Work Piece in the Dry Etching Test

Configuration Layer		Thickness
Material substrate	Si	500 μm
Si etching mask	SiO_2	200 nm
EB resist	ZEP520A	320 nm

to reactant gases. Both anisotropic and isotropic etching are possible with this equipment if suitable etchant gases are used. Trifluoromethane (CHF_3) is used for anisotropic SiO_2 etching. DRIE is carried out in Si substrate etching to keep the vertical sidewall of the etched grooves. In this process, a mixture of sulfur hexafluoride (SF_6) and argon (Ar) gases was used in the etching step, while a mixture of octafluorocyclobutane (C_4F_8) and Ar gases was used in the passivation step. There are many parameters to be controlled in etching and passivation.

5.5.4.1 Electron Beam Exposure

To examine the capability of EB exposure to produce fine lines, several radial grating patterns were drawn on a work piece. The patterns are long radial trapezoids, which have line widths varying from 500 nm to 5 μm. Each line pattern has a length of 500 μm. The designed grating pattern is shown in Figure 5.48.

EB exposure was carried out at acceleration voltage of 30 kV, using the high-resolution electron beam lithography system previously described. The EB resist of the black drawn area was exposed. Figure 5.49 shows an SEM image of the designed grating patterns after EB exposure and development. They are exposed as designed, and all the grating lines are patterned exactly. To study the drawing capability in detail, each grating pattern is shown in close-up in Figures 5.50–5.55. In the case of the 2 to 5 μm lines, the exposed line width and the spacing

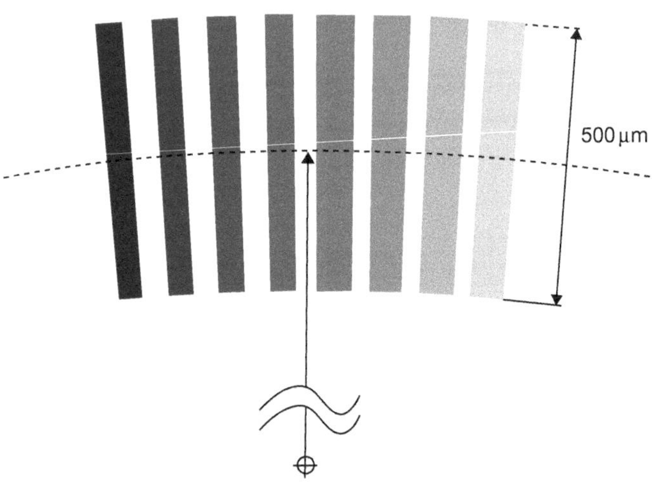

Figure 5.48 Designed grating patterns

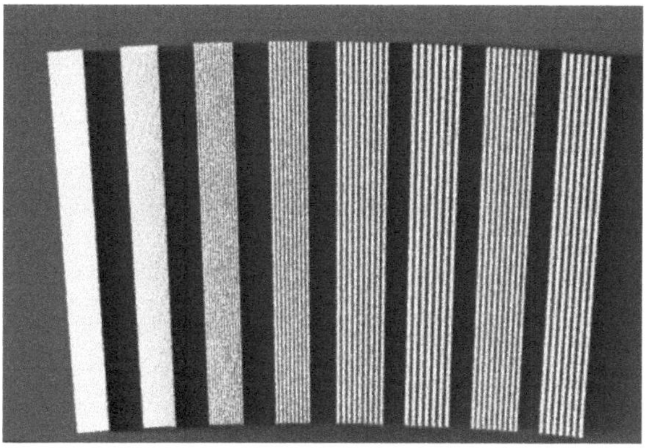

Figure 5.49 SEM image of the designed grating patterns after EB exposure and development

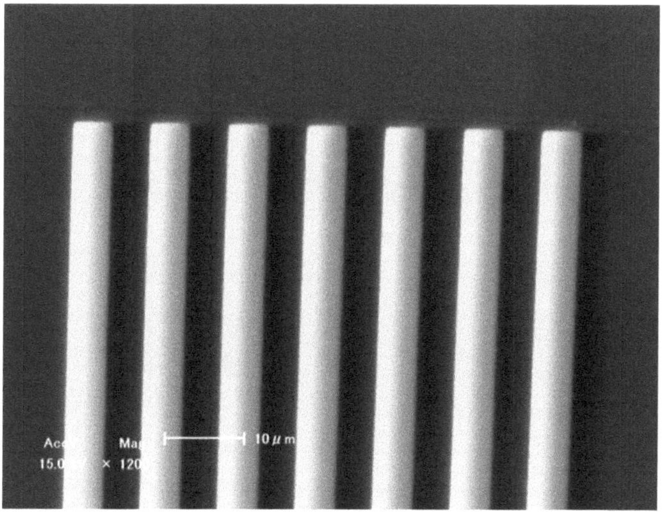

Figure 5.50 SEM image of the designed grating pattern with 5 μm line width

between exposed lines are almost exactly the same. However, at 1 μm, the exposed line width is a little wider than the spacing. This phenomenon becomes more prominent at 500 nm. If the wide exposed area is etched according to the observed width in next etching process, the etched groove width will also get wider than the spacing. The inhomogeneity of the line width and the spacing leads to inaccurate gratings. In the preparing the electron gun, the electron beam axis alignment and polarization must be adjusted correctly, which requires skillful operation and the

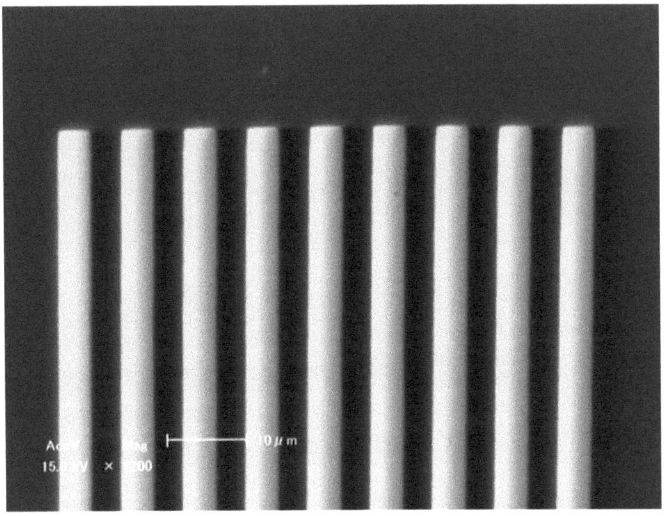

Figure 5.51 SEM image of the designed grating pattern with 4 μm line width

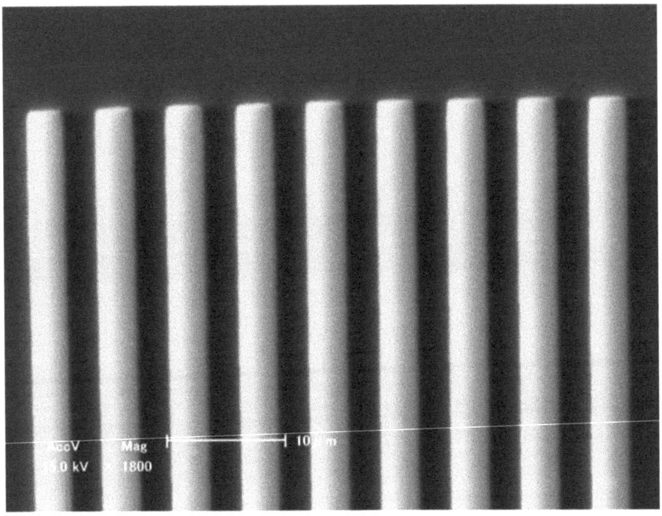

Figure 5.52 SEM image of the designed grating pattern with 3 μm line width

operator needs experience. Beam refining is key for clear line drawing. These EB resist patterns are used as a mask for etching the SiO_2 layer.

5.5.4.2 Dry Etching

Etching the SiO_2 layer was carried out using a dry etching process. The etchant was CHF_3 gas, and the etching time was examined, because this

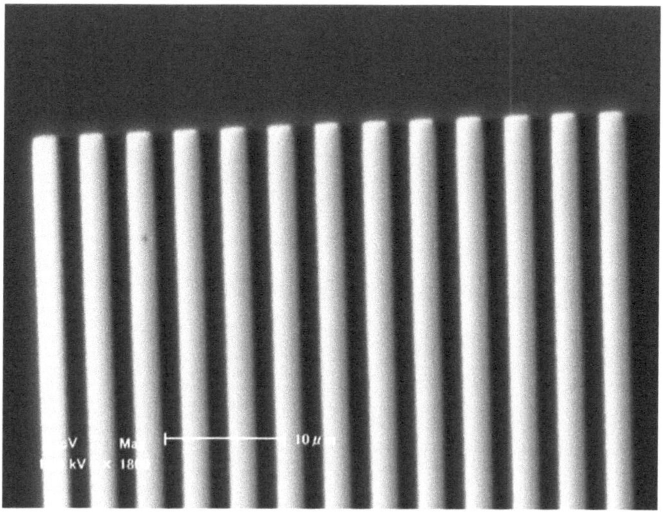

Figure 5.53 SEM image of the designed grating pattern with 2 μm line width

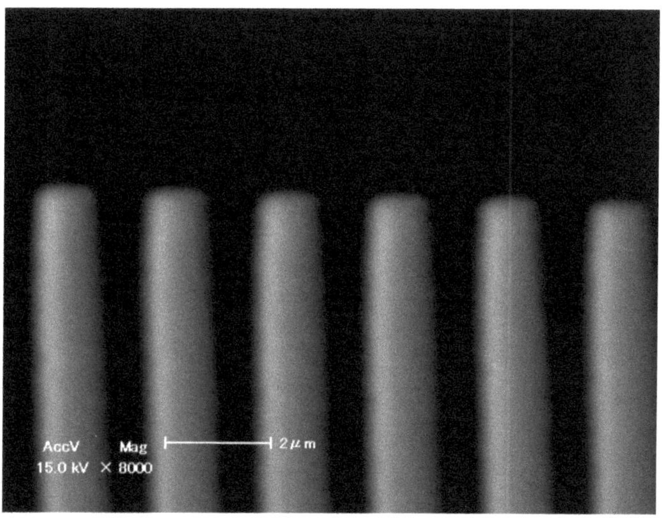

Figure 5.54 SEM image of the designed grating pattern with 1 μm line width

strongly influences the quality of the grooves after Si substrate etching. The anticipated effect is illustrated in Figure 5.56. If the SiO_2 etching time is too long, the etched line width of the SiO_2 layer will become wider than the resist line width by undercut, due to the high selectivity of CHF_3 between SiO_2 and Si. If an undercut SiO_2 mask is etched, the following Si substrate etching results in highly undercut grooves. On the other hand, if

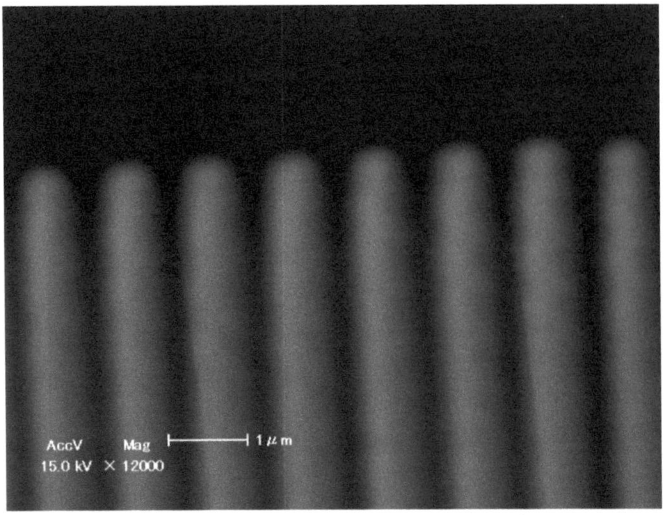

Figure 5.55 SEM image of the designed grating pattern with 500 nm line width

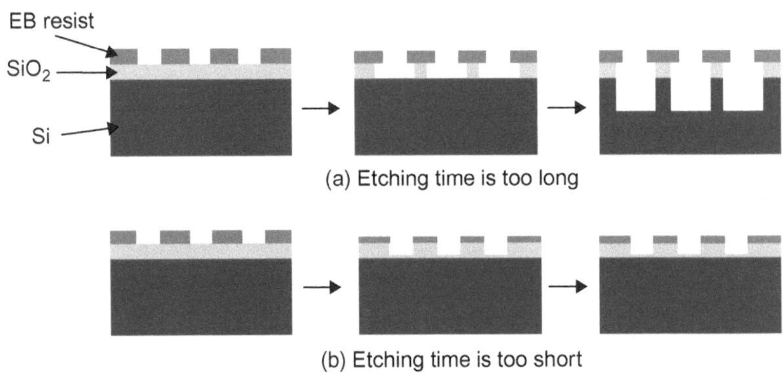

Figure 5.56 Influence of long and short SiO_2 etching time

the etching time of the SiO_2 layer is too short, the SiO_2 layer will not be etched completely. In this case, etching of the Si substrate would not progress any further. As a result, only quite shallow trenches within the SiO_2 layer will remain. The optimum etching condition lies between these limits.

The Si substrate was also etched after SiO_2 layer etching, using the same dry etcher with SF_6 gas for etching and C_4F_8 gas for passivation in a DRIE process. The conditions for DRIE are the same as during the optimization test. The parameters are listed in Table 5.4.

The SiO_2 etching time is first set at 83 s. Figure 5.57 shows the SEM image of the outer end of the grating pattern, which has a line width of

Table 5.4 Etching and Passivation Parameters

Si Etching			**Passivation**		
Gas flow rate (sccm)			Gas flow rate (sccm)		
SF_6	Ar	C_4F_8	SF_6	Ar	C_4F_8
50	20	0.5	0.5	20	50
Bias (W)		50	Bias (W)		20
ICP (W)		700	ICP (W)		550
Pressure (Pa)		4	Pressure (Pa)		4
Time (s)		4	Time (s)		4

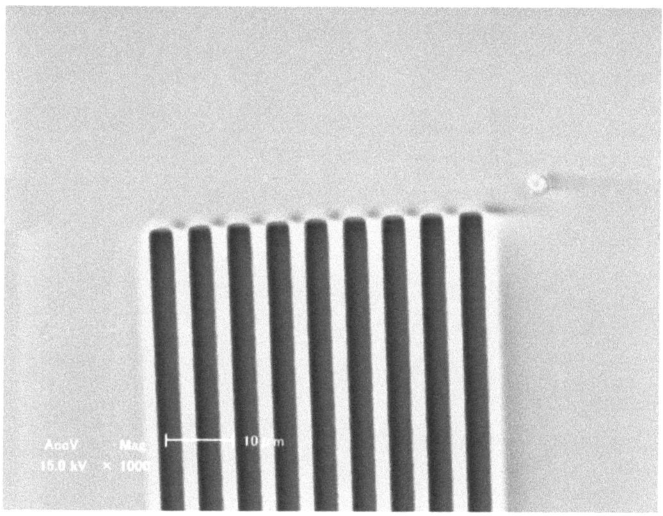

Figure 5.57 SEM image of the outer end of the grating pattern with 3 μm in line width after SiO_2 etching of 83 s

3 μm. Figure 5.58 shows a close-up of Figure 5.57. The etched line is obviously wider than the spacing, resembling typical over-etching. Figure 5.59 shows the cross-section of the etched grooves. The typical undercut phenomenon can be seen in both the SiO_2 layer and the Si substrate. Even if the DRIE process is adopted, the undercut occurs in the Si substrate due to the narrow SiO_2 mask. That means, strictly speaking, that Si substrate etching depends on the opening width of the mask. It is predicted that the parameters for the DRIE process must be optimized in respect to the opening width.

The SiO_2 etching time was then shortened to 63 s. Figure 5.60 shows an SEM image of the outer end of the grating pattern, which has a line

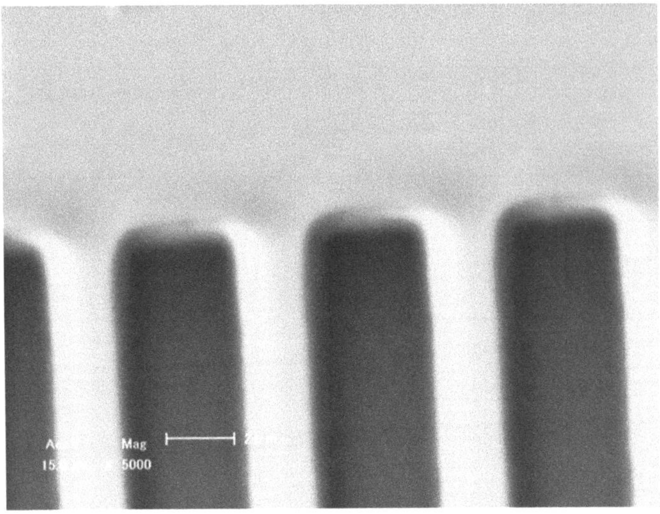

Figure 5.58 Close-up of SEM image in Figure 5.57

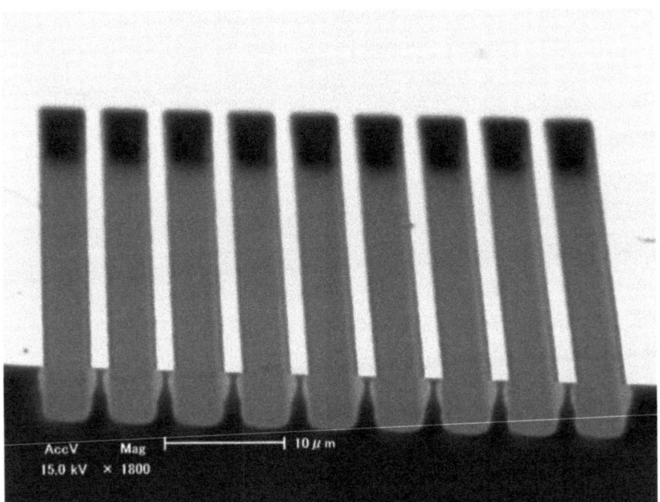

Figure 5.59 Cross-section in etched grooves of Figure 5.57

width of 3 μm, and Figure 5.61 shows a close-up of it. The difference between the width of the etched line and the spacing is smaller than in Figure 5.58, so the SiO_2 etching undercut has improved by decreasing the etching time. The optimum etching time is when the whole etched SiO_2 layer just vanishes away. Therefore, the SiO_2 etching time was shortened further in the next test.

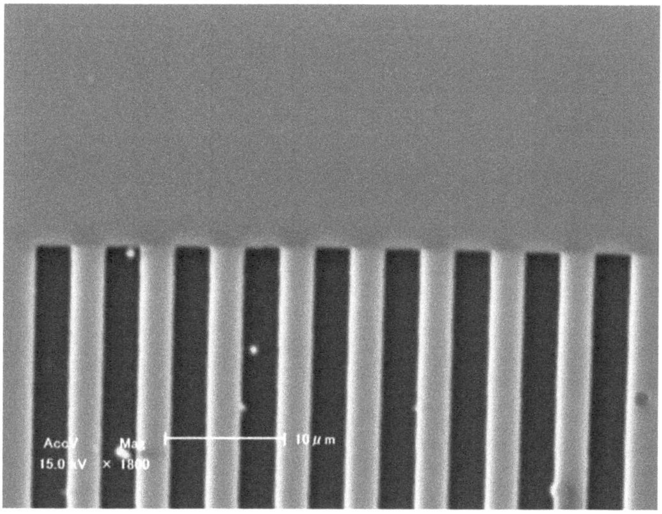

Figure 5.60 SEM image of the outer end of the grating pattern with 3 μm in line width after SiO_2 etching of 63 s

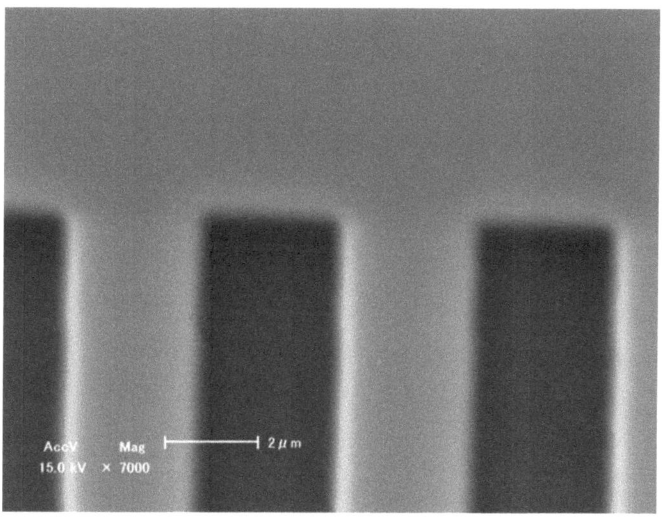

Figure 5.61 Close-up of SEM image in Figure 5.60

In the third test, the SiO_2 etching time was further shortened to 58 s. The SEM image of the outer end of the grating patterns of 2 μm and 3 μm width is shown in Figure 5.62, and Figure 5.63 indicates that the difference between line width and the spacing is almost zero. Quite good homogeneity is achieved under this condition. The cross-section of the

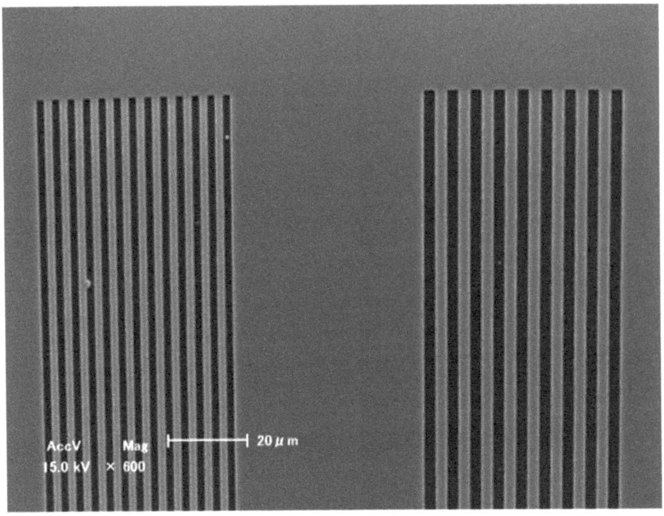

Figure 5.62 SEM image of the outer end of the grating patterns with 2 μm and 3 μm in line width after SiO_2 etching of 58 s

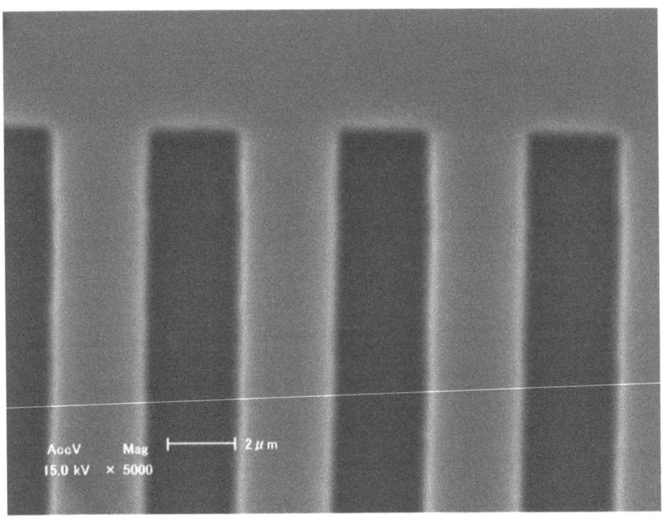

Figure 5.63 Close-up of SEM image on right-handside of Figure 5.62

etched grooves of Figure 5.63 is shown in Figure 5.64. The depth of etched grooves seems to be very small. While the thickness of the SiO_2 layer of this work piece is about 200 nm, the etched depth of the groove is obviously greater than 200 nm. That means that the whole SiO_2 layer

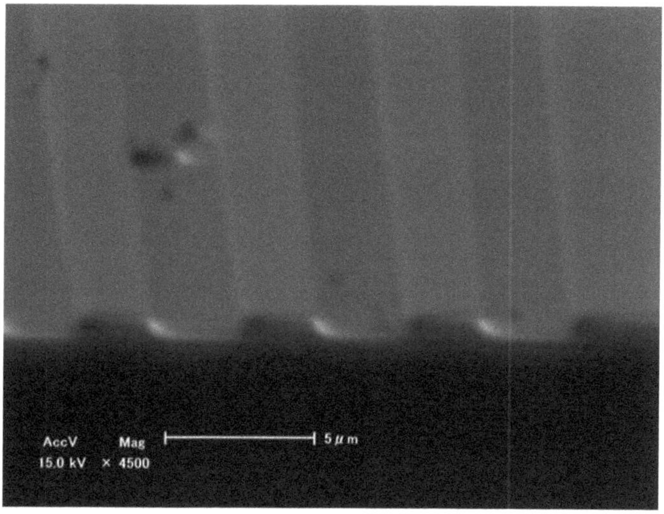

Figure 5.64 Cross-section of etched grooves of Figure 5.63

has been completely etched away, and the Si substrate has also been etched a little. From this evidence, this condition is considered to be almost optimum. In order to confirm this, the SiO_2 etching time was shortened further.

For the final test, the SiO_2 etching time was set to 53 s, which was 5 s shorter than the test above. The SEM images of the outer end of the 3 μm grating pattern are shown in Figures 5.65 and 5.66. In these SEM images, the contrast between the etched line and the spacing is very poor. That indicates that the etched groove is very shallow, as is shown in the cross-sections in Figures 5.67 and 5.68. It is very hard to observe a trench depth in this etched region. It is obvious that the etching time of the SiO_2 layer is too short.

From the above four cases, the optimum etching time of the SiO_2 layer is shown to be about 58 s. This case study shows the influence of the etching time of the SiO_2 layer, but this is just one of many parameters that affect the etching process. In the following Si etching process, the gas flow rates of SF_6, Ar, and C_4F_8, the applied voltages for bias and ICP, the pressure and cycle time are the main parameters, and varying and combining these gives an enormous number of options. Atmospheric conditions, such as room temperature and humidity, also have an impact on the etching results. In the EB exposure process just described, the EB resist type and dose rates are

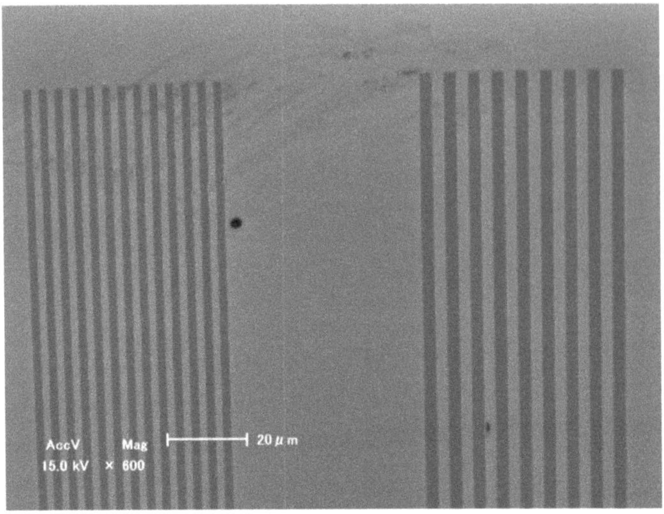

Figure 5.65 SEM image of the outer end of the grating patterns with 2 μm and 3 μm in line width after SiO_2 etching of 53 s

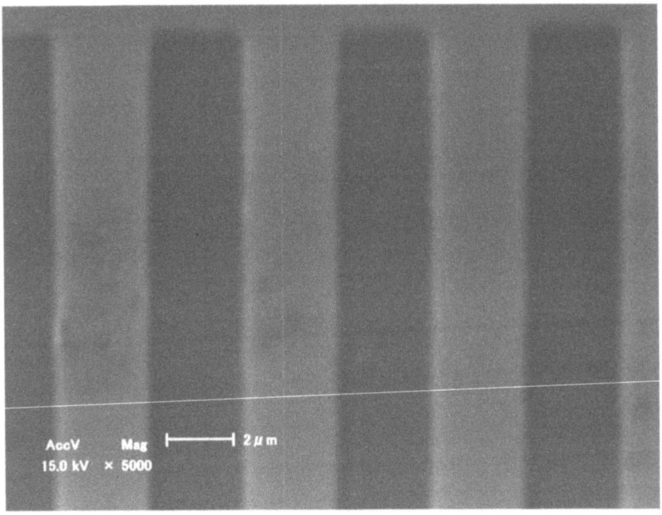

Figure 5.66 Close-up of SEM image on right-handside of Figure 5.65

significant parameters. It takes quite a long time and needs a lot of trials to optimize the parameters, but once they are determined, the dry etching process provides a useful and effective manufacturing method for mass production.

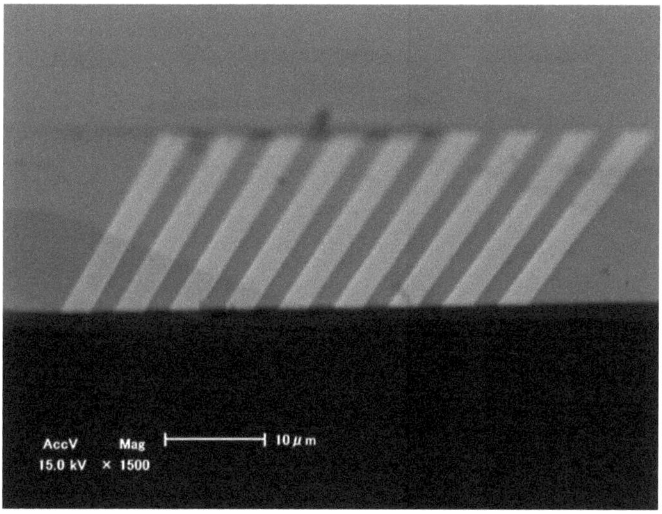

Figure 5.67 Cross-section of etched grooves of Figure 5.66

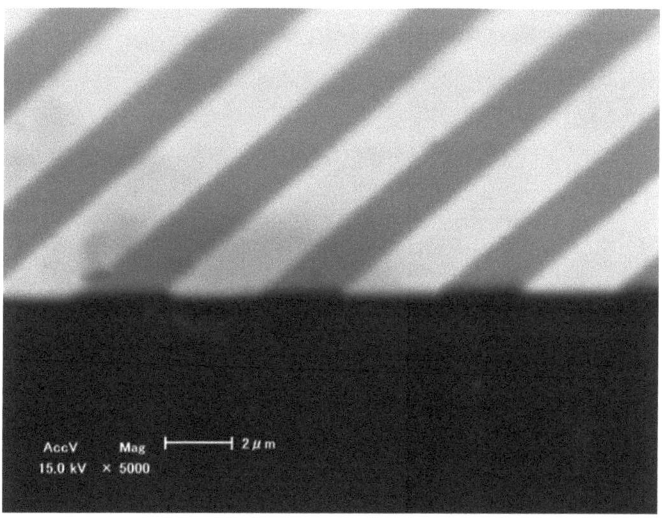

Figure 5.68 Close-up of SEM image in Figure 5.67

5.5.5 Application for Micro-Machining

An example of DRIE over a large area is shown in Figure 5.69. This is a 2.5D tiny-tipped stylus with large body for measurement of the micro-machine element. The wafer surface is polished by the CMP process to achieve high planarity over the whole area. The etched area is

Figure 5.69 DRIE example of large area planarized by CMP

Figure 5.70 Close-up SEM image of the 2.5D tiny stylus tip after DRIE in 300 cycles

3.5 × 3.5 mm. The result of the stylus pattern, whose tip radius of curvature is 4 μm, after 300 cycles of the etching-passivation step, is shown in Figure 5.70. It seems that the passivation process might have been a little too strong, but the sidewall remains as desired and the stylus tip keeps the same shape along in the depth direction. The etched depth is about 72 μm, so the aspect ratio is about 9. The undercut phenomenon is hardly observed, as desired. Since the sacrificial layer (Cr) remains thick enough, if the DRIE process is continued under the same conditions, a thick and rigid stylus is expected. By combining the CMP and DRIE

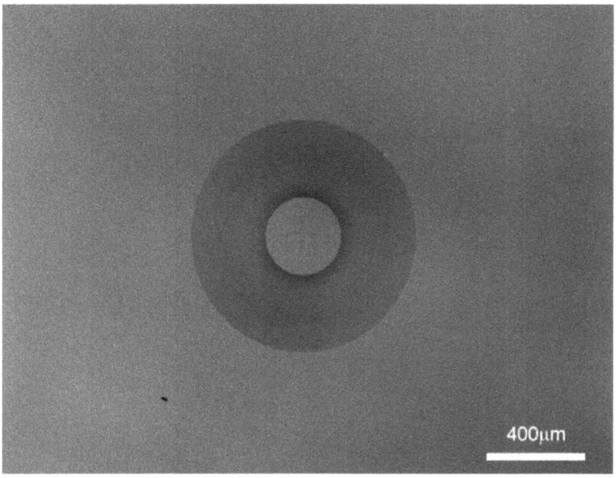

Figure 5.71 CLSM image of the whole radial grating pattern with 10,000 gratings

processes, an area as large as a few millimeter squared can be etched, and the desired structures obtained.

Another example of high resolution EB exposure is shown in Figure 5.71. The designed pattern is 10,000 radial gratings. The whole etched image of the radial pattern is captured by the confocal laser scanning microscope (CLSM). All essential features of the grating pattern are arranged within the field of 1 mm^2. The outer diameter is 950 μm, and the inner diameter is 320 μm. Each of the radial gratings has the same trapezoidal shape, and is 315 μm long in the radial direction. EB exposure was carried out using a high-resolution electron beam lithography system. The system is based on an electron gun as exposure system (JBX-9300FS, JOEL Ltd.). After EB exposure and development, dry etching was carried out using a dry etcher with an RIE by utilizing ICP. In the RIE process, hydrogen bromide (HBr) gas was used to etch the Si substrate, and the depth of etched grooves was intended to be about 30 nm. The whole shape of the grating pattern seems to have been etched successfully, but the radial gratings cannot be seen due to the limitation of low resolution of CLSM. Figure 5.72 shows the SEM image of the outer end of the part of radial gratings, and the inner ends are shown in Figure 5.73. The etched line width at the outer end is about 150 nm and at the inner end is about 50 nm as intended. The pattern has also been manufactured almost homogeneously with satisfactory pitch accuracy. From these results, a grating pattern with line widths up to 50 nm can be manufactured successfully and accurately.

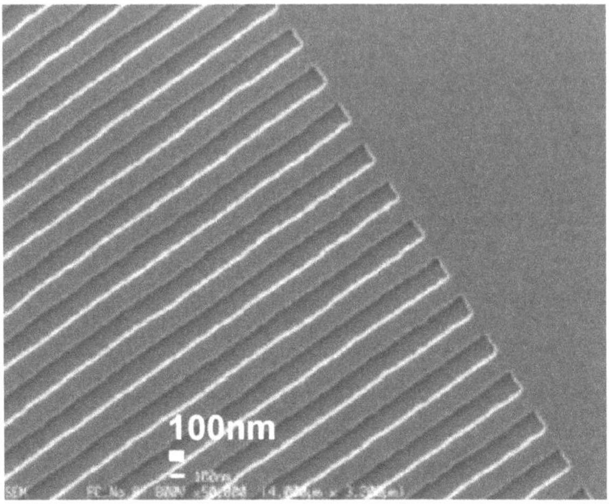

Figure 5.72 SEM image of the outer end of gratings

Figure 5.73 SEM image of the inner end of gratings

Further applications are proposed using the combination of wet and dry etching processes. These are known as surface micro-machining, and involve the deposition and etching of multilayer thin films. Alternating deposition of two different materials is followed by pattern lithography and wet etching. The simplest possible structure is a 3D bridge shape. By applying three-layer processes, a micro-gear which is able to rotate around its own axis can be fabricated.

5.5.6 Summary

The basic characteristics of dry etching and the influence on it of a range of parameters have been examined by electron beam (EB) lithography. Dry etching applications for surface micro- and nano-machining have also been introduced in combination with the CMP process. The new field of surface micro-machining has been introduced.

Currently, lithographical techniques are capable of resolving a few tens of nanometer in line-and-space. Although it uses high-cost facilities, dry etching has a wide range of potential applications, especially in semiconductor devices and MEMS fabrication. Both LSIs and MEMS sensors are widespread in our ordinary life and further development will be of significant general benefit.

REFERENCES

1. S. Kondo, et al., Proc. IEDM (2003) 151.
2. S. Tokitoh, et al., IITC (2004) 130.
3. A. Fukuda, A. Kodera, et al., Removal rate simulation of dissolution-type electrochemical mechanical polishing, JJAP 49 (2010) 076701.
4. M.T. Tsai, et al., IEDM 01 (2001) 80–83.
5. <http://asm.com/pln.asp/>
6. I. Noji, et al., Application of electro-chemical polishing in DI water to Cu damascene wiring planarization process, Tech. Rep. IEICE.
7. H. Zhu, A. Luiz, Chemical mechanical polishing (CMP) anisotropy in sapphire, Appl. Surf. Sci. 236 (2004) 120–130.
8. C.P. Khattak, F. Schmid, Growth of the world's largest sapphire crystals, J. Cryst. Growth 225 (2001) 572–579.
9. Y. Wang, S. Liu, Effects of surface treatment on sapphire substrates, J. Cryst. Growth 274 (2005) 241–245.
10. K. Fathima, A. Senthil Kumar, M. Rahman, H.S. Lim, A study on wear mechanism and wear reduction strategies in grinding wheels used for ELID grinding, Wear 254 (2003) 1247–1255.
11. J.H. Liu, Z.J. Pei, G.R. Fisher, ELID grinding of silicon wafers: A literature review, Int. J. Mach. Tool Manufact. 47 (2007) 529–536.
12. H.S. Lim, K. Fathima, A. Senthil Kumar, M. Rahman, A fundamental study on the mechanism of electrolytic in-process dressing (ELID) grinding, Int. J. Mach. Tools Manufact. 42 (2002) 935–943.
13. S. Malkin, Grinding Technology, Theory and Applications of Machining with Abrasives, Ellis Horwood, Chichester, 1989.
14. J.E. Mayer Jr, G.P. Fang, Effect of grit depth of cut on strength of ground ceramics, Annals CIRP 43 (1994) 309–312.
15. A.V. Gopal, P. Venkateswara Rao, A new chip thickness model for performance assessment of silicon carbide grinding, Int. J. Adv. Manufact. Technol. 24 (2004) 816–820.
16. K. Fathima, A. Senthil Kumar, M. Rahman, H.S. Lim, A study on wear mechanism and wear reduction strategies in grinding wheels used for ELID grinding, Wear 254 (2003) 1247–1255.

17. R.L. Hecker, S. Liang, Predictive modeling of surface roughness in grinding, Int. J. Mach. Tool Manufact. 43 (2003) 755–761.
18. R.L. Hecker, S.Y. Liang, Grinding force and power modeling based on chip thickness analysis, Int. J. Adv. Manuf. Technol. 33 (2007) 449–459.
19. S. Agarwal, P. Venkateswara Rao, A probabilistic approach to predict surface roughness in ceramic grinding, Int. J. Mach. Tool Manufact. 45 (2005) 609–616.
20. H. Zhu, A. Luiz, Chemical mechanical polishing (CMP) anisotropy in sapphire, Appl. Surf. Sci. 236 (2004) 120–130.
21. C.P. Khattak, F. Schmid, Growth of the world's largest sapphire crystals, J. Cryst. Growth 225 (2001) 572–579.
22. Y. Wang, S. Liu, Effects of surface treatment on sapphire substrates, J. Cryst. Growth 274 (2005) 241–245.
23. E. Dorre, H. Hubner, Alumina: Processing, Properties, and Applications, Springer, 1984.
24. K. Fathima, K.-A. Senthil, M. Rahman, H.S. Lim, A study on wear mechanism and wear reduction strategies in grinding wheels used for ELID grinding, Wear 254 (2003) 1247–1255.
25. J.H. Liu, Z.J. Pei, ELID grinding of silicon wafers: A literature review, Int. J. Mach. Tool Manufact. 47 (2007) 529–536.
26. H.S. Lim, K. Fathima, K.A. Senthil, M. Rahman, A fundamental study on the mechanism of electrolytic in-process dressing (ELID) grinding, Int. J. Mach. Tools Manufact. 42 (2002) 935–943.
27. B.P. Bandyopadhyay, H. Qhmori, Efficient and stable grinding of ceramics by electrolytic in-process dressing (ELID), J. Mater. Process. Tech. 66 (1997) 18–24.
28. I.D. Marinescu, Handbook of Advanced Ceramics Machining, Taylor & Francis Group, 2007.
29. H. Ohmori, I. Takahashi, B.P. Bandyopadhyay, Ultra-precision grinding of structural ceramics by electrolytic in-process dressing (ELID) grinding, J. Mater. Process. Tech. 57 (1996) 272–277.
30. B.P. Bandyopadhyay, H. Ohmori, The effect of ELID grinding on the flexure strength of silicon nitride, Int. J. Mach. Tools Manufact. 39 (1999) 839–853.
31. J.Q. Gu, et al., A study on the electrolytic in-process dressing (ELID) for silica glass in a ductile mode, J. Non-Cryst. Solids 354 (2008) 1340–1398.
32. D. Dornfeld, H. Cai, An investigation of grinding and wheel loading using acoustic emission, Trans. ASME 106 (1984) 28–33.
33. I. Inasaki, K. Okamura, Monitoring of dressing and grinding process with acoustic emission, Annals of the CIRP 34 (1989) 277–280.
34. J.F. Gomes de Oliveira, D.A. Dornfeld, Application of AE contacting in reliable grinding monitoring, CIRP Ann. – Manufact. Technol. 50(1) (2001) 217–220.
35. D.J. Stephenson, X. Sun, C. Zervos, A study on ELID ultra precision grinding of optical glass with acoustic emission, Int. J. Mach. Tools Manufact. 46 (2006) 1053–1063.
36. C. Zhang, H. Ohmori, W. Li, Small-hole machining of ceramic material with electrolytic interval-dressing (ELID-II) grinding, J. Mater. Process. Tech. 105 (1999) 284–293.
37. H. Chen, J.C.M. Li, Anodic metal matrix removal rate in electrolytic in-process dressing, J. Appl. Phys. 87(6) (2000) 3151–3158.
38. T. Doi, Details of Semiconductor CMP Technology, Kogyo-chosakai Publishing Co., Tokyo, Japan, (2001) (in Japanese).
39. S. Kondo, S. Tokitou, B.U. Yoon, Proceedings of VMIC Conference, Sep. 19–20 (2002) p. 87–94
40. T. Doi, et al., Design and performance of a controlled atmosphere polisher for silicon crystal polishing, Electrochem. Solid State Lett. 7(8) (2004) G158–G160.

41. T. Doi, et al., A new bell-jar type controlled atmosphere polishing (CAP) machine and its characteristics, J. JSPE 70(5) (2004) 726–731, in Japanese.
42. D. DeNardis, T. Doi, et al., Impact of gaseous additives on copper CMP in neutral and alkaline solutions using a CAP system, J. of ECS 152 (2005) 11.
43. L.M. Cook, J. Non-Cryst. Solids 120 (1990) 152.
44. B. Bhushan (Ed.), Handbook of Nanotechnology, Springer-Verlag, 2004.
45. C.M. Waits, B. Morgan, M. Kastantin, R Ghodssi, Microfabrication of 3D silicon MEMS structures using gray-scale lithography and deep reactive ion etching, Sensors Actuator A 119 (2005) 245–253.
46. T. Bourouina, T. Masuzawa, H. Fujita, The MEMSNAS process: microloading effect for micromachining 3D structures of nearly all shapes, J. Microelectromech. Syst. 13 (2) (2004) 190–199.
47. S. Kurokawa, S. Imamura, Nano manufacturing of a small-tipped stylus for measurement of micro-machine elements with concave surfaces, Proc. of the 35th International MATADOR Conference (2007) 7–10.
48. S. Kurokawa, Y. Umezaki, M. Hoga, T. Doi, O. Ohnishi, M. Uneda, et al., Gear metrology and nanomanufacturing application for micro gear measurement, Proc. of the 1st International Conference on Manufacturing Process Technology (2011) 1–6.
49. R.W. Johnstone, M. Parameswaran, An Introduction to Surface-Micromachining, Kluwer Academic Publishers, 2004.

CHAPTER 6

Progress of the Semiconductor and Silicon Industries – Growing Semiconductor Markets and Production Areas

6.1 A PARADIGM SHIFT IN THE SEMICONDUCTOR INDUSTRY

The European-American economy, which focuses on financial investment, is now in a difficult state due to the failure of subprime mortgages and the collapsing economy of Greece, and similar problems in other European nations. Furthermore, the center of the global economy is drastically shifting to the BRICs (Brazil, Russia, India and China), and the VISTA (Vietnam, Indonesia, South Africa, Turkey, Argentina), where the number of people in the middle income bracket is on the rise, and both demand and markets are growing. BRICs GDP is expected to grow in the future, as first identified by an analyst in Goldman Sachs. Among them, China in particular was predicted to climb to the top of GDP tables, overtaking the US. This forecast indicated that Japan would come second to China in 2010, and growth of the world economy would be driven by BRICs, in particular China and India.

The credit crisis and recession in 2008 have had a significant impact on the semiconductor industry. However, since the latter half of 2009, sales of digital electric appliances, such as flat panel televisions (FPD), have been increasing thanks to "the policy to support people in the local area to purchase electric appliances by giving a subsidy," which was advocated by the Chinese government and contributed to the recovery of the semiconductor industry. There are still some risks, though. As a result, according to a survey conducted by the author, the number of FPD TVs shipped in 2010 was expected to be 2,380, a 58.5% increase over the

DOI: 10.1016/B978-1-4377-7859-5.00006-5

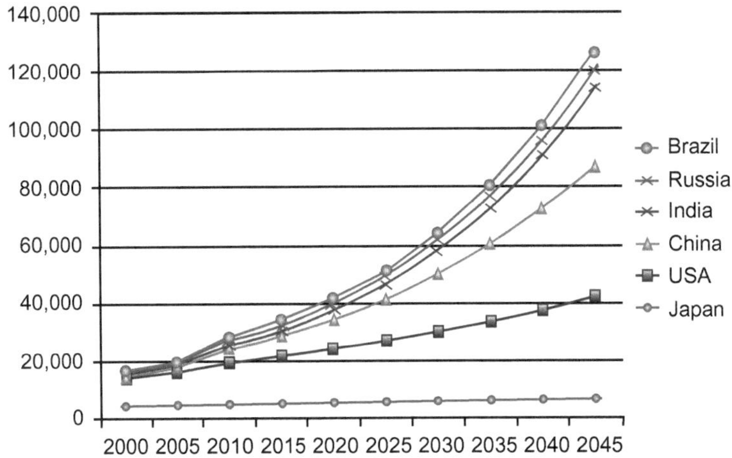

Figure 6.1 GDP forecast for BRICs and US/Japan

previous year. The future of the electronics industry will follow the same path as that of the world economy. Following China, the middle class is likely to expand in a new economic zone like VISTA, and the electronics industry is expected to grow in this region. Rio de Janeiro in Brazil, a member of the BRICs, has been selected as the venue for the Olympics in 2016, and an unprecedentedly large market will be created in this region for electronics (Figure 6.1).

After the IT bubble, digital electric appliances like FPDTV and cell phone spread widely and contributed to the growth of the world economy. In the midst of this trend, the subprime mortgage problem suddenly hit the world economy. In the US, President Obama took office on January, 21 2009, and the nation has been and is still trying to recover with a large scale economy policy using $825 billion for solar cell batteries, and promoting the growth of alternative energy resources like biofuels. This policy of considering the environment is called the "Green New Deal." President Obama's idea is correct, and the growth of the economies in the nations mentioned above will result in an increase in energy consumption. As a result, it is apparent that energy resources will be depleted and development of new sources – called "Green Technology" – as an alternative to fossil fuels will be essential.

Economists Kondratef and Shumpete have analyzed past changes in the economy, and found that stable economies have been generated by inventions and discoveries that created innovations in industry. It also is known that changes in the semiconductor industry have a tendency to

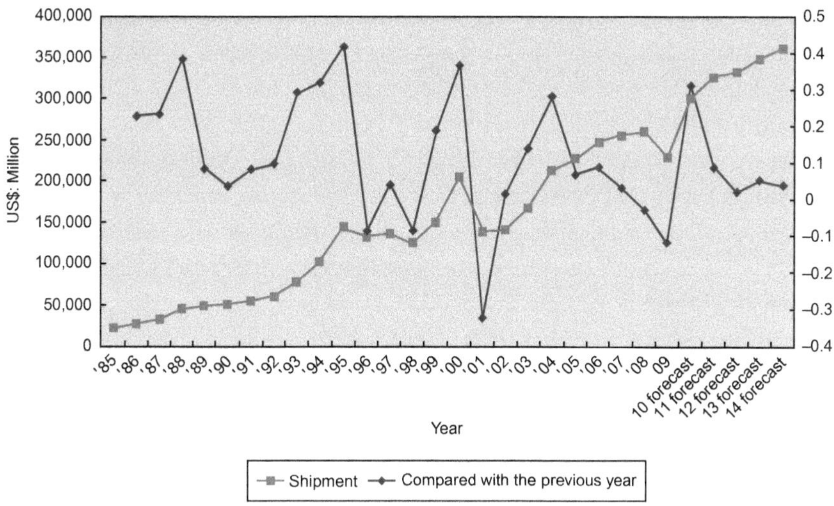

Figure 6.2 World semiconductor market trends

follow a four-year cycle, called the "silicon cycle." It is our opinion that development of an operating system and CPU are important for computer development, and this requires four years to reach production. However, the application area has diversified to computers, cell phones, digital appliances and cars. Computers are no longer the center of market growth, and, as a result, the "silicon cycle" has changed subtly. In addition, "Moore's Rule," which states that semiconductor integration will double in 18–24 months (initially proposed by Gordon Moore of Intel in the US), is less applicable. This is partly because of the emergence of new applications, and partly because microprocessing is approaching physical limits. It is likely that, in the evolution of energy and digital electric appliances, the development of LED and power devices that do not require integration will become increasingly important, and key technology will change.

Statistics for the world semiconductor market are shown in Figure 6.2. The market grew overall from 1986 to 2007, during which time there were three periods of decrease. In 2008 the market dropped to $255.6 billion, down 2.8% from the previous year. In 2009 there was a big drop of 11.8%, to $228.8 billion. In addition, there have been five periods of decline in the past. The semiconductor industry also shows complex economy circulation, including a small size silicon cycle (4 years) and a middle size cycle (8-10 years), like the collapse of the IT bubble in 2000 and the subprime mortgage crisis in 2008.

The semiconductor industry is expected to achieve a value of $300 billion, which is up by 31.08% thanks to favorable sales of digital electrical appliances in Asia. It is clear that smart phones, like the iPhone, and LED FPD TV are driving the industry. However, the shrinking semiconductor market caused by the subprime mortgage problem has changed semiconductor-based companies; they are tending to reorganize and consolidate their production lines. Companies making up the semiconductor industry consisting of IDM (perpendicular integrated production), foundry (consignment production) and fabless (having no plant) are co-existing. Huge amounts of investment are needed to design, develop and produce high density and ultra-micronized devices, which effectively limits these activities to giant corporations like Intel in the US and Samsung in Korea, whose sales of semiconductors are over $10 billion.

On the other hand, fabless companies, who just design semiconductors, have been gaining orders from IDM companies, since the latter have difficulty with the levels of investment now required. Foundry companies produce the semiconductors for the fabless companies, and try to increase their production capacity through the introduction of new technology. In this environment only those who can make new investments can survive. The advent of foundry companies has made it possible for fabless companies, who do not have production facilities, to produce and sell semiconductors. In this environment, Qualcomm in the US, a fabless company, has achieved semiconductor sales ranking within the top ten companies. Production takes place in countries where large-scale investment is possible and there are cost advantages in terms of electricity, infrastructure and tax.

6.2 CORRELATION BETWEEN THE ELECTRONICS AND SINGLE CRYSTAL SILICON WAFER INDUSTRY

High-quality 300 mm diameter wafers are currently being introduced as substrates for semiconductor production. Silicon wafers were first introduced in the 1940s when the vacuum tube was in current use. The transistor was invented in 1947, and a silicon transistor was announced by TI (Texas Instruments) in the US in 1956, followed by the MOS (Metal Oxide Semiconductor) in 1966. In 1967, production of silicon semiconductors exceeded germanium semiconductors in value and accounted for 58% of semiconductor element production. Intel developed the 1 k DRAM (Dynamic Random Access Memory) in 1970. They used a 50 mm diameter wafer with 8 μm of machining line and space. During this time, silicon semiconductors were in their glory and the trend of "Moore's Rule" began.

At this point, the trend to increase the diameter of the silicon wafer began, in order to increase the number of silicon tips produced per wafer, and hence increase productivity. A wafer 75 mm in diameter with 7 μm line and space was introduced in 1974, followed by a 100 mm diameter wafer with 5 μm line and space in 1977. It took six more years to move from a 125 mm to a 150 mm wafer, because 3 μm line and space was the minimum machining width. Time and financial investment were required to develop processes and facilities because of the need for steppers in lithography and dry etching facilities.

Six further years were needed to move from 150 mm to 200 mm wafers, and then 10 years to move from a 200 mm to a 300 mm wafer, due to difficulties in developing a 100 nm miniaturization process, and facilities that could cope with a 300 mm wafer. There are pros and cons in introducing a 450 mm wafer, but this is inevitable when considering cost reductions in mass production.

6.3 ANALYSIS OF SINGLE CRYSTAL SILICON WAFER MARKET FOR THE SEMICONDUCTOR INDUSTRY

Figure 6.3 shows the performance of the single crystal silicon wafer market from 2001 when the IT bubble burst and production value dropped by 23.1% to \580.5 billion JPYEN. Favorable growth was seen until 2006 the market was \1 trillion and $239.8 billion yen which was more than double

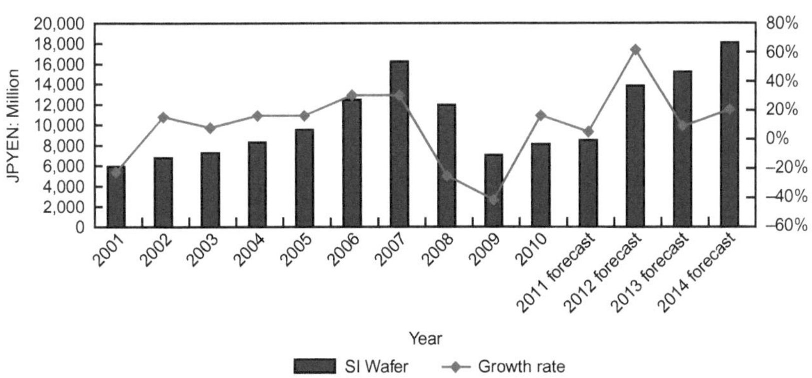

Figure 6.3 The world market for silicon wafers for semiconductor manufacture (Source: SEMI)

the figure in 2001. In 2007 a huge market was created. It grew by 30.1% to \1 trillion and 613.5.2 billion yen. In 2008 it fell by 26% to \1 trillion and 195.0 billion yen, and in 2009 it showed a further drop of 42% to \687.4 billion yen due to an influence of Lehman shock. In 2010 it grew by 17.2% to 806.0 billion yen.

Figure 6.4 shows the single crystal wafer market by diameter. Production of 300 mm diameter wafer was 85.2 million m^2 in 2001. This increased after 2002, and by 2006 it was 18 times as big as 2001, at 1.532 million m^2. In 2009 it grew to ten times bigger than 2006 and was 3.517 million m^2, and for 2010 a 32% increase of 4.642 million m^2 is expected. In the latter half of 2008 the market for 200 mm wafers was surpassed by that for 300 mm wafers; a turning point for the larger diameter. In the future, it is likely that 200 mm silicon wafer lines will be reduced, and it is predicted that in 2014, 300 mm silicon wafers will have a 75% market share.

Silicon wafers are produced by growing single crystalline silicon in the CZ single crystal pulling up process. This single crystal is then cut into blocks and sliced to wafers by a wire saw, followed by lapping and double sided polishing to produce mirror-faced wafers with high planarity. For this process, various facilities are required, such as a single crystal pulling facility, a wire saw, a lapping machine, a polishing machine, a cleaning facility, a diffusion facility, an etching facility, and a wafer carrying machine. All these facilities are needed to deal with large wafers.

Investment needs to be made earlier by semiconductor makers, and if they delay in producing large diameter wafers, profits will suffer. This can be seen in the fact that repeated mergers and acquisitions among silicon wafer companies has resulted in five big companies. Kyushu Electronic Metal Co. merged with Osaka Titanium Co. and later with Sumitomo

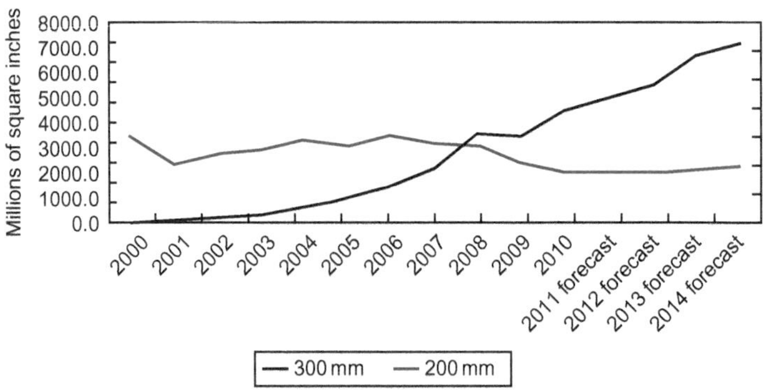

Figure 6.4 World shipments by wafer size (Source: SEMI)

Metal to set up Sumitomo Sitix. This was merged with by Mitsubishi Material Silicon, formerly Nihon Silicon, to form Mitsubishi Sumitomo Silicon. The company was rebranded to SUMCO in 2005, and in 2006 it acquired Komatsu Electronic Metal as a consolidated subsidiary. Komatsu Electronic Metal was rebranded as SUMCO TECHIV in 2007 and became a subsidiary of SUMCO in 2008. Various mergers and acquisitions were conducted like this, resulting in the existence of only three main silicon wafer companies in Japan: Shin-Etsu Semiconductor, SUMCO, and Covalent.

Shin-Etsu Semiconductor Co. specializes in the semiconductor silicon wafer business and has Shin-Etsu Chemistry Co. as a parent company. They have been making positive investments in the industry since the sudden rise of the semiconductor industry, have not experienced merger and acquisition, and achieved production of \254.5 billion JPYEN, and a 37% market share in 2009, as shown in Figure 6.5. They have a production capacity of 1.3 million 300 mm silicon wafers. SUMCO announced that their monthly production would reach 2 million wafers per month for 300 mm silicon wafers at the end of 2010. Their sales value was \203.8 billion JPYEN in 2009, with 30% market share, but they are \96.5 billion JPYEN in debt. The same thing happened to Siltronic AG. in Germany, when it set up Wacker NSCE as a joint venture between Nittetsu Electronics Co. and Wacker Siltronic, and later Siltronic gained the entire share. They produce 300 mm wafers for Samsung Electronics in Korea, having 12% market share and achieving \82.5 billion JPYEN in sales. MEMC Co. in the US produced 350 thousand wafers per month in 2009, and plan to raise that figure to 700,000 by 2010. Their sales

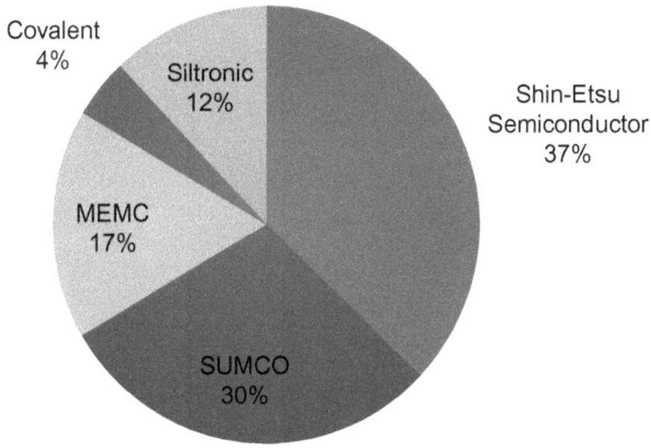

Figure 6.5 Share of semiconductor wafer market in 2009

turnover in 2009 was \116 billion JPYEN, and they have a 17% market share. The reduced price of silicon wafers has resulted in big differences in sales.

The single crystal silicon wafer industry has progressed to the manufacture of larger diameter wafers in accordance with rules of economy, and become an oligopolistic market because of the necessity for continuous investment in facilities. In addition, as the wafer diameter gets larger, it takes more time for the market to prepare for it, and a business scale that can withstand continuous investment is required, which has forced companies to reorganize. It is expected that a market for wafers over 450 mm will emerge, and the silicon wafer industry will change in the future. The electronics application market has been moving to car electronics, energy, robotics and medical applications, which requires diversification of the semiconductor substrates. Various substrates like SOI (silicon on insulator), GaN, sapphire and SiC will be needed for this range of devices. These substrates are hard to machine and further development of production technology is needed to realize cost reductions.

REFERENCES

1. Y. Takeno, J. Satoh, Fundamentals and Mechanism for Electronic Material, Shuwa System Co Ltd., 2010, pp. 10–11.
2. World Semiconductor Trade Statistics, WSTS Forecast Summary, 2010.
3. SEMI , Semiconductor Silicon Wafer Shipment, 2010.
4. Y. Takeno, T. Doi, T. Kasai, Semiconductor industry about progress circumstance (Sixth news), JSPE Spring Meeting proceedings 2009, pp. 443–444.
5. Y. Atagawa, The History of the World Economy seen From Long Waves, Minervashobo, pp. 23-53, 2005.

CHAPTER 7

Summary – The Future of CMP/Polishing Technologies

Almost 20 years have passed since the arrival of the planarization technology called "CMP" for fabricating device wafers. For the researchers specializing in precision processing, the arrival of this processing technology was truly shocking in many ways. Device surfaces formed over the silicon wafers used to be a sacred, untouchable area for mechanical processing engineers, who might have felt as mystified by the processing involved as if it dealt with another world.

Planarization CMP technology was born after many techniques had been tried, and as a last resort. For the first time, a mechanical processing method drew attention. Although CMP is a polishing method, it wasn't mechanical processing specialists who came up with the idea. Silicon wafers are required to be damage-free and clean, with a planar, mirror surface. Currently, silicon surfaces are finished with this polishing method without significant problems. It is not an exaggeration to say that there is no other practical processing method that can produce better surfaces than polishing. As a natural consequence, the planarizing of device wafers has shifted to today's CMP technology.

However, CMP was believed to pose contamination problems. Although the polishing produces smooth mirror surfaces, its working processes were viewed as being very dirty because of the polishing abrasives, which were thought to cause dust generation. This meant that they could not be introduced into the device fabrication process, which should be conducted in a clean room using pure water for cleaning.

Such contamination problems proved unfounded, as silicon wafers are currently finished by polishing/CMP with abrasives, and cleaned. However, in a sense, it could be said that device wafer surfaces are contaminated, because various materials are adhered to them, and such adhered materials are partially removed. This is the process of patterning to make surface topography.

Advances in CMP/Polishing Technologies
DOI: 10.1016/B978-1-4377-7859-5.00007-7

CMP applied to the planarization of device wafers still has some problems to be solved.

Abrasive polishing is a mechanical process to remove excess materials from the wafer surface by letting abrasives cut into the surface to be polished. When the hardness of the work piece changes, the removal amount naturally also changes. When a work piece composed of different materials is polished, soft material is polished rapidly, producing a concave result, while the hard material remains projected. In other words, when it comes to composite materials, generation of concavity and convexity is an inevitable weak point of the polishing process.

Device wafers are, in a way, a typical composite material, composed of hard materials, such as SiO_2, W, TiN and TaN, and soft materials such as Al, Cu and Au. For the surfaces of Al base material with W particles included, it is difficult to make the surface planar by polishing. Furthermore, even if the hardness is the same, there are many materials whose workability is very different. Apart from the processing efficiency, polishing is very effective for processing hard, brittle materials. Soft metals can also be polished to a reasonably acceptable quality. However, it is extremely difficult to planarize these different materials simultaneously, and CMP offers the best solution for this problem. Particles in slurry, chemical solutions and additives can be adequately designed for CMP, and in particular, additives are carefully designed to keep the removal rates of many work pieces more or less the same; in other words, to make the selected ratio of the processing equal.

Further difficulties lie in the uniform planarity, or so-called "global planarization," required for multiple chips and across the wafer surfaces. However, only uniform removal of the projected areas from a large-sized wafer is difficult. For example, although accurate wafer thickness is obtained in the lapping process, it is possible that the edge of the wafer will droop during the final polishing process. This implies that there is no method yet established to let abrasives make uniformly relative motions at a uniform pressure across the entire wafer surface. To make it more difficult, the amount of removal allowed here is at submicron levels, which implies that projected areas only should be preferentially removed from the wafer surface. In other words, extra removal is not permitted here, which also poses a serious challenge.

Given that the work pieces are subjected to polishing, there are still several technical improvements to be made. For the processing efficiency, influential factors are; abrasives, slurry (polishing reagent), pad materials,

Table 7.1 The Achievement Level of Each CMP Technical Field, Including Process/ Equipment Technology, Cleaning Technology, Measurement/Assessment Technology and Device Fabrication

Field	The Achievement Level for Fiscal Year 2011
Processing / equipment technology	ϕ 300 mm are the mainstream of mass production. The mass production trial starts for ϕ 450 mm. Edge exclusion (EE): 3 → 2 mm, setting removal rate variation <2.5% (1 σ). Throughput >50/h (interlayer interconnection) Special note : Practical implementation of low-k and dual-Damascene Cu-CMP process, low polishing pressure due to Cu/low–k, direct low–k–CMP, application to TSV, etc. Cu–CMP : Polish uniformity WIWNU < ~2%, WTWNU < ~2%, (1 σ). Dishing <30 ~ 10 nm (20 μm isolated pattern). Erosion <20 nm (L/S : 20 μm/0.25 μm). Cu removal rate >1000 nm, barrier polishing: Cu/barrier film 2 step polishing (consider about the nanotopography problem, dishing erosion <50 nm)
Cleaning/ washing technology	In the case of Cu-CMP: ϕ 300 mm : Metal impurities concentration $<1 \times 10^{10}$ atoms/cm^2. ϕ 300 mm : Particle size (include scratch) Surface 80/wafer (>0.225 μm) Back side 200/wafer(>0.16 μm)
Measurement/ evaluation technology	Practical realization of APC (Advanced Process Control). Integrated management of in-process and in-line monitor, interlocking movement with other process, CIM–ization. Accuracy of in-process monitor <1 % (practical value on the oxidation film). Particle detection performance >0.1 6 μm (practical value on the oxidation film). In-process monitor (mainly used for EPD), in–line monitor (feed back the information about uniformity, planarity, deflection of wafer to the process).
Device–ization technology	Apply Cu-CMP to hp 32-45 nm generation device. Number of wiring layers 10–12 layer, wiring width: 0.05 ~ 0.8 μm, permittivity of interlayer dielectric: k = 2.3 ~ 2.7 (dishing) + (erosion) <30 nm (Metal 1), wiring height < 5 ~ 10%.

polishing pressure, removal rate and polishing pass. Studies of these influential factors continue, and steadily improve CMP efficiency. The most important issue at this point is further global planarization across the entire device wafer and detection of the end point of polishing.

Furthermore, there is another problem to be solved; by continuing polishing, polishing efficiency varies, causing deterioration to the polishing pad. As a countermeasure, conditioning is carried out after completion of every polishing, which, however, not only shortens the life of the pad but also makes it difficult to detect the end of the polishing by using the polishing time. Hence the development of high efficiency, long-life polishing pads is very necessary.

CMP technology has been adopted and is already established as a planarization method in the multilayer wiring process. Table 7.1 shows current (year 2011) achievements in the following four fields: CMP machine manufacturing technology, wafer cleaning technology, measurement & evaluation technology, and device technology, compiled by the JSPE Planarization CMP Committee (Chairman Doi). Actual achievement has reached a very high precision level in comparison to 20 or 10 years ago.

Lastly, even though semiconductor technology may change in the future, the planarization process will continue to play an indispensable role in it. Changes in the technology will provide challenging tasks for the researchers and engineers specializing in abrasive machining and high-precision processing. Only some 20 years have passed since wafers started to be planarized and dramatic changes are expected in the future. CMP is expected also to be applied to cutting edge sensors, MEMS, three-dimensional LSI, and HDD or FPD glasses. It is hoped that developments in new technologies continue, and CMP application fields expand further without limiting quantitative expansion. With a flexible mind-set, now is the time to rethink the planarization process from the bottom up to find what is required and what measures should be taken for better manufacturing.

INDEX

F

G

H

I

J

K

L

M

T

U

Kolibri 360 GmbH
Ferdinand-Jühlke-Straße 7
99095 Erfurt, Deutschland
produktsicherheit@kolibri360.de